The Wraparound Universe

The Wraparound Universe

Jean-Pierre Luminet

translated by Eric Novak

A K Peters, Ltd.
Wellesley, Massachusetts

Editorial, Sales, and Customer Service Office

A K Peters, Ltd.
888 Worcester Street, Suite 230
Wellesley, MA 02482
www.akpeters.com

Library of Congress Cataloging-in-Publication Data

Luminet, Jean-Pierre.
[Univers chiffonné. English]
The wraparound universe / Jean-Pierre Luminet ; translated by Eric Novak.
p. cm.
Includes bibliographical references and index.
ISBN 978-1-56881-309-7 (alk. paper)
1. Cosmology. I. Title.

QB981.L913 2007
523.1–dc22

2007033467

Cover image: Ghosts in the Weeks space. (See page 114. Image courtesy of Jeffrey Weeks.)

Printed in Canada
12 11 10 09 08 10 9 8 7 6 5 4 3 2 1

Contents

Space cries out in play: "I do not know!"

—Stéphane Mallarmé, *Toast funèbre*

Preface to the English Edition

The aim of this book is to describe a particular approach to cosmology, known as *cosmic topology*, that seeks to discover the overall size and shape of the universe, one of the biggest questions in cosmology for more than 25 centuries.

Along with a small group of collaborators, I have tenaciously defended this approach for more than a decade against the skeptical—and, as it seemed then, "reasonable"—opinions of the majority of researchers in the field. In fact, one of the most surprising possible conclusions of our wraparound universe models is that of a physical space that is smaller than the apparent one that we explore through telescope observations of celestial objects. As a consequence, some portion of distant light sources must be nothing more than "ghosts," their light reflected back to us from the depths of a strangely shaped space, which is folded in on itself.

In 2001, when the first French edition of this book was originally published, the theoretical tools needed to study cosmic topology were more or less complete. I gave explanations and discussions of these tools that were understandable to a general audience, and these can be found without change in the present edition. On the other hand, I did not hide from the reader the fact that observational proof was still lacking. I sketched the different methods and instruments that would, in the future, be capable of distinguishing the various possibilities, and at the same time would resolve innumerable theoretical questions on the nature of space. In particular, in Chapter 38, I explained how, with NASA's planned MAP satellite, a precise map of the cosmic background radiation—allowing us to see the pattern of light that emerged for the first time in the history of the universe—would allow one to decode the original vibrations of space, and therefore to discover, perhaps, the shape of the "cosmic drum." In Chapter 23, I even expressed a personal aesthetic preference for a particular model, that of a *spherical dodecahedral space.*

It was thus in a state of expectation mixed with excitement that I gave the manuscript for *The Wraparound Universe* to my publisher. I had no idea that my wishes would be realized so quickly, and in such a spectacular fashion. The MAP satellite (since renamed WMAP, the Wilkinson Microwave Anisotropy Probe), launched on June 30 2001, has, in fact, replaced and improved upon the data collected by its smaller-scale predecessors. Beginning with the first experimental results, published in February 2003, it became apparent that certain anomalies in the data allowed us to test our hypotheses on the shape of the universe, and we thought that perhaps our dreams were coming true. The article that we subsequently wrote made the cover of the prestigious journal *Nature* on October 9, 2003, representing the archetype of scientific work: a new observation, a proposed explanation by a specific theoretical model—more precisely, the spherical dodecahedral model for the universe, which the journalists, in search of striking images, immediately popularized under the name of "the soccer ball universe"—and precise predictions coming from this model, which could be refuted or confirmed by more detailed observations.

In addition to the detailed corrections and improvements suggested by particularly diligent readers, it therefore seemed to me indispensable to provide an update for the present English translation, in the form of an afterword. This afterword takes stock of the progress made since 2001 in the field of cosmic topology—a subject which previously was little-known, but which since has been propelled to the front of the cosmological scene.

The news coming in every month from the faraway cosmos will demand further additions. There is no doubt that in the coming years the dazzling advances made in observational astronomy will uncover additional celestial surprises, and new discoveries will refine our theories, while invalidating some. However that may be, these discoveries will continue to cause us to ponder the place that we humans, small beings made of dust, occupy in this immense cosmos, whose shapes we brazenly attempt to discover.

Acknowledgments

Christine Ehm reworked my manuscript with rare clearsightedness; and Jean-Philippe Uzan, Roland Lehoucq, and Didier Pelat have made me privy to their accurate critiques.

The Fondation des Treilles in Tourtour, France welcomed me many times and allowed me to bring to a close this extended project, providing me with long stretches of serene work that it would otherwise have been impossible for me to find. This long labor has also been sustained at every instant by Olivier Bétourné, who armed himself with patience and understanding in spite of my numerous postponements. For the second edition, I thank the readers of the first edition who discovered some typos and gave me some pertinent critiques, in particular Georges Melki. For the present English-language edition, I acknowledge Eric Novak, for the excellence of his translation. To one and all I express my deep gratitude.

Reading Guide

The description of science began in Greek antiquity. In those times natural philosophers such as Democritus, Heraclitus, and Plato were both scholars and masters of language. In their studies of nature, physical reasoning and poetic expression went hand in hand. A high point of the genre was reached by Lucretius. This Latin poet of the first century BCE, a follower of Epicurus and the atomistic philosophy, left a crucial work, *On the Nature of the Universe* (*De natura rerum*), which admirably combined an epistemological design with a concern for form. Moreover, Lucretius constructed his work as fittingly as possible for its content.[1]

Throughout history, this manner of communicating science has seen some success, as well as some virulent criticism. Some purist thinkers have proposed that no literary expression can give a proper account of the subtleness and complexity of scientific thought. As for me, I have always thought that there is no contradiction between the scientist's work and the writer's art. Of course, I know the limits of language, of analogies and metaphors. In the use of vocabulary, for example, the physical and mathematical sciences have borrowed a great number of terms from everyday language, to which they have given very precise meanings which often are estranged from their everyday connotations: space, field, wave, chaos, etc. These shifts in the meaning of scientific words often cause confusion, or even lead to seeming nonsense, either for the general public or in the writings of intellectuals who do not have adequate scientific training.[2]

However, I have always been intrigued by the idea that the very form of a book could reflect, in one way or another, its content—through its size, its layout, its

[1]The poem, as much in its general organization as in its syntax and its alphabet, makes use of a cycle of creation and destruction that mimics one of the essential aspects of atomistic philosophy.

[2]See, for example, [Sokal and Bricmont 03].

organization, its literary construction, and its rhythm. The first book I wrote for a general audience [Luminet 92], which was about black holes, had the structure of a detective novel. It focused on two main characters, gravity and light. The latter had been murdered, and this uncommon crime led to a sort of police investigation. The inquest began with an autopsy, described in the part of the book entitled "Exquisite Corpses," and continued until the puzzle was solved in the final part, named "Light Regained." I must admit that this construction passed by unnoticed, save by a very few readers. Therefore, I am tempted to explain at the very beginning the particular construction of the present work and to propose a guide for reading it.

This book concerns the topology of the Universe, a subject with many ramifications, which will oblige us to make forays into mathematics, physics, and astrophysics, as well as into the history of ideas, philosophical concepts, and even artistic references. The reader may have some difficulty following the main theme if it were presented as a single, linear block. I have therefore opted for a tree-like construction, by giving the book the structure of a graph. In mathematics, a graph is a visual collection of elements mutually connected to each other by arcs, arrows, and loops. Now, as we shall see later with the riddle of the bridges of Königsberg, the mathematical theory of graphs is an offshoot of topology; it allows us to solve complex problems through the use of simple graphical representations.

The first part of the book makes up the main trunk of the graph. It borrows largely from popular lectures I have often given on the subject. The logical sequence of ideas and concepts progresses smoothly from the first naive questions about the shape of space to the most sophisticated observations about cosmic topology: the models of wrapped universes and their experimental ramifications.

The chapters in the second part, entitled "Folds in the Universe," may be seen as the branches of the graph, providing clarifications and in-depth treatments of various points raised along the principal route. Just as a graph may contain loops and crossings and may be traced in many ways, the exploration of these ramifications can be attempted at many moments while reading. The simplest way is to digress as soon as an exit point is indicated in the principal part (by an ← arrow and the page number of the relevant chapter); it is then up to the reader whether to return to the trunk or to wander about from branch to branch in the subject's tree. The reader is also free not to explore any branches before having completely climbed the main trunk.

In Syracuse, in the Greek Sicily of the third century BCE, King Hiero called upon Archimedes to turn his art from purely intellectual issues to concrete objects, to render his reasoning accessible to the senses and tangible to the common man. I would be happy if this book, dedicated to a subject never before treated at this level of exposition, succeeded in this goal.

I
The Shape of Space

And you, gentle Space,
where are the steppes of thy breasts, that I may dream there?

—Jules Laforgue, *Complainte du Temps et de sa commère l'Espace*

1
The Universal Mollusk

What is the shape of space? This is one of the questions that has most intrigued me for the last 30 years, since a summer afternoon that has remained etched in my memory. I was a teenager, reading a popular encyclopedia of astronomy in my sun-drenched garden. At the end of the book, some more technical pages introduced the ideas of general relativity and curved space. I understood nothing, but I was fascinated. A certain Albert Einstein had shown that space and time were not as simple as was suggested by our geometric intuition. One statement above all piqued my curiosity: it said that, in a gravitational field, the space-time continuum was no longer Euclidean, but became a referential mollusk. A very clear image formed right away in my mind. To the flavor of the words was added their metaphorical value: in my imagination the space-time mollusk gave birth to a picturesque vision of an immense cosmic snail, its skin streaked through with light, variegated in bends and curves. From then on, I have never stopped seeking to clarify this strange assertion—what is this universal mollusk?—and I have never again looked the same way upon the beautiful starry skies of my native Provence. It was no longer the myriad of stars, flowing through the Milky Way like rivers of diamonds, that intrigued me, but that which surrounded them, space. It was no longer the contents but the container that caused questions to rush forward: does this impalpable space, which contains the stars, have a structure? Is it flat, dented, curved, folded, smooth, rough, grainy? Is it finite or infinite? Does it have boundaries, holes, handles? And then, what does it mean, exactly, to say that space has a shape?

It was at this time that I decided to interest myself not so much in the Universe as it is, but in the Universe as it *may be*, while remaining within the limits of reasonable physics (without which, all fantasies would be permitted and my hypotheses would no longer correspond to scientific work). It took me many years before I began to put rational content into this poetically resonant mollusk. As understood by common sense, it is normal to think of space as some kind of empty, shapeless receptacle, which welcomes material bodies. Some philosophers have tried to flesh out this receptacle by picturing it more like a material, an ethereal substance that would simultaneously contain and interpenetrate physical objects. For a scientist, questions about space are not pertinent unless some mathematics gives them meaning. It is in fact only today that mathematics, properly applied, can pose proper questions about space and, if need be, help us to resolve them. Although it may not teach us that space necessarily exists as an objective ← 250 reality (a question that still lies in the realm of philosophy), mathematics at least gives space an existence as a structure, defined by its symmetries and by its shapes.

The study of shapes is a science in itself, which is connected to geometry. Geometers have the habit of juggling with spaces that defy common sense. These include spaces that possess 11 or 80 dimensions; spaces that are completely smooth or, on the contrary, discontinuous; spaces riddled with holes; and spaces that have boundaries or that end brutally at angular points. However, these abstract spaces have almost no relation to physical space. Then again, if you think about it, even the most elementary geometrical objects, although used in everyday language, are abstractions: a point being an element without extent, a line a collection of points, a plane being formed by lines, and a solid volume being an aggregate of planes—none of this exists in the real Universe. However, this real Universe, which contains material objects, living creatures, stars, galaxies, waves, and radiation—what can be said of it by the physicist, the mathematician, or the astronomer? Was there a beginning? Is it eternal or ephemeral? Is space finite or infinite?

When I first began to wonder about nature, about the grandeur and origin of the Universe, I did not know that these were some of the questions that humankind has asked itself since long ago. In nearly every culture, philosophers, scholars, or artists have supplied various explanations, which have evolved over the course of history. If modern cosmology and the big-bang solutions that flow from it see such public success at present, it is because they try to respond to these ← 173 questions by combining mathematical reasoning, physical models, and astronomical observations.

After a century of remarkable advances, cosmologists are on the verge of resolving some of these enigmas. I shall do my best in particular to show how, with a little geometry and telescopes of greater perfection, we hope over the next few years to succeed in measuring the size and shape of space.

Up until now, I have sometimes used the word *Universe* in place of the word *space*. This identification is common in everyday usage. Nevertheless, it is incorrect. It is therefore useful to outline the distinctions between mathematical space, physical space, space-time, and the Universe:

- A *mathematical space* is a collection of objects upon which a structure is defined (e.g., a collection of points, between which one defines distances). There are as many mathematical spaces as there are possible structures, that is, an infinite number, which will never be exhausted by the human imagination. Until the nineteenth century, the only known mathematical space was Euclidean space, whose geometric structure derived from the postulates given by Euclid in antiquity, and whose elementary rules we all have learned in school.

- *Physical space*, on the contrary, is unique. If one believes the definition in the dictionary, it is the limitless expanse that contains material objects. In everyday language, it is the space of conquest, through which the Starship Enterprise sails in *Star Trek*. But that this space possesses a certain number of dimensions, three, as it happens, that it is finite or infinite, that it is flat or curved, etc., is far from evident! The reason for this is that our perceptive space is a priori distinct from physical space.[1] The physicist tries to precisely describe the indefinite expanse that is space by means of a geometric model. As we shall see, there are many possible models; the description obtained depends notably on the degree of precision with which the physical space is analyzed.

- *Space-time* is a theoretical entity that unites a three-dimensional geometrical space with a one-dimensional temporal continuum. It is, therefore, a four-dimensional space supplied with a number of possible (generally non-Euclidean) structures, which serves as a framework for the theories of relativity. 161→

- *The Universe*, by definition, is the ensemble of everything that exists. In general relativity, it is modelled by a complex combination that unites the container with the contents, that is to say, space and time, but also energy

[1]The semicircular canals of our internal ear, which detect the rotational acceleration of the head in three perpendicular planes, help to construct a mental space of locally Euclidean structure; see for example [Berthoz 02].

in all its forms (matter, light, vacuum energy). The relativistic Universe is physical space, woven through time and carved by matter. Because of this, certain cosmological assertions that identify the Universe with its spatial component only—for example, the assertion that the Universe is flat—can ←245 cause some confusion for the reader.

- Finally, there is the notion of the *observable Universe*, that I shall from now on call *universe* with a lower case u. In contrast to the Universe, space-time and matter in their entirety, which seemingly has neither center nor frontier, neither interior nor exterior, the observable universe is centered on the terrestrial observer and has a border. Every astronomical investigation actually plunges us into the past, since the information comes to us through the intermediaries of radiation (electromagnetic or gravitational) or of particles (neutrinos, cosmic rays), travelling at a finite speed. However, this return to the past necessarily reaches a limit, corresponding to an epoch when no radiation source had yet formed. The observable universe is therefore only a portion of space-time, circumscribed by a *cosmological horizon*: it is the interior of a sphere centered on us, whose present-day radius is approximately 50 billion light years. Over the course of cosmic expansion, the radius of this spherical horizon grows as a function of time, at a speed that depends on the chosen cosmological model.

Let us keep these different meanings in mind. It is probable that, in the rest of this book, I will now and again use the word *universe* in place of the word *space*, simply for ease of discourse, and only in those places where there is no risk of error in the interpretation that follows.

A world is a circumscribed portion of the universe, which contains stars and earth and all other visible things, cut off from the infinite, and terminating in an exterior which may either revolve or be at rest, and be round or triangular or of any other shape whatever.

—Epicurus

2
The Size of Space

The history of ideas about the shape of space stretches over 25 centuries. It is full of surprises, developments, and, sometimes, steps backward; it provides us with deep insights as well as picturesque anecdotes. However, it has long been confined to a single question: is space finite or infinite? In different cultures and different epochs the response has oscillated, in a hesitant waltz, between these two extreme visions of the world, that arise as much from philosophy and aesthetics as they do 143→ from astronomy.

An episode from antiquity will help illuminate the type of reasoning used by the ancient sages. In a treatise entitled *The Sand Reckoner*, the geometer Archimedes discussed the size of the Universe according to the hypotheses of the Greek astronomers. To do this, he proposed to write the number of grains of sand that would fill the volume of the world. For most of the ancients, the world was not limited to the Earth, but extended to the entire Universe, encompassing the Sun and the planets, out to the fixed stars. The latter were thought to be situated all at the same distance, golden nails of some sort set in an ultimate sphere encircling the entirety of space. Archimedes thus wanted to calculate the volume of the world by counting the grains of sand that one would have to amass to fill it up entirely. By taking as the diameter of the world the one assumed by the geocentric system of Aristotle and Eudoxus, Archimedes obtained a number of grains equal, in modern notation, to 10^{51}. However, his contemporary Aristarchus of

Samos had conjectured a much more extensive world than that of his predecessors. He had, in effect, proposed a heliocentric world system, in which the Sun was immobile at the center of the Universe, with the Earth turning in a circle around it. In these conditions, the stars needed to be extremely far away. If they ← 192 were not, through a perspective effect called "parallax," their position in the firmament would change over the course of the year. By adopting the diameter of the world given by Aristarchus, Archimedes found 10^{63} grains of sand. Finding this number to be unreasonably large, he concluded that Aristarchus had overestimated the size of the cosmos and that he was mistaken in his cosmological hypotheses.

Modern astronomy has taught us that not only was Aristarchus correct as far as the movement of the Earth, but that he had also foreseen the immensity of the cosmos. The Universe is incredibly greater in extent than anything that had been imagined by the scholars of antiquity. The total number of atoms contained in the observable universe is estimated to be approximately 10^{79}. This value is very easy to calculate: knowing that the universe is made up essentially of hydrogen atoms, it suffices to multiply Avogadro's constant (the number of hydrogen atoms contained in one gram, about 10^{24}) by the mass of the Sun (about 10^{33} grams), by the average number of stars in a galaxy (around 100 billion, or 10^{11}), and by the number of galaxies in the universe (also 10^{11}).[1] Moreover, the hydrogen atoms do not touch each other in the manner of Archimedes's grains of sand; there is only, on average, a dozen or so per cubic meter of space. The volume of the observable universe is therefore 10^{78} cubic meters, which corresponds to a sphere having a radius of approximately 10 billion light years.

Archimedes, like most of the astronomers in antiquity, believed that the Universe was finite. When we speak of a finite Universe, many people have in mind a bubble of matter embedded in a sea of nothingness. They imagine a zone with matter, including some stars, that extends to a certain frontier, beyond which there would be nothing more than empty space. This picture is erroneous, in more than one way. First, there is probably no physical space without matter or energy; it is impossible to prove this, of course, because one cannot remove all forms of energy from the Universe in order to verify that space indeed disappears with them. Nevertheless, the theories of general relativity and quantum mechanics, the twin pillars of modern physics, reinforce this idea. In addition, space may perfectly well be finite without having a frontier. This is what we learned with the ← 161 nineteenth-century discovery of new geometries and non-Euclidean spaces with puzzling properties.

[1]The notation in powers of ten allows us to reduce multiplication to simple addition: $10^{24} \times 10^{33} \times 10^{11} \times 10^{11} = 10^{24+33+11+11} = 10^{79}$.

One may better grasp the notion of a space that is both finite and without boundary by considering an analogy given by George Lemaître—the father of the big bang—in one of his lectures entitled "The Size of Space." We may simplify the reasoning by passing from three to two dimensions; in other words we reduce the volume of space to a surface, for example, that of the Earth. This surface has a finite extent with a measurable surface area, just as ordinary space has a volume. It is a mosaic made up of unequal, juxtaposed pieces, e.g., countries and oceans, each piece having a precise frontier. However, if one progressively combines all the countries, all the continents, and all the oceans, the frontiers are increasingly pushed back, until they disappear completely, because the union of all the different parts has been realized. The surface of the Earth has no frontiers. Why should it not be the same for the entirety of space in three dimensions? For this analogy to work, we must forget that we can detach ourselves from the terrestrial surface and move in the third dimension. Indeed, we cannot escape ordinary space through some fourth dimension and examine it from the exterior.

It is customary to think that physics only accesses some minute part of reality. As far as cosmology is concerned, when one contemplates the Universe, one might think that the observable universe—that which is brought to our view or to our telescopes—is necessarily a small fraction of the *real* Universe. However, this idea could be false. We shall see that the wraparound models of the Universe, equipped with their exotic topology, offer a paradoxical alternative where the real space is smaller than the observable space! Some cosmic objects which are apparently situated ten billion light years away may fill a space measuring only three billion light years across. How is this possible?

There an immense room, through a hundred windows
Of limpid crystal, takes in the light of day
And in a beautiful decor of fine mosaics
It holds the face of the universe.

—Giambattista Marino, *L'Adone*

3
The Hall of Mirrors

Anyone who visits the Hall of Mirrors in the Chateau of Versailles, or even the most modest hall of mirrors at a carnival, is captivated by the trick played by the mirrors. Everyone marvels at the optical illusion caused by the mirror images, appearing like phantoms.

The mirrors do more than just reflect images, they also contain certain secrets about the infinite. Everyone knows that by covering the walls of a room with mirrors one gives the illusion that it is much larger. Let us place ourselves between two parallel mirrors—a classic situation in a hair salon. We see ourselves alternately from the front and from the back, repeatedly and nearly to infinity. Nearly, because in practice the infinite is not attained: the smallest lack of parallelism between the mirrors transforms the system into a great kaleidoscope. A kaleidoscope is made from a cylinder within which a set of small mirrors is arranged at various angles, which produce multiple combinations of images. In a hair salon, the imperfection in parallelism distributes the mirror images along a circle until, after a certain number of reflections, they ultimately disappear from the mirror.

Let us now imagine a room panelled with mirrors on all four walls, and place ourselves somewhere within the room; a kaleidoscopic effect will be produced in the closest corner. Moreover, the repeated reflections of each pair of opposing mirrors ceaselessly reproduce the effect, creating the illusion of an infinite network extending in a plane. This paving of an infinite plane by a repeating design is called, in mathematics, a *tessellation* (*tessella* being the name for a mosaic tile).

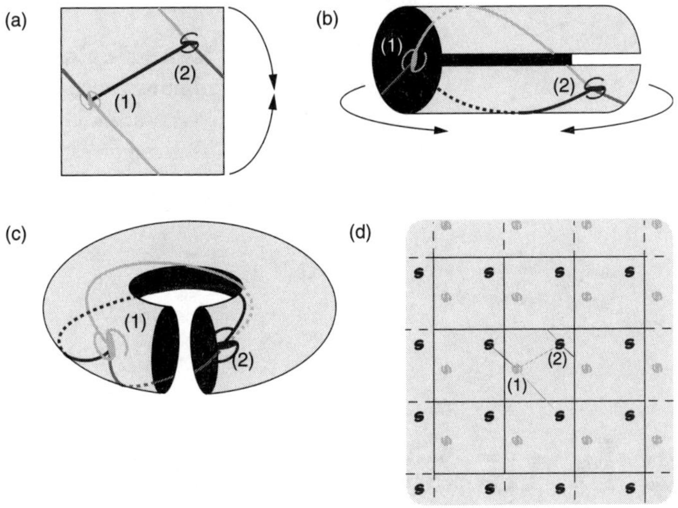

Figure 3.1. Mirror games. A very simple, so-called toric universe in two dimensions (c) shows how an observer situated in galaxy (1) can see multiple images of galaxy (2). This model of a "wraparound" universe is constructed by starting with a square (a), whose opposite borders have been glued together (b): everything that leaves on one side reappears immediately on the opposite side, at the corresponding point. The light from galaxy (2) reaches galaxy (1) by several distinct trajectories, because of which the observer in galaxy (1) sees numerous images of galaxy (2), spread in all directions across the sky. Although the space of the torus is finite, a being who lives there has the illusion of seeing a space that, if not infinite (in practice, there are horizons that limit the view), at least seems larger than it really is. This fictional space (d) looks like a network constructed from a fundamental cell that endlessly repeats each of the objects within the cell.

Finally, let us consider a room panelled with mirrors on all six surfaces (now including the floor and the ceiling). If we go into the room, the interplay of multiple reflections will immediately cause us to have the impression of seeing infinitely far in every direction, as if we were suspended at the top of a bottomless pit, ready to be swallowed, at the least movement, in one direction or another.[1]

Cosmic space, which is seemingly gigantic, might be lulling us with a similar illusion. Of course, it possesses neither walls nor mirrors, and the ghost images would be created not by the reflection of light from the surface of the Universe, but by a multiplication of the light trajectories following the folds of a wraparound universe (see Figure 3.1).

[1]The architect Serge Salat has designed astonishing installations where the spectator loses all spatial landmarks, and has the impression of entering a crystalline space that repeats out to infinity in every direction. See [Salat 02].

Call Space to the rescue, the old gouty vulture
who leaves behind himself like trails of slime
the white ribbon of the roads and the great arcs
of the horizon, like immense rounded slugs!

—Filippo Marinetti, *L'aeroplano del papa*

4

The Four Scales of Geometry

The shapes found in nature are limited by certain constraints. First and foremost are those imposed by the three-dimensional character of space.[1] Space is not a passive background; it has a structure which influences the form of everything that exists. Its architecture dictates the rules according to which every material form that resides therein must behave.

The real architecture of space and the constraints that it imposes are still unknown to us. If he could study the buoyancy and pressure, as well as the laws of fluid dynamics and the other constitutive properties of his natural space, doubtless the fish would have a different perception of his environment! In fact, our species has this capacity: by studying the large range of abstract spaces proposed in mathematics, by studying their curvatures, their transformations, and their expansions, humankind succeeds at better understanding the universe. Certainly, most minds

[1] I speak here of the usual three dimensions of length, width, and height, acknowledging that some recent theories postulate the existence of supplementary spatial dimensions, which would only be perceptible at short distances.

are incapable of picturing a non-Euclidean space, but geometers can give them coherent mathematical descriptions. Renaissance painters worked extensively on the representation of space in order to improve their technique. Faced with the difficulty of projecting three-dimensional space onto a planar surface, they invented perspective, thus mathematizing perceived space by means of the laws of geometry. This particular type of geometry was given the name "projective" in the seventeenth century by Girard Desargues. Leon Battista Alberti, Piero della Francesca, Paolo Uccello, and Albrecht Dürer were the true geometers of their time.

Which mathematical space is capable of correctly representing real space? The problem is much more complicated than it seems. The microscopic and macroscopic worlds differ profoundly from the immediate space of our surroundings. The question of the geometric representation of space presents itself at four different levels, or four "scales," as the physicists say: microscopic, local, macroscopic, and global. What exactly does this mean? Let us return to the image of the universal mollusk. At the macroscopic level, the surface of the mollusk seems softly curved, supple, and regular. If we look at it through a magnifying glass, the curvature becomes imperceptible, and at the local scale each small neighborhood of a point on the surface of the mollusk could be identified with part of the plane. If one now uses a microscope, one discovers numerous irregularities and overall roughness on the surface of the mollusk. Finally, by moving as far away as necessary, one can see the global, overall morphology of the animal. Is it a slug, a snail, or a squid? Any observations that were made at the other resolutions teach us nothing about this question.

It is somewhat the same with the structure of space. At the local scale, e.g., on distances between 10^{-18} meters (a billionth of a billionth of a meter, the length currently accessible to experiment) and 10^{11} meters (100 million kilometers, on the order of the Earth-Sun distance), the geometry of space is described very well by that of ordinary three-dimensional Euclidean space, without curvature. This in turn implies that this mathematical structure serves as a natural framework for those physical theories which, like Newton's classical mechanics and special relativity, allow for the correct explanation of nearly all natural phenomena. Euclidean space gets its name from the Greek geometer Euclid who, in the third century BCE, collected the entirety of the mathematical knowledge of antiquity into the thirteen volumes of his *Elements.* Euclid's work deals with the shape of geometric figures, their interrelations, their dimensions, and their proportions. For a long time, Euclidean space was the only space known to mathematicians, to the extent that the eighteenth-century philosopher Immanuel Kant maintained that Euclid's postulates were inherent to human thought. However, he wrote this at just that moment when the Euclidean era was reaching its end. 143→

At the macroscopic scale,[2] roughly speaking between 10^{11} and 10^{25} meters, the geometry of space is better described by a non-Euclidean geometry, or more precisely by a continuous Riemannian manifold (a three-dimensional generalization of a surface with variable curvature), curved to a greater or lesser degree by the massive bodies that it contains.[3] In Einstein's general theory of relativity, the Universe is better described by a supple and elastic fabric; space, coupled with time, is a harmonious anatomy made up of curves, holes, and bumps, giving rise to the phenomenon of gravitation.

At the infinitesimal extremity of scale, at distances smaller than 10^{-18} meters, we enter into the realm of microscopic space, as yet unexplored by experimental physics. Its most intimate structure is inaccessible to present-day methods of investigation, which include powerful electron microscopes and particle accelerators of very high energy. There are only some speculative theories that propose geometric models for this scale. Microscopic space could display distinctive geometric properties. What is it really made of? Do grains of space exist, like the grains of energy that exist in quantum physics? At the beginning of the twentieth century, the chemist Dimitri Mendeleiev, famous for having designed the periodic table of elements, put forth the idea that space was made up of particles one million times smaller than hydrogen atoms, and that atoms were nothing but combinations of these particles. Inventive theorists like Paul Dirac and John Wheeler have developed this idea and have likened space to a collection of grains or soap bubbles. In this picture, space no longer has the passivity of a simple coordinate system. It is an active participant giving birth to the material world, a magic substance whose curvature, granularity, and excitations determine the mass and electric charge of particles, as well as all the interaction fields to which they are subject. For example, space could be shaken by fluctuations that constantly modify its shape and render it extremely complex: unstable, discontinuous, and chaotic. Space could even reveal extra hidden dimensions.

These highly speculative subjects are presently being developed and make up one of the great challenges for the physics of tomorrow. I will not discuss these in the present work, because what interests us here, first and foremost, is the widest perspective from which the universal fabric may be viewed. And there, just as many surprises lay in wait. It is not yet known if, at the cosmological scale, space is infinite, with a negative or null curvature, or if it is finite with a positive

[2] In everyday language, the terms "microscopic" and "macroscopic" that I use here to describe the geometry of space do not have the same meaning. A bacterium is microscopic because it cannot be seen with the naked eye, while a flower is macroscopic. However, according to my definitions, a bacterium and a flower are both connected to the local scale of space.

[3] In the neighborhood of exceptionally massive or dense bodies like black holes, the effects of the curvature can be detected over distances of a few meters.

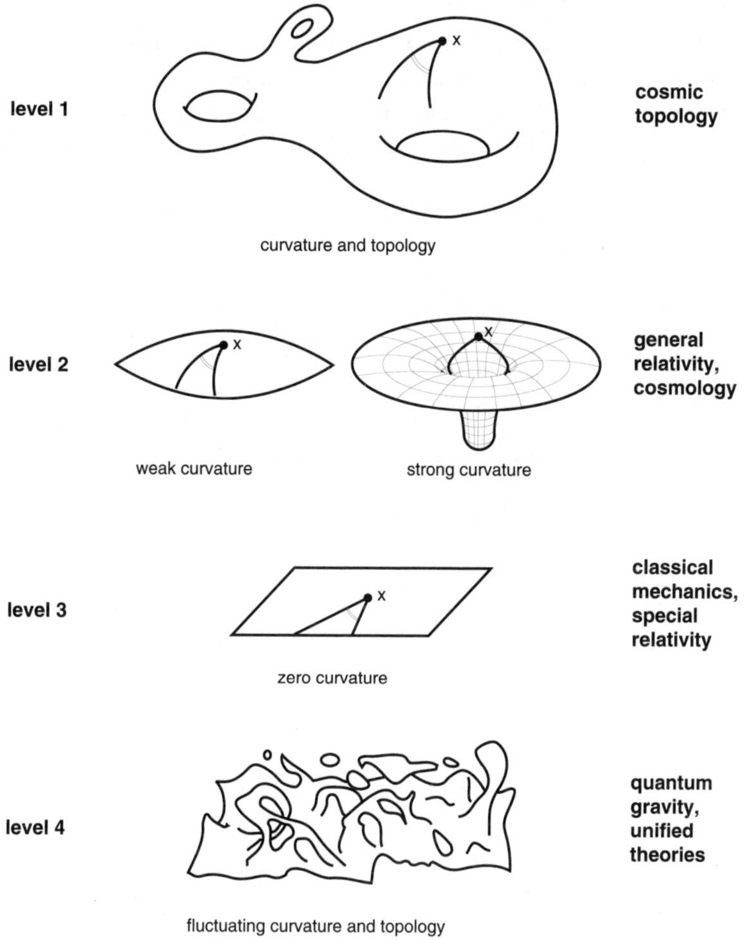

Figure 4.1. The four levels of geometry. The levels of geometry can be defined through what is seen by an observer whose eye gets closer and closer to a point in space. We examine the shape of space with a telescope of variable magnification, centered on a chosen point x. At magnification 1, the field of view includes the global geometry of space, whose shape could be extremely complicated (level 1). When the magnification grows, a smaller portion of the space appears in the field of view, which looks like a more or less curved disc; this is the space of macroscopic geometry (level 2). When the magnification grows even more, the disc is reduced to a smaller and smaller neighborhood of the point, its curvature becomes imperceptible and it seems flat; this is the space of local geometry (level 3). At nearly infinite magnification, we see the intimate microscopic structure of space, which may once again become very complicated (level 4). In the column on the right, I have indicated the various theories of fundamental physics that, at the present time, seem to be in the best position to describe correctly the corresponding level of magnification.

curvature, like a multidimensional sphere. The most intriguing possibility to consider is that space could be *wrapped*. For example, a Euclidean or negatively curved space could nevertheless be folded in on itself such that its volume remains finite. To deal with these global aspects of space, a new discipline is necessary, mixing mathematical advances and subtle cosmological observations: cosmic topology—the subject of this book.

When, in a given bedroom, you change the position of the bed,
can you say you are changing rooms, or else what?

—Georges Perec, *Espèces d'espaces*

5 Absolute or Relative Space?

Speculations on the nature of space, in continuous development from antiquity to the present day, can be classified into two large categories: those of absolute space and those of relational space, depending on whether space is thought to feel the influence of the matter that it contains. Both types of speculation have alternately had the favor of physicists. At the beginning of the eighteenth century, for example, the absolute space advocated by Newton triumphed over the relational space supported by Descartes and Leibniz. At the beginning of the twentieth century, the pendulum swung the other way. With the theory of general relativity, our conception of space and time has experienced a profound refashioning. At first glance, the change seems minor, appreciable to at most only a handful of specialists: Newton's law of universal attraction was replaced by another theory of gravitation, Einstein's general relativity. But since the latter has direct implications for the nature of space and time, the landscape of the Universe changes completely.

In the Newtonian vision of the world, space is a background, absolute and given a priori, in the sense that space intervals are the same for all the observers that measure them, whatever their position and speed may be. This space is

Figure 5.1. The Newtonian Universe. This view of Newtonian cosmic space allows one to visualize the structure of infinite Euclidean space, woven through with light rays following rectilinear trajectories. Planets, stars, and galaxies move along orbits curved by the universal attractive force of gravity. (Image from the film *Infiniment Courbe*, © Laure Delesalle.)

described by the simplest possible geometric structure (and the only one known to Newton): that of a three-dimensional Euclidean space of infinite extent. Time is also absolute, represented by a one-dimensional mathematical space: a line oriented from the past toward the future. It is in this preestablished framework of space and time, fixed once and for all, that matter moves. To explain the motion of bodies, however, it is necessary to introduce a supplementary concept, that of force. Here is why: a body in motion, if it is not subject to any force, must follow a straight line in space. Now, we observe that the trajectories of bodies subject to gravity are not straight: a comet passing near the Sun, for example, deviates from a straight line, and the planets trace out ellipses. The problems of celestial mechanics therefore reside entirely in some kind of physical action: we must introduce forces which pull the system away from its natural trajectory. Newton's idea was to imagine that all massive bodies, whether an apple falling from a tree, the Moon orbiting around the Earth, or the Earth revolving around the Sun, were experiencing a physical action of the same nature: a universal, attractive gravitational force. According to him, every massive body exerts an attractive action-at-a-distance on other bodies. This explanation of gravity in terms of force is added to the absolute structure of space and time. In the same way, all of the laws of classical physics, including those developed later, in the nineteenth century, to describe electric and magnetic phenomena, are defined within this frozen system: every body, whether celestial or terrestrial, moves under the action of forces within a relatively meager geometric framework.

The much richer vision of the world laid out by Einstein, as summarized by the two theories of relativity, is radically different. These theories are most easily explained in terms of pure geometry .

Special relativity, which deals with the motion of bodies in the absence of gravity, is still content with a geometry without curvature, but space and time are now combined into a four dimensional structure: space-time. The equations which describe, for example, phenomena as curious as the apparent dilation of time intervals and the contraction of lengths, reduce essentially to the Pythagorean theorem generalized to four dimensions, up to one subtlety: the time coordinate t is treated differently than the spatial coordinates (x, y, z), with the square of the time interval carrying an additional minus sign: $s^2 = x^2 + y^2 + z^2 - c^2t^2$, where c is the speed of light. We say that space-time has a "signature" $+ + + -$ (three spatial dimensions, one time dimension) and that its geometry is pseudo-Euclidean (the purely spatial part reduces to a Euclidean space).

General relativity deals with gravity. This no longer is described as a force acting at a distance, but as a local manifestation of the non-Euclidean geometry of space-time. Einstein chose in effect to eliminate the physical action for explaining gravity, paying a price which translated into a needed enrichment of the underlying geometry. This leads one to consider non-Euclidean spaces, in which the straight lines are no longer the usual ones. The deviation in the trajectory of a comet is interpreted as a consequence of the fact that the space-time geometry is curved by the presence of the Sun. The trajectory followed is still a line, but now in a curved geometry; a taut wire would follow this path.

Relativistic space-time is thus endowed with curvature. When one measures distances in this curved universe (using a mathematical construction called the metric), the Pythagorean Theorem turns into a more complicated formula. The signature is still $+ + + -$ (three spatial coordinates and one time coordinate), but 195→
the spatial part is a curved, so-called *Riemannian* space.

Anyone can make his own little model of elastic space-time. However, in order to facilitate the model's visualization and to be able to easily manipulate it in ordinary space, we must once more restrict ourselves to two dimensions; a piece of fishnet stocking, stretched and fixed on a rigid framework, will do perfectly.

Figure 5.2. The relativistic Universe. General relativity offers a new vision of the Universe in terms of a supple space-time, woven through by light and curved by matter. According to Einstein, celestial mechanics looks like a tennis match that is played not on a hard ground, as with Newton, but on a trampoline; the game is much more complicated because the players and, to a lesser degree, the ball, deform the terrain by their movements and displacements. In general relativity, the shape of space is more variegated than the Euclidean "dull plane."

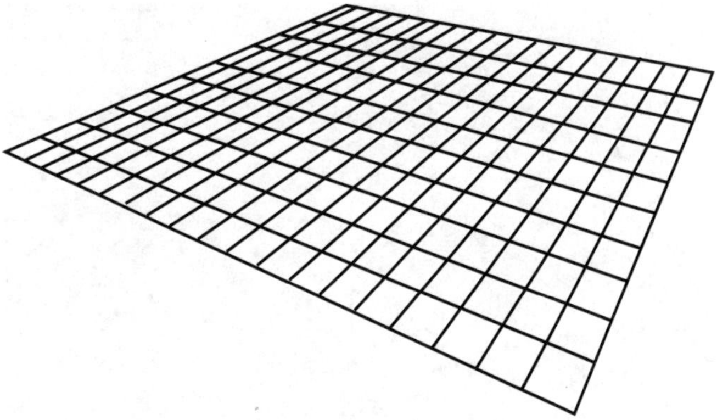

Figure 5.3. Flat elastic fabric. In the absence of mass, which is to say the absence of gravity, the fabric of space-time has no curvature. The trajectories of light rays trace out a rectilinear grid, which defines the Euclidean geometry of space.

I constructed this object for the first time a decade or so ago, to be used in a program for Italian television in which I was to introduce my book on black holes, freshly translated into the language of Dante. My publisher was quite worried and had begged me to find a spectacular visual metaphor, so that the viewers would not immediately flee. Every visual analogy is defective in some way, but this is a game whose rules must be known and respected. Therefore, I went to a lingerie merchant in Rome. I asked him for a large pair of fishnet stockings, opened the packet and, to his horror, cut off a section of stocking with a pair of scissors, pulling it every which way to test the elasticity. The shopkeeper was not far from calling the *carabinieri* who was passing in the street to denounce this sadist .... Satisfied, I then went to a toy store where I bought balls and marbles of different masses, sizes, densities, and colors. That evening, my demonstration produced its little effect. Since then, I have regularly used this device in my public lectures and it never has failed to produce the desired effect. I therefore encourage you to construct this artifact of such astonishing pedagogical virtues.

For the first step, let us imagine the fabric in mid-air without any object placed on top. It has no shape, no structure. More precisely, it possesses the simplest geometric structure there is, that of the Euclidean plane. Here we see the Newtonian conception of space. The rectangular mesh materializes the trajectories of light rays. Its weave is a universal system of coordinates that defines the architecture of space.

We recall that Newton postulated a rigid, absolute space, one that is not influenced by the material bodies found within. The container is inflexible, whatever

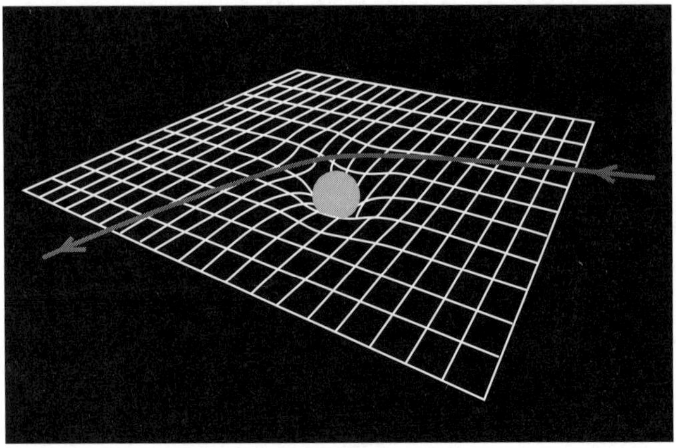

Figure 5.4. Curved elastic fabric. In the relativistic framework, gravity is an illusory force created by the curvature of the Universe. A massive star like the Sun imprints a dent into the elastic fabric of space-time. The trajectories of bodies naturally follow the curvature of the fabric.

the contents may be. Let us therefore place the elastic frame on a table, to impose inflexibility,[1] and place a ball in the center, to represent a massive body such as the Sun. Nothing in particular happens. If we add a small marble representing a planet and then push this marble, it will move in a straight line—in other words, it will follow the rectangular frame of the support. However, this is not what happens in nature; the planet should rotate around the Sun, following an ellipse. To explain this deviation with respect to its natural movement, we must call on a central force that keeps the marble from moving in a straight line. Moreover, this force acts from a distance (without direct contact) and instantaneously.

In Einstein's picture, the elastic fabric stays flat only if it is absent of matter. Let us therefore lift the frame from the table, with the Sun-ball still placed in the middle. What happens? The mass produces a hollow, a deformation of the fabric which is revealed by the fact that the rays of light themselves—the web of the fishnet—are curved around it. We see that the grid of the fabric embraces the curvature, and thus that space-time is still woven through with light ray trajectories. This leads to the idea that in the neighborhood of large masses, e.g., in the vicinity of the Sun, light ray trajectories are deflected by gravity. This curvature, just like the gravitational field, varies from point to point, so that it can be negative,

[1] A barbecue grill would in this case serve better, since it would not need a rigid support to stay inflexible. However, it would not be very easy to transport.

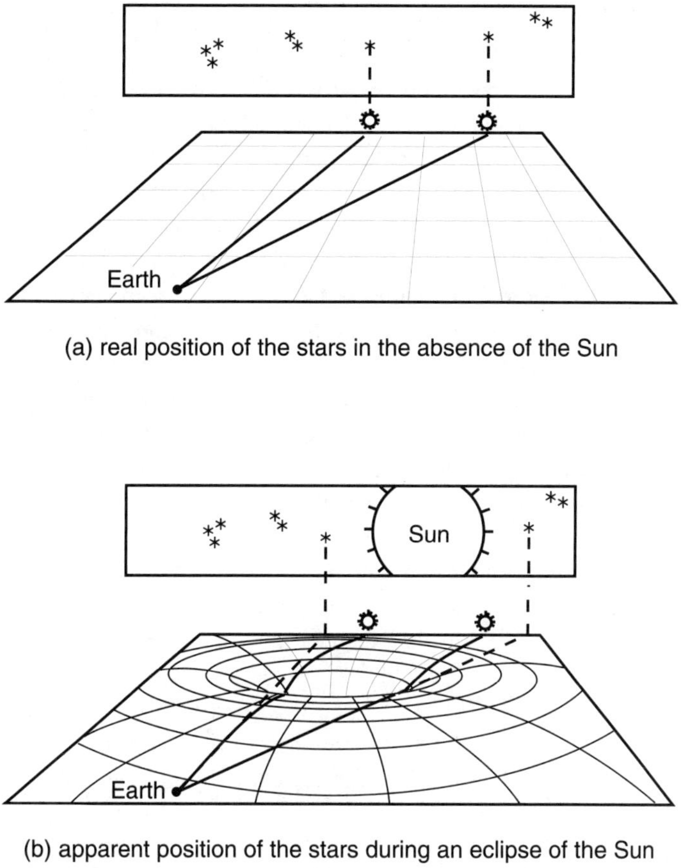

Figure 5.5. The deflection of light rays by the Sun. When the Sun is placed in front of a distant field of stars (visible only during a sufficiently long total eclipse), it curves the space-time in its vicinity and deflects the light rays coming from behind it. The apparent direction of the stars (b) is shifted with respect to their real position (a).

positive, or null. Thus the curvature of the fabric tends toward zero as one moves away from the central ball.[2]

In this framework, gravitation is not a force acting at a distance; it is only a local manifestation of the curvature. If we throw a little ball representing a planet onto the fabric, it will follow the natural slope of the fabric, without any need to be affected by a force of attraction (in real conditions, the phenomenon would be distorted through frictional forces, which would cause the marble to fall directly into the hole). A natural slope in a curved space is called a *geodesic.* It is the equivalent of a straight line in flat space, in other words, the line of shortest possible length between two points. Now, the geodesics of a space-time curved by a central spherical mass are viewed, in three dimensions, as ellipses. Moreover, these are ellipses whose major axes turn slowly over the course of time. General relativity thus explains, as a bonus, a phenomenon that had been observed by astronomers as early as the middle of the nineteenth century and that the Newtonian theory had left unexplained: the advance of Mercury's perihelion.

General relativity consequently forces us to revise the very concepts of time and space. The Universe no longer has an immutable Euclidean spatial structure, woven through by an independent time; it is a space-time deformed by the presence of matter and energy. In the neighborhood of a mass like the Sun, a star, or a galaxy, there is a curvature which measures the deviation of the geometric properties of the Universe with respect to the properties of Euclidean space.

One can manipulate the fishnet stockings further in order to explore the deeper content of the theory. We have seen that by placing a star, that is to say a ball of a certain mass, inside, we obtain a certain curvature. If, instead, we place a star having exactly the same size but a larger mass or, in other words, a higher density, the deformation of the fabric is much greater. This property of gravity is integrated into the equations for the gravitational field formulated by Einstein in 1915, which encoded the way in which the distribution and motion
←195 of the material bodies of the Universe determine the curvature of space-time.

[2]The experiment with the elastic fabric is only a highly simplified analogy, which one should take with a grain of salt; it is terrestrial weight, a gravitational field supplied from the exterior of the two-dimensional space of the tissue, that causes the marbles to deform their support, and not the gravity field of the marbles themselves. In fact, to properly represent space-time, we would need a soft three-dimensional "sponge" whose shape changes over the course of time. All the same, at the risk of displeasing the purists, the metaphor of the fabric is not too deceitful; so-called *embedding* diagrams are a mathematical technique allowing one to visualize the shape of a space by embedding it in a fictional space of higher dimension. This is what our brains do automatically when visualizing surfaces. This technique, applied to the representation of a space-time deformed by a spherical mass, gives a shape that basically reproduces that of the elastic fabric with a hollow produced by the weight of a ball—see chapter XII of my book *Black Holes.*

His eye is like the four-leaf clover of space.

—Blaise Cendrars, *L'Eubage, aux antipodes de l'Unité*

6 Celestial Mirages

Our piece of fishnet also allows us to understand easily why a space curved by gravity introduces optical deformations, and thus real celestial mirages. In certain situations, in fact, curved space multiplies the trajectories by which rays of light reach us. Let us place our eye at one of the corners of the fishnet and observe a faraway star that is located at the opposite corner. Now, along the line of sight, we have placed a massive body. Rays of light coming from the faraway star in the background of this intermediate body are, therefore, going to have curved trajectories and will reach our eye arriving from several different directions. (It is obviously necessary to imagine the phenomenon occurring in three dimensions.) The massive intermediate body could quite simply be the Sun. The first experimental test of the theory of general relativity was in fact provided by a solar eclipse, in 1919. For the first time, the British astronomer Arthur Eddington observed that the stars situated in the background, but near the border of the Sun, were deflected slightly from their usual position—that is to say their position without the Sun along the line of sight—by an angular distance that agreed with Einstein's theoretical calculation.

More generally, the massive object in the foreground could be some other entity, e.g., a star, a black hole, an entire galaxy, or a galaxy cluster. The sources situated in the background may be stars, very distant galaxies, or quasars. In ideal conditions—a distant luminous source, a massive intermediate galaxy, and a

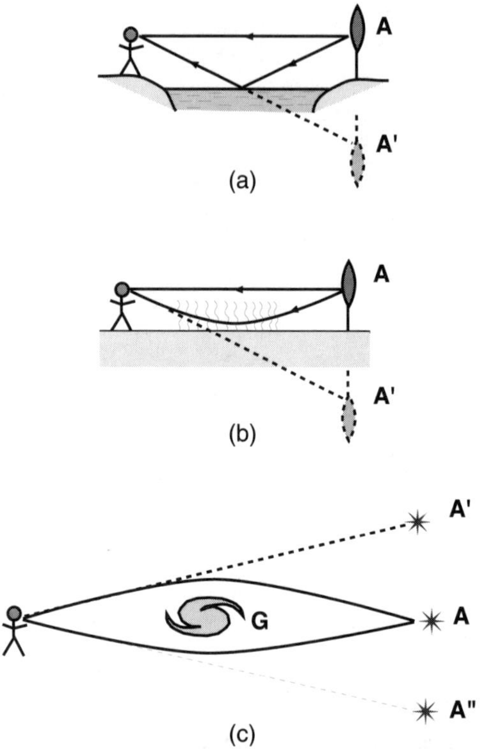

Figure 6.1. The principle behind mirages. (a) Light reflection from an aquatic surface situated between the observer and the object A causes two light rays issuing from A to meet once again; there is a doubling of the image. (b) In a terrestrial mirage, the light rays issuing from A are refracted and deflected while crossing overheated layers of air near the ground. Once again, there is a multiplication of images. (c) In a gravitational mirage, a massive body G situated between the observer and the luminous source A deflects, deforms, and multiplies the images of A.

terrestrial observer all perfectly aligned—the observer should see an image of the source that is deformed by the foreground galaxy into a ring of light. This is what is known as a *gravitational mirage.*

The space-time curvature created by an object along the line of sight, which functions as a gravitational lens, causes optical illusions for all objects situated in the background, because the curvature increases the number of distinct trajectories taken by light rays. It was not until 1979 that powerful modern telescopes could detect these gravitational mirages, not in the ideal appearance of a ring, but in that of a multiple quasar.

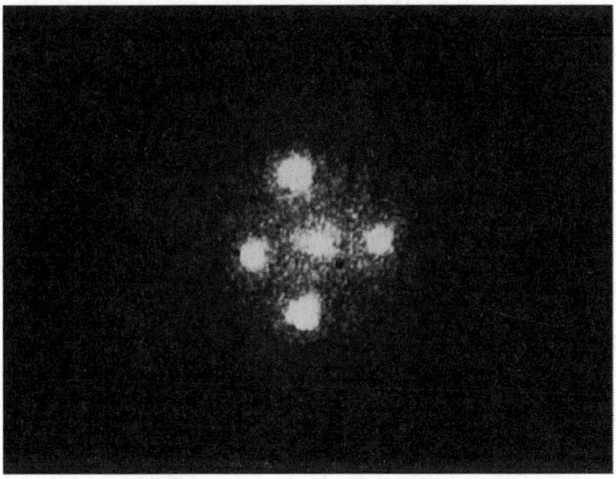

Figure 6.2. The Einstein Cross. This gravitational mirage was photographed by the Hubble space telescope. Four copies of a quasar, situated eight billion light years away, are observed because of a lens galaxy (at the center) which is twenty times closer. (Image from STScI/NASA.)

Figure 6.3. Gravitational arcs. The most spectacular mirages lead to the formation of luminous arcs. These arcs are the highly distorted images of distant galaxies, situated in the background of the galaxy cluster Abell 2218 (in yellow in the photograph). The gravitational effects of the cluster bend the light rays and distort the images, ultimately giving the images their characteristic arc shape. (See Plate I. Image from STScI/NASA.)

We now leave the fabric behind to admire an astronomical image taken with a telescope that shows the reality of curved space-time and its gravitational mirages (Figure 6.2). This superb cosmic object, called the Einstein Cross, shows four images of a single light source, a very distant light source situated in the background of a nearby galaxy. Why four images? Along the line of sight there is a massive lens, visible at the center of the cross, which is a spiral galaxy. There is therefore a multiplication of the images coming from the background. How can one be sure that they all come from the same quasar? Quasars are variable luminosity objects; their light curves—showing the variation of their luminosity over the course of time—are a veritable identity card. To be assured that the four images all come from the same quasar, it suffices to follow their light curves and to notice that they are identical, up to a small shift in time, corresponding to the different trajectory times for the light rays following the various contours of the curved space.

Today, several dozen gravitational mirages have been detected, all with different characteristics. One should like to see some perfect rings of light, but for this one would need a spherical lens placed exactly along the line of sight. However, in nature, the situation never is ideal; we must content ourselves with ring fragments. In certain cases which approach perfect symmetry, we observe gravitational arcs (Figure 6.3).

We shall see further on how *topological mirages*, created by the topology of a wraparound Universe, may produce image multiplications which are much more surprising than even the gravitational mirages.

Where a dark solar body, infinite and leaden,
was swallowing flames and suns, without becoming any brighter.

—Jean Paul Richter, *Der Komet*

7
Black Hole

The notion of the black hole is the ultimate consequence of the flexibility of space-time. Still equipped with our elastic fabric, let us imagine that we place a marble on top of it that is so dense that it forms a bottomless well, to the point that nothing, as soon as it has crossed the edge of the well, can escape. Is it a real hole? It is in any case an "infinitely" deep well, in the sense that it would take infinite energy to get out again, if one went wandering in the interior of a critical zone marked off by the edge of the well. In this situation, let's throw a marble in the direction of the neighborhood of the well, but not directly toward it, and give it enough speed; the marble is deflected, since it follows the curvature of space, but it possesses enough energy to escape from the well. Nevertheless, there is a critical angle below which the marble will be captured no matter its speed, even if we give it the speed of light, in other words even if, instead of a marble, we fire a photon, a corpuscle of light. The photon will also be captured by the hole. The relativistic definition of the black hole flows from this demonstration: it is a confined region of space-time from which matter and light cannot escape. The black hole is a natural and inescapable corollary of the theory of relativity.

Since the black hole causes extreme deformations of space-time, it also creates the strongest possible deflections of light rays passing in its vicinity and gives rise to spectacular optical illusions. One aspect of scientific research that has always particularly interested me is when a problem of visualization is posed. The

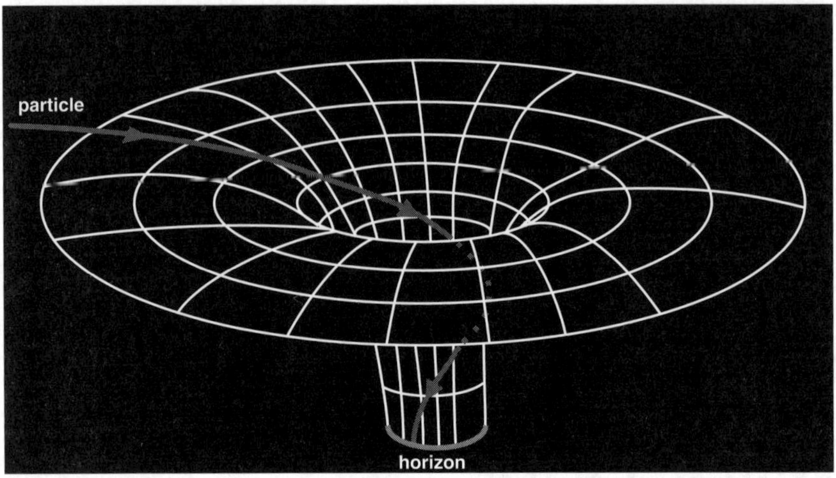

Figure 7.1. The well of the black hole. The black hole forms a gravitational well in the elastic fabric of space-time that is so deep that every particle and every light ray that crosses its edge is captured. The surface of the black hole, called the event horizon, has no material consistency: it is a purely geometric border around a zone of no return.

typical example is, of course, that of the black hole, a fascinating concept that is impossible, by definition, to visualize directly. By its very nature, the black hole is invisible: it lets neither matter nor light escape. One can only understand all of its properties by solving the appropriate equations from general relativity, and this understanding is unfortunately reserved solely for the specialists. If, despite all this, one wants to give a representation, one can, for example, imagine that this black hole is surrounded by hot gas, which lights up in a certain way and reveals the presence of the invisible. This question constituted one of my first research directions. At the end of the 1970s, I used a computer to calculate a virtual photograph of a black hole.

Today, it is possible to observe black holes indirectly, through the effects that they have on their environment: their powerful gravitational fields suck in the gaseous substance of neighboring stars into strange vortices, called *accretion disks*. These phenomena have been detected in some double-star systems that emit x-ray radiation (with black holes of about ten solar masses) and in the centers of numerous galaxies (with black holes whose mass adds up to between one million and one billion solar masses).

Figure 7.2 shows how a black hole surrounded by a disk of gas would look. The images experience extraordinary optical deformations, due to the deflection of light rays in the vicinity of the black hole. This is all explained perfectly by

(a)

(b)

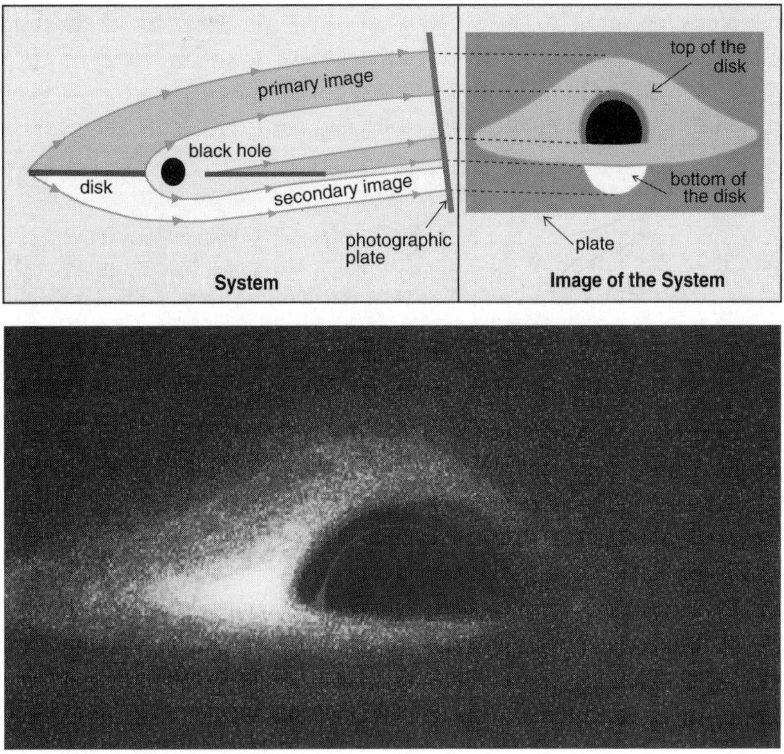

Figure 7.2. The black hole photographed. (a) A black hole, surrounded by a thin disk of gas and viewed from very far away in a direction weakly inclined with respect to the plane of the disk. General relativity allows the calculation of the deflection of light rays due to the strong curvature of the space-time in the vicinity of the black hole; it allows one to see all of the top of the disk and part of the bottom. (b) By accounting also for the physical properties of the gaseous disk, in particular rotation (which gives the left right asymmetry of the light flux), temperature, and emissivity, this simulation of the photographic appearance of a black hole surrounded by a disk of luminous gas was completed on the computer at the Meudon observatory in France in 1979.

general relativity. Let us take a black hole and a thin disk of gas viewed from the side, either by a distant observer or a photographic plate, if one wants to immortalize the setup. In an ordinary situation, meaning in Euclidean space, the curvature is weak. This is the case for the solar system when one observes the planet Saturn surrounded by its magnificent rings, with a viewpoint situated slightly above the plane. Of course, some part of the ring is hidden behind the planet, but one can mentally reconstruct their elliptic outlines quite easily. Around a black hole, everything behaves differently, because of the optical deformations due to the curvature of space-time. Strikingly, we see the top of the disk in its totality, whatever the angle from which we view it may be. The back part of the disk is not hidden by the black hole, since the images that come from it are lifted to some extent by the spatial curvature and reach the distant observer. Much more astonishing, one also sees the bottom of the gaseous disk. In fact, the light rays that normally propagate downwards, in a direction opposite to that of the observer, climb back to the top and provide a *secondary image*, a highly deformed picture of the bottom of the disk. In theory, there is a tertiary image that gives an extremely distorted view of the top after the light rays have completed three half-turns, then an image of order four that gives a view of the bottom that is even more squashed, and so on to infinity.

To describe this final image, no caption fits better than these verses by Gérard de Nerval, written a century and a half ago:

> In seeking the eye of God, I saw nought but an orbit
> Vast, black, and bottomless, from which the night which there lives
> Shines on the world and continually thickens
>
> A strange rainbow surrounds this somber well,
> Threshold of the ancient chaos whose offspring is shadow,
> A spiral engulfing Worlds and Days!

(Gérard de Nerval, "Le Christ aux Oliviers")

So much for the exterior of the black hole. What happens in the interior? Let us return to our elastic fabric. If we dig a sufficiently deep well to simulate a black hole, what will happen at the bottom? Is there a knot that blocks space-time (what in mathematics is named a singularity, where the curvature becomes infinite) or rather is the material pierced? In the latter case, is the opening a gaping hole, or does it look like a passage to somewhere else? How do these space-time distortions act on the elastic material? Only the mathematical solutions of general relativity can guide us. The most spectacular possibility, although it is quite speculative, is a strange geometric structure known as a *wormhole*: a tunnel space that would connect the bottom of the black hole to another region of the Universe called a *white hole*, since it behaves like the opposite of a black hole. According to some models, this region could even be located in another Universe. This mathematical solution of the equations of relativity opens singularly novel perspectives for space-time voyages, since they would ultimately offer space-time shortcuts permitting voyages of several light years in a few hours, all the while never exceeding the speed of light. Some trajectories could even go back in time!

However, let us not get carried away. The occurrence of this phenomenon, rendered commonplace by science fiction stories, is studied quite seriously in our

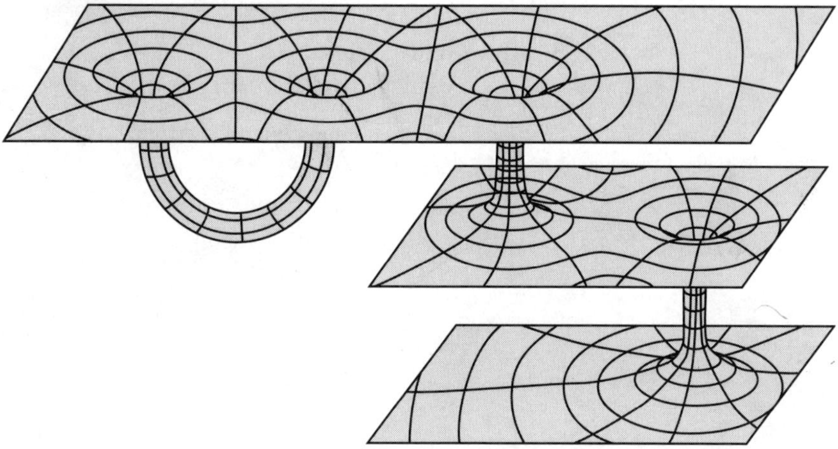

Figure 7.3. Wormholes. Wormholes could interconnect distant regions of the Universe and permit interstellar voyages in space and time. They could also connect an infinite series of parallel universes.

laboratories; the great majority of specialists estimate that in ordinary conditions, e.g., for stellar and galactic black holes, this type of space-time anomaly could not form. In fact, a wormhole is fundamentally unstable: as soon as it is created in idealized conditions, the tunnel would be destroyed by the slightest particle or faintest light ray penetrating into the hole. The models of physics deal with mathematical equations, but few of these solutions are actually pertinent for physics, that is to say for the description of the real universe.[1]

[1]Nevertheless, on the scale of microscopic black holes, governed by the laws of quantum physics, such a voyage would be possible and some elementary particles could furtively take these transitory tunnels in order to move back in time.

How many worlds, maimed and failed, have disappeared,
reforming and disappearing at each instant....

—Denis Diderot, *Lettre sur les aveugles*

8 Foam, Strings, and Loops

The preceding considerations on the shape of space have not taken us out of the intermediate scale, between 10^{-18} and 10^{25} meters, with the upper limit of this interval corresponding to about one tenth of the radius of the observable universe.

I shall briefly discuss the shape of space in the domain situated at the small end of this interval, that is to say, at the microscopic level. The distance of 10^{-18} meters is that which we have presently succeeded at exploring in the laboratory.[1] In reality, the experimental physicist uses particle accelerators to explore high energies, but energies are connected to distances: the greater the energy that one imparts to the particles, allowing them to experience collisions with other particles at speeds approaching that of light, the more finely one probes the spatial extent of these particles, and the more precisely one tests the detailed structure of space. At this scale, it seems that physical space can still be described by a continuous mathematical space—a *differentiable manifold.* But some researchers, such as Alain Connes in France, believe that the experimental results obtained at CERN already imply a more complex structure, in terms of a so-called non-commutative

[1]The future collider at CERN, the European Organization for Nuclear Research, will allow us to explore distances that are ten times smaller.

geometry. In this new approach, space is no longer defined in terms of points, characterized by precise coordinates that commute with each other, but in terms of interacting fields. What is certain is that if one considers distances that are much smaller, space must reveal an extremely complicated structure, necessitating a completely new geometric landscape.

Space cannot be understood without gravity. Its description at the subatomic level, where general relativity and quantum mechanics intersect, thus assumes the development of a quantum theory of gravity. This will be the major challenge for fundamental research in the twenty-first century. In the quest to understand quantum gravity, physicists have been generous with analogies and metaphors.

One of them, proposed in the 1960s by John Wheeler, is that of space-time foam. If we look at the ocean from an airplane flying at high altitude, we have the impression that it is smooth and flat, or at least weakly curved; we would thus describe it by a mathematically simple space: a section of the Euclidean plane or a portion of a spherical surface with uniform curvature. If we now move closer, we obtain a better resolution and we see that waves appear; the surface of the ocean remains connected, that is to say, all of one piece, and continuous, but it now reveals a curvature which varies from place to place, varying between troughs and waves. The ocean is dented. Finally, if we descend to the level of the waves, that is, if we scrutinize the surface of the ocean with maximal resolution, we can describe it only with the aid of an extremely complicated mathematical space, whose form varies constantly over the course of time. It is a bubbling foam that ceaselessly changes its spatial structure in an unpredictable way. The space is no longer even connected, since droplets now continually detach themselves.

To understand the structure of space at very small scales, and at the same time to unify the fundamental interactions according to theories such as quantum gravity, some physicists postulate that microscopic space could be similar to this ocean. If one had access to a resolution sufficient to scrutinize distances on the order of 10^{-35} meters, a characteristic length called the *Planck length*, the granular and fluctuating nature of space-time would be brought to light. However, the Planck length is as small compared to the size of an atomic nucleus as the atomic nucleus is compared to us.

This approach has gone barely farther than the stage of analogy, since the necessary mathematical tools, like the understanding of dynamical topology, are not sufficiently developed. Also, in the 1980s, another hypothesis supplanted that of space-time foam: string theory, which stipulates that the fundamental constituents of matter are not point-like particles but open or closed strings on the scale of the Planck length, whose vibrational modes define particle properties. In this framework, space-time becomes a derived concept that only makes sense at a scale larger than that of the strings. String theory, which comes in five different

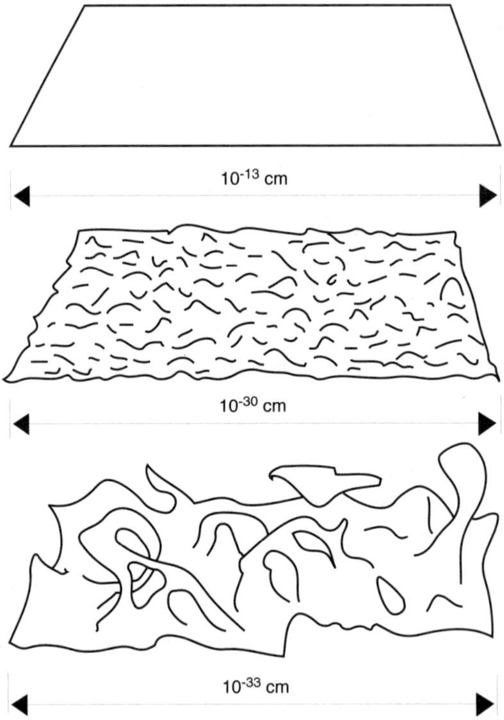

Figure 8.1. Space-time foam. At the microscopic level, space-time is well described by a smooth continuum, deprived of roughness like the surface of the ocean seen from high altitude. Approaching the quantum level, the fluctuations of space-time geometry should become perceptible, like waves on the ocean. At the truly quantum level (10^{-33} cm), the fluctuations would completely dominate the geometry, which would then become extremely tormented, in perpetual transformation like foam on the waves.

varieties, aroused such keen interest that in the 1990s, certain theorists believed it was capable of giving a *theory of everything*. However, the mathematical difficulties involved are formidable, and it is not certain that they will be resolved in the near future.

Recently, the five different string theories have given birth to a larger supposed theory, of which the string theories would only be limits; two-dimensional surfaces and other spaces of higher dimension, and not only one-dimensional lines, could vibrate, such as membranes—from whence the name *M-theory*, given to this unifying hypothesis.[2] Also in the 1990s, Abhay Ashtekar proposed another

[2]For some researchers, the letter *M* could just as well stand for "Meta" or "Mystery."

formalism, the theory of loops, in which the space-time continuum is again an approximation that holds at large distances, while at short distances the geometry would be a sort of weave, a complex interconnection of threads. We should also mention the twistor theory of Roger Penrose and the non-commutative geometry of Alain Connes, which are also rich in possibilities.

All of these approaches are not necessarily mutually contradictory, but their individual formalisms are so complex that it takes nearly a lifetime of research to penetrate and untangle any one of them, so much so that it leaves little time to try to establish a link with the formalisms of other approaches. Perhaps there is a major step to be taken by unifying these formalisms into one overarching theory, from which all the others could be derived.

In addition, most of these innovative conceptions lead, through constraints imposed by mathematical self-consistency, to models of space-time with more than three spatial dimensions. They bring completely new light to the birth of space and time during the big bang: quantum cosmogenesis. A bubbling ocean of energy is associated with the quantum vacuum. This energy generates certain fluctuations that are much larger than others on extremely short time-scales, by virtue of the time-energy Heisenberg uncertainty relation that, in quantum mechanics, constrains variations in energy to values that are as large as the time-scales on which they vary are short. In this quantum ocean, the biggest fluctuations of energy could spontaneously eject drops of foam which, once detached, would evolve according to their own laws, in the form of entirely separate universes. In such a scheme, developed in particular by Andre Linde, our universe would no longer be unique; it would make up part of a *multiverse*. Our Universe, whose structure and history astronomers laboriously try to reconstruct by making astronomical observations and building cosmological models, would thus only be one very particular bubble, which would have detached itself from the quantum vacuum about fifteen billion years ago—in time measured by its own proper clock. As for the multiverse, itself free of time and space, it would be a foam of chaotic universes ceaselessly regenerated, giving birth to bubbles of universes with perpetually changing interconnections, each bubble having different properties, including the number of spatial dimensions. The fascinating prospects of such scenarios could be developed in another volume, a logical complement to this one.

Lively, the nebulae were taking off, while forming a shrewd space.

—Raymond Queneau, *Petite cosmogonie portative*

9 From Relativity to the Big Bang

The rest of this book will be dedicated to the description of the other extremity of the spatial scale, that which deals with the global shape of space. To enter into such an imposing and complex subject while using everyday language is not without risk; as Paul Valéry wrote, "The great words: infinity, the absolute, nature ... these are the cardboard weights that the literary Hercules lifts, brandishes, and puts back down" [Valéry 60, 1429]. Let us nevertheless attempt the adventure and enter the realm of cosmology—the science that studies the structure and evolution of the Universe as a whole.

One day, after a seminar that I gave on the subject, an attendee approached me and asked me with a bewildered air:, "Okay, so, you do *cosmetology*?" At that moment, I admit I was surprised, if not offended, and I instinctively passed my hand through my hair to check that I had not used too much hair spray. Some time later, the company L'Oréal asked me to present to its personnel a quick overview of cosmic evolution. In front of a carefully made up and manicured audience, I had the following illumination: *cosmetology* and *cosmology* both derive from the Greek *cosmos*, which means "order, ornament, beauty."

The big questions posed by cosmology in fact concern the spatial and temporal organization of the cosmos. Investigations into the finite or infinite character of space, the beginning of time, the future destiny of the Universe, the constitution of cosmic matter, and the nature of the void have challenged the imagination of humankind for thousands of years. Today they have received some partial 173→ answers, thanks to a conjunction of theoretical and experimental progress.

However, having become a science in its own right only in the twentieth century, cosmology still has its adversaries, as much among physicists, some of whom believe that it relies more on myth than on experiment, as among philosophers, some of whom believe that there is no "good definition" of the Universe. However, professional cosmologists are quite resolved in exploring the world, even 169→ without a definition.

At the cosmological scale, the landscape of space and time is governed by gravitation. As such, models for the Universe are built on the fundamental theory that describes this interaction: general relativity. Relativistic cosmological models are particular solutions of the gravitational field equations that are capable of representing, not only a small region of space-time locally curved by a star, but the Universe as a whole, filled with its billions of galaxies, with its radiation, and with 195→ all of its other energetic constituents.

The equations of relativity are quite complex. Mathematically, they possess a great number of solutions; but not all of them are possible at the physical level. Some of them, for example, violate the basic laws that the physicist is nowhere close to being ready to abandon, like causality; others can be eliminated because they are in flagrant contradiction with observations. Alas, the solutions still remain numerous, and the task remains arduous. Cosmologists thus begin by studying the simplest models, those that assume that there are no privileged positions in space. This *homogeneity* presupposes that physical features in the Universe (for example, the average distribution of matter) are the same at every point in space, although they can change over the course of time.

Furthermore, cosmologists assume that space is *isotropic* (*iso* means "the same" in Greek, and *tropos* means "turn"), which is to say that there is no privileged direction. They thus eliminate universes that would be in rotation about a particular axis, as well as universes that would expand or contract at different speeds in the different directions of space.

These two symmetries, homogeneity and isotropy, do not necessarily come as a pair. An unending grassy field, whose blades all grow in the same direction, or a universe where all galaxies would have parallel axes of rotation, is homogeneous without being isotropic; an ovoid surface is inhomogeneous (its curvature varies) but at the two ends, the surrounding space is isotropic (the curvature is the same in all directions).

Let us examine the fundamental combined hypothesis according to which space is both homogeneous and isotropic. Isotropy is justified, at least in our local patch of space, by observation of the background radiation, the cooled relic from the big bang, characterized by a background temperature that has proven to take ←231 the same value, up to a precision of nearly one in a hundred thousand, in every direction around the observer.

Homogeneity, on the other hand, may seem to be a less realistic hypothesis. In the first place, to check the homogeneity of space, one would need to be able to move instantaneously through the Universe, which is impossible. The immobile observer sees the Universe at different epochs of the past, and thus can not experimentally prove homogeneity. Even neglecting this temporal factor, we observe that at the small scale, that is to say at the spatial scales probed by investigations of galaxies, space presents no trace of regularity. The galaxies are grouped together in clusters, and the clusters in superclusters, separated by great zones that are apparently empty (Figure 9.1).

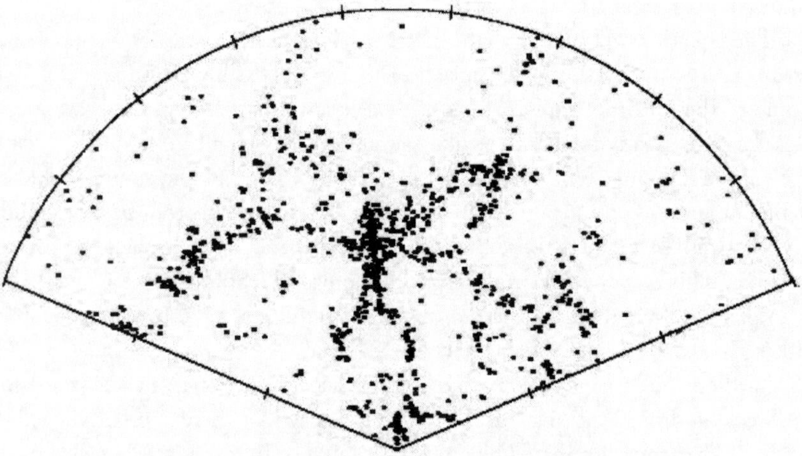

Figure 9.1. The cosmic Great Wall. Galaxies are grouped in clusters, and galaxy clusters are assembled into larger structures, the superclusters. The slice of space represented here includes more than a thousand galaxies, grouped in filaments organized around empty bubbles. A strange structure, in the form of a stick figure viewed face on, with legs and arms spread out, stands out 300 million light years from Earth. The arms of the stick figure stretch out for more than 500 million light years, but measure no more than 15 million light years in thickness—from whence the name "Great Wall." (Image adapted from M. Geller et al.)

Nevertheless, on a scale larger than a few hundred million light years, corresponding to volumes sufficiently large to encompass several galaxy clusters, it seems that matter is spread out in a statistically uniform manner.

More generally, it is reasonable to presume that physical space is equivalent at every one of its points. This hypothesis, known as the *cosmological principle*, does not result only from the need to simplify things. It constitutes an extrapolation to the scale of the universe of the Copernican idea that we are not at the center of the world: our position is insignificant, and there is no reason to think that other positions would be any different; as a consequence, all positions are equivalent. Furthermore, if space is homogeneous and isotropic around us, it is isotropic at every point.

Homogeneity and isotropy translate mathematically into the fact that space must possess a constant average curvature. In fact, since curvature is produced by matter, the fact that the latter is distributed uniformly implies that the curvature on a large scale must itself be uniform. To better visualize this, let us return to the metaphor of the elastic fabric. In the case of the solar system, which is a minute portion of the Universe, we only needed one marble standing in for the Sun. This central body produces a hollow and a variable curvature that tends toward zero when one moves away to infinity. All around, the grains of sand that model planets have masses that are so small that they do not affect the curvature of the fabric. If we now wanted to represent the Universe as a whole, we would

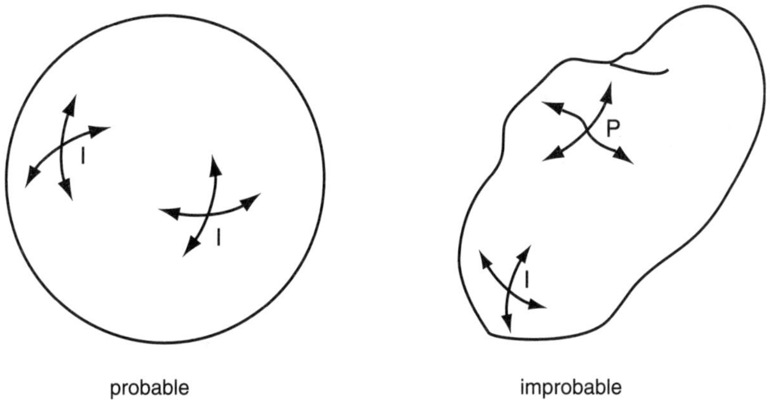

Figure 9.2. Illustration of the cosmological principle. On the surface of a sphere, all points are centers of isotropy I: the curvature is the same measured in every direction. On the surface of a "potato" only exceptional points are centers of isotropy. The cosmological principle forbids a privileged position of this type for any terrestrial observer: if the Universe appears isotropic to us, it must be so at every one of its points. It is, therefore, completely improbable that its geometry is anisotropic like the surface of a potato.

have to stretch out an immense surface, on which a great number of marbles representing galaxies and possessing roughly the same mass would be placed in a regular fashion. The distribution of these masses being uniform, we can easily imagine that the fabric would acquire a constant curvature.

Let us therefore accept this hypothesis of a homogeneous and isotropic Universe, with constant curvature. Three different cases then arise for cosmologists: this constant curvature can be positive, negative, or zero.

In 1917, Einstein constructed the first model of the Universe founded on his theory of relativity. He opted for a non-Euclidean geometry with positive curvature, allowing for a precise representation of a space that was both finite and ←154 limitless: the *hypersphere*, discovered by Bernhard Riemann a half-century earlier, which had the advantage of evading any edge paradox.

Nevertheless, by interesting himself only in the spatial aspect of the Universe with respect to its curvature, Einstein ignored its temporal aspect and committed, as he subsequently recognized, the greatest blunder of his life: he passed over cosmological dynamics. For Einstein's model assumed, in addition to the homogeneity of space, another hypothesis which seemed natural, namely that the circumference of the Universe did not vary over the course of time. Now, the deep philosophy of relativity is that space and time are, in a certain manner, inseparable. Applied to cosmology, this principle should logically lead to models for the Universe where the spatial structure is inseparable from its history. Actually, in parallel to Einstein's revolution, progress in observations led astronomers to announce the discovery of a strange phenomenon: other galaxies seemed to be systematically moving away from ours, at speeds which were larger the farther away they were. The spectrum of light received from them is shifted toward the ←179 red end, which is to say toward larger wavelengths, which can be interpreted to a first approximation as the effect of a generalized receding motion.

Einstein's static model, when confronted with experimental reality, therefore had to be abandoned in favor of dynamical models of the Universe that evolve over the course of time. Such models had, in fact, just been discovered by the Russian Alexander Friedmann and the Belgian Georges Lemaître. Both took notice of the fact that, since the very essence of general relativity was to couple the geometric properties of the Universe to its material contents, cosmic space, molded by matter, should naturally have varied over the course of time, that is to say, expanded or contracted globally.

Lemaître, in 1927, was the first to make the link between relativistic models and cosmological observations: it is space which is expanding at each of its points, and not galaxies fleeing from the terrestrial observer. The elastic fabric of space, in addition to being curved by the distribution of massive bodies, expands as a whole. The distance between two arbitrary points that are sufficiently distant to escape particular movements of non-cosmological character grows with time. Let us imagine some microbes distributed over the surface of a balloon that is inflating: each microbe might notice that its neighbors move away from it; it would thus have the impression, but only the impression, of being at the center 176→ of a universe where all the other galaxies are running away!

If these solutions of the relativistic equations naturally take into account the observed expansion of space, it remains to extrapolate the global evolution of the Universe, on the one hand toward the past (every cosmology is thus a cosmogony), and on the other hand toward the future.

As for the past, a film of the expansion played backwards leads inescapably to an earlier epoch where the expansion began, when the Universe was found in a highly concentrated and extremely hot state. Lemaître, inspired by the radioactive processes discovered 30 years earlier, imagined that the entirety of cosmic matter, before the beginning of expansion, must have been found in the form of a

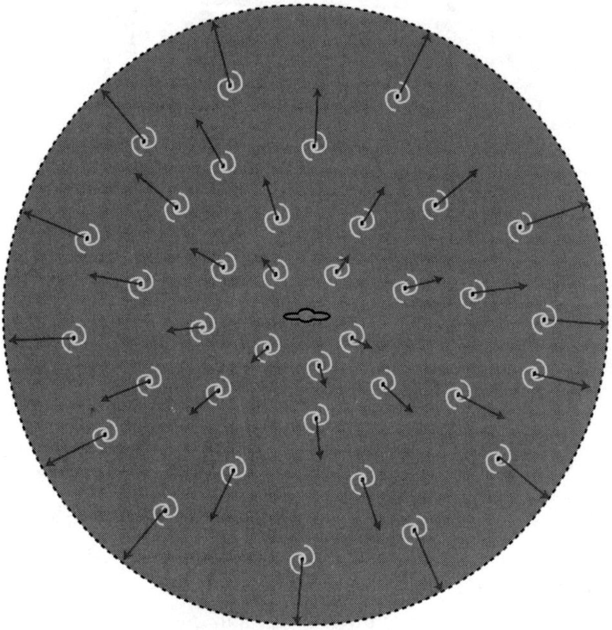

Figure 9.3. Does the recession of other galaxies imply an explosion? Around us, we observe that other galaxies systematically move away at radial velocities that are larger (as shown by the arrows), the farther away they are. Does this mean that we are at the center of a great explosion, as the figure suggests? No, because by virtue of the cosmological principle, at every point in space the same type of observation could be made. Space expands at every one of its points.

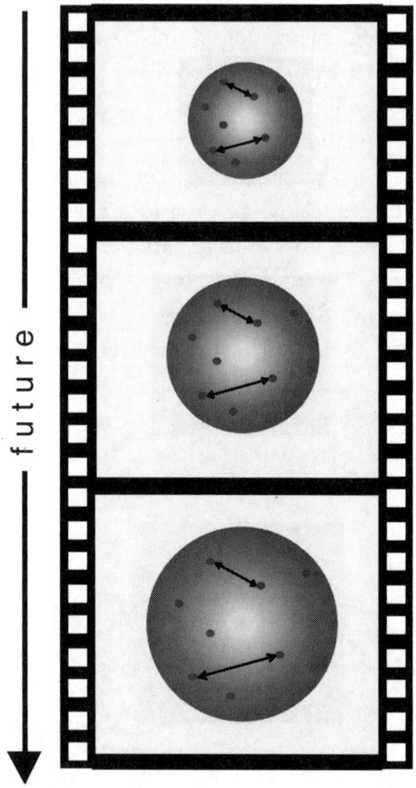

Figure 9.4. Snapshots of the expansion. The expansion of space is interpreted as a growth in the distance between two arbitrary positions in space (occupied, for example, by galaxies). These three instants during the expansion show a volume of space that expands over the course of time. The concept of the big bang rests on such a film running backward: in going back through time, the mutual distances between all points in space tend toward zero.

compact, ultra-dense nuclear fluid in a nucleus comparable to ordinary atomic nuclei, but much larger, and whose fundamental instability would have caused the expansion: the *primeval atom*. Lemaître's vision, revised and corrected by later progress in nuclear physics, gave birth to the famous big-bang models. The big bang corresponds to a singularity, in the mathematical sense of the term, where the curvature of space becomes infinite. As an extrapolation of general relativity, the big bang thus escapes the known laws of physics. A better description of this singular state, which could be obtained by a theory of quantum gravity, for example, constitutes one of the great challenges for research in the twenty-first century.

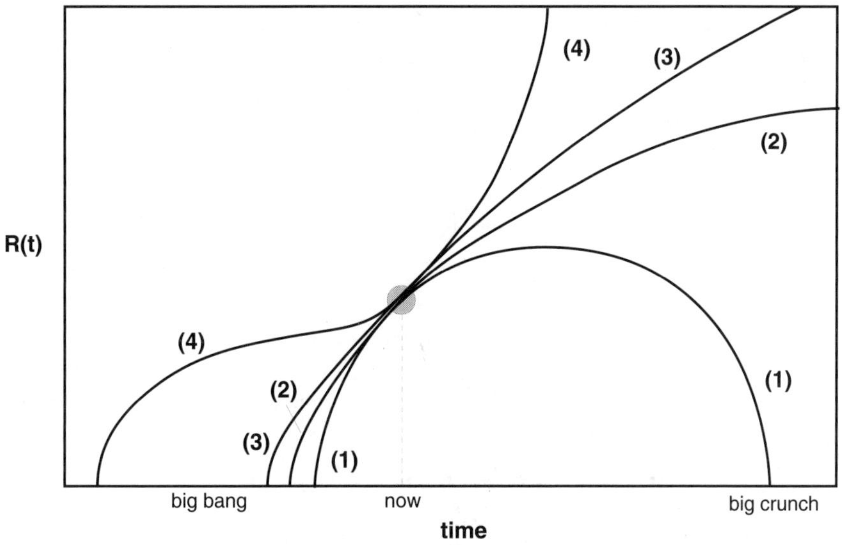

Figure 9.5. Big-bang models. Homogeneous and isotropic cosmological models, called Friedmann-Lemaître models, are classified by their curvature and their dynamics. Their essential characteristic is that they are solutions where space varies over the course of time. We represent (along the y-axis) a spatial scale factor, that is to say, the distance between two arbitrary sufficiently distant points, whose separation one measures over the course of time (the x-axis). The viable models of the big bang are the closed (1) and open (2) Friedmann-Lemaître models, the Einstein-de Sitter model (3)—a particular case with strictly zero curvature—and the Lemaître model of accelerated expansion, (4). The small circle represents the present epoch. To determine on which branch we really are, and therefore to extrapolate the ultimate destiny of the universe and its distant past, we must determine the density and the nature of cosmic energy. Present measurements tend to favor the model of accelerated expansion, (4).

As for the future of the Universe, two scenarios are possible: either the expansion slows down extensively, as space expands up to a maximal volume, then gives way to an inverse motion of contraction and reheating, finishing in a final squeeze of the Universe: the "big crunch"; or the present expansion continues forever, ceaselessly diluting and cooling the Universe. This second possibility itself subdivides into two, according to whether the expansion slows down without ever stopping, or, on the contrary, accelerates. All of these scenarios are predicted by Friedmann-Lemaître solutions and depend on the value of certain cosmological parameters, such as the curvature of space and the total density of energy. I will outline later on what the models predict exactly, and the partial answers provided by the most recent observations.

Since their invention, big-bang models have been the object of endless disputes between their supporters and their detractors. The historical controversies between Einstein, the father of general relativity, Friedmann and Lemaître, the fathers of the big bang, and Fred Hoyle, the famous adversary of big-bang cosmology, are the joy of epistemologists. Today, a large enough consensus has emerged on the validity of general big-bang models; rather, disagreements concern different variants, for example, the model of inflation (a theory coming from high energy particle physics according to which, in its primitive phase, space ←188 would have blown up fantastically quickly), the age of the Universe, the uniform character of the distribution of galaxies, and the question of the finite or infinite nature of space. In these impassioned debates, many have a hard time keeping a cool head. Scientific journalists, the media, and a fortiori the reading public rarely have all the technical data for the problem at hand. Regularly, articles with sensational titles are published that once again put into debate the supposedly established big-bang model. Now, on the one hand, there is an entire palette of big-bang models that differ in their detailed parameters, and on the other hand, *not one* serious experimental datum that until now has contradicted the basic scheme.

I am reminded of a story from my native Provence. A couple who pass their time arguing regularly take as their arbiter the old man of the village, who plays the role of the sage. Each time that the husband comes to see him to complain of the offenses of his wife, the wise man nods his head and says, “Yes, you’re right.” Each time that the wife comes to see him to complain of the offenses of her husband, the wise man nods and says, “Yes, you’re right.” One day the wise man’s own wife jumps from behind the curtain where she was hidden and, exasperated, yells, “Come now, you’re crazy, you can’t just tell both of them that they’re right!” Then the wise man nods his head and says, “Yes, you’re right.”

We shall dream until the end
Of an excellence of the subjunctive
Which shall curve space-time
Until it bites its tail.

—Jean Rousselot, *Poèmes en espoir de cause*

10
What is the Curvature of the Universe?

In an arbitrary space, the curvature is a complicated mathematical object called a *tensor*, described by a large number of components. In four dimensions, all information on the curvature of space-time is thus contained in the *Ricci tensor*, which has sixteen components and can be represented by a table, or matrix, with four rows and four columns, like the squares of a small checkerboard. In the case of homogeneous and isotropic universes, the problem simplifies considerably, since the curvature of space is constant. It is thus expressed by a single number, of which it suffices to know the sign.[1] The Friedmann-Lemaître models thus distinguish three families, with curvatures which are respectively negative, zero, or positive.

To understand the classification of models for the Universe by their curvature, let us return once again to the case of surfaces, meaning spaces with two dimensions. The simplest surface is the plane, with zero curvature. Here, Euclid's geometry reigns: through a given point there is a unique parallel to a given

[1]If R is the radius of curvature of space, its curvature is k/R^2, where $k = -1, 0$ or $+1$ according to whether the curvature is negative, zero, or positive. Over the course of time, R varies but k does not. The geometry of space is therefore fixed once and for all by the sign of its curvature.

line (Euclid's fifth postulate), the sum of angles in a triangle is always equal to 180 degrees, the ratio of the circumference of a circle to its radius is always equal to 2π, and so on. Another simple surface is the sphere, whose curvature is constant and positive. There are no straight lines on a sphere, but great circles (those that lie in a plane passing through the center of the sphere) play the same role: when one moves straight ahead, one follows a great circle; the shortest route between two points is always a portion of a great circle that passes through these two points. More generally, on an arbitrary surface, the curves having this property are called *geodesics*. Euclid's fifth postulate does not apply to spheres, since geodesics always have two points of intersection, such as the north and south poles for longitudes. The other properties of the plane are not found either: the sum of angles in a triangle is greater than 180 degrees and the ratio of the circumference to the radius of a circle is less than 2π. The geometry of constant positive curvature therefore differs from Euclidean geometry. Finally, there are also surfaces of constant negative curvature, discovered in 1829 by the Russian mathematician Nikolai Lobachevsky. Euclid's fifth postulate, once again, fails to hold: through a point exterior to a line, one can trace an infinite number of parallels, the sum of the angles in a triangle is smaller than 180 degrees, and so on. These surfaces are difficult to visualize in our mental space. Let us content ourselves here with imagining that, at every point, they are identical to the hollow of a horse's saddle or to the horn of a trumpet.

Like surfaces, three-dimensional spaces of constant curvature are classified in three families. Besides "ordinary" (Euclidean) space, with zero curvature, one encounters spaces of constant positive curvature, called spherical spaces, whose geometric properties generalize those of the sphere, and spaces with constant negative curvature, called hyperbolic spaces. None of the latter can be visualized, since they cannot be contained in a Euclidean universe: one must not consider them as objects *within* an ordinary space, but *in the place* of ordinary space.

Since the Friedmann-Lemaître models are spatially homogeneous and isotropic, they require that at cosmic scales physical space belongs to one of these three families, without, a priori, privileging one or the other. At any rate, the sign of the spatial curvature remains the same over the course of cosmic evolution (either expansion or contraction), even if the radius of curvature varies (grows or diminishes). One of the pressing questions of cosmology is therefore to know in which of these cases we really find ourselves. Can the curvature of space be measured? One could do so directly with a geometric experiment: by tracing out a vast triangle, determining its angles and comparing their sum to 180 degrees; or by drawing a gigantic circle, evaluating its circumference and its radius, and comparing their ratio to 2π. Nevertheless, the difficulties of such an undertaking are considerable. The German mathematician Carl Friedrich Gauss attempted such

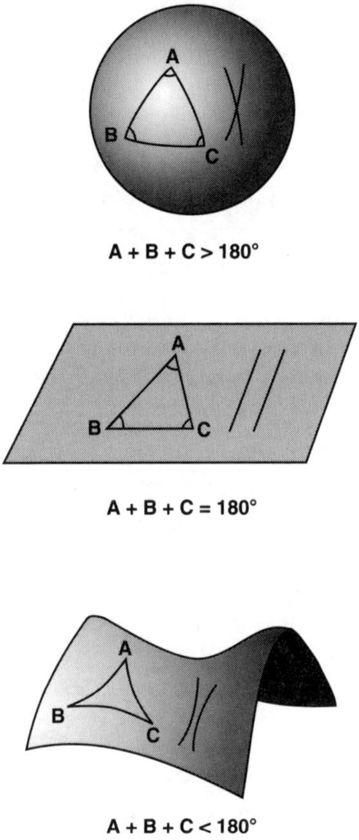

Figure 10.1. Three types of curved surfaces. The surface of the sphere has positive curvature, the plane has zero curvature, and a horse saddle has negative curvature. On the sphere, the sum of the angles in a triangle is greater than 180 degrees; in the plane, it is exactly equal to 180 degrees; and on the saddle, it is less than 180 degrees. Three-dimensional spaces can be curved just like surfaces, but in a more complicated fashion. In principle, one could detect the curvature of the Universe by measuring the angles formed by a gigantic cosmic triangle.

an undertaking in the first half of the nineteenth century, by experimenting on a triangle of 100 km on each side whose corners coincided with mountain peaks, but his efforts were in vain. At this scale, the deformations of space ascribable to curvature are much to weak to manifest themselves. However, they must play an important role in cosmology, so much so that astronomers hope, by looking far enough into space, to find evidence for them. It is now possible to experiment on 161→ cosmic triangles whose sizes are comparable to that of the observable universe.

A more indirect approach consists in deducing the curvature from the equations of general relativity. These, in fact, indicate that the sign of the curvature depends on the average density of matter and energy spread throughout space, as well as on a term called the *cosmological constant.* From the mathematical point of view, the cosmological constant constitutes a logical generalization of the equations of relativity; from the physical point of view, it is interpreted as a vacuum energy. It thus behaves like a supplementary contribution to the energy density and must be included in the equations in order to evaluate the curvature of ←207 space.

The Friedmann-Lemaître models teach us that, depending on whether the average density of matter-energy is larger or smaller than a certain critical value, the curvature is positive or negative, and null at the exact crossover point. This critical value, equal to 10^{-29}g/cm^3 (up to a factor which depends on the present expansion rate of the Universe), is equivalent to between two and ten hydrogen ←184 atoms per cubic meter.[2] For comparison, the best vacuum obtained in a modern laboratory still contains two hundred billion air molecules per cubic meter. It is thus striking to appreciate the change in perspective. While at the human scale the critical density seems to be an extraordinarily small value, on the cosmic scale it takes on gigantic importance, to the point of deciding the ultimate fate of the Universe, since it corresponds to the threshold past which it will end up collapsing under its own gravity.

For convenience, cosmologists use the *density parameter*, defined as the ratio of the real density (including all forms of matter and energy) to the critical density. Depending on whether this ratio is smaller than, equal to, or greater than one, the curvature is negative, zero, or positive, respectively. This link between the curvature and the density allows us experimental access to the macroscopic geometry of space. At least this is true in principle, for when astronomers want to take inventory of all the energy contained in the Universe, they are confronted by a certain number of complications. Without dedicating themselves to that fool's task, evoked in tales and legends the whole world over, of counting the stars one by one, they nevertheless pursue a statistical count of visible sources: galaxies, galaxy clusters, and so on. The most difficult part is to identify a region of the Universe that constitutes a good sample, that is to say, a region whose quantity of matter is properly representative of the rest. After having evaluated this mass, they calculate that the density of visible matter represents only about one hundredth of the critical density.

---

[2] This also means that a cube of the universe that is 10,000 kilometers on one side weighs, on average, 10 milligrams.

The density of visible matter necessarily constitutes only a lower bound on the real density. Not everything is detectable by our telescopes, as perfect as they may be. There is *dark matter*: e.g., black holes, small stars which have too little mass to start nuclear reactions and are therefore sub-luminous, vast clouds of interstellar gas formed from cold hydrogen, and massive particles produced during the big bang. We make use of certain indirect methods to detect the traces of this dark matter, not through its electromagnetic radiation, but by the gravitational effects that it causes on the motion of stars and galaxies. Several independent results lead us to believe that the Universe contains at least ten times as much dark matter as it does visible matter.

The study of gravitational mirages is one way to indirectly bring dark matter 217→ to light. The principle is simple: the greater the total mass (both visible and invisible) of the gravitational lens, the stronger the optical deformation of background objects. By analyzing the precise geometric structure of a mirage, for example the angular separation between multiple images, one can thus deduce the mass of the lens. For a cluster of galaxies responsible for such a mirage, the mass determined in this way systematically proves to be ten times greater than that obtained by only counting the visible galaxies of the cluster. In other words, at least 90% of the mass of the Universe is found in the form of dark matter. This is the same ratio as that which exists between the submerged and exposed parts of an iceberg drifting in the Atlantic.

Until 1996, most determinations of the density of the Universe based on this type of technique were converging toward a common minimal value for the density parameter, in the range between 0.1 and 0.3 (10 to 30 percent of the critical value). This therefore led to an experimental preference for a hyperbolic Universe, with negative curvature and in perpetual expansion.

The situation changed brusquely with the implementation of new methods for determining the curvature. In fact, other components for the dark matter are possible that would not necessarily be detectable in gravitational mirages; massive elementary particles, produced during the big bang, that would bathe the universe with a uniform background, the contribution of a non-zero cosmological constant, or that of a mysterious field of negative pressure called *quintessence*, could make the real density of the Universe climb up to a notably larger value. This would be a veritable windfall for numerous cosmologists who, for various reasons, retained a preference for models of "flat" space, meaning those with zero curvature.

Certain theories about the very first instants of the Universe, such as that of inflation, suggest in fact that the Universe would have experienced, in the first few fractions of a second of its evolution, a frenetically high rate of expansion, to the point that at the end of this inflationary phase space would have become basically flat and would remain so from then on. Since it is fairly well established that the density of matter in all its forms combined represents at most a third of the critical density, this would mean that cosmic energy finds itself primarily in an additional form, perhaps due to the quantum vacuum, the dynamic role of this form of dark energy (either a cosmological constant or quintessence) being to accelerate the cosmic expansion. Some teams have observed supernovas (exploding stars) in distant galaxies and from this have deduced that the density parameter is equal to 1, of which a contribution of 0.3 comes from ordinary matter-energy and 0.7 from the cosmological constant.

In the spring of 2000, balloon borne telescopes, dedicated to the analysis of certain detailed characteristics of the cosmic microwave background, furnished data allowing one (assuming some debatable hypotheses) to fix the density parameter in a narrow enough range of values, not varying much more than 15% on either side from the critical value of 1. In other words, space should be nearly flat, in perpetual expansion, and this expansion should accelerate over the course of time.

Large daily newspapers the world over, including the *New York Times*, *Le Monde*, the *Herald Tribune*, and *La Stampa*, had a field day with this news. Any amount of triumphalism is allowed in this type of exercise, encouraged by the thundering declarations of some big names in cosmology. One of them, during a press conference held at NASA's headquarters in Washington, did not hesitate in announcing: "This is a moment which we will all remember. It will remain marked in the annals of history." Declarations of this type occur every year, supported by organizations that have become masters in the art of scientific marketing, aiming to sell their programs. This somewhat diminishes the historical impact of major scientific events.[3]

Outside of the effects of these sensational announcements, it seems essential to recall that the precise measurement of the curvature today remains out of our reach, by reason of our ignorance concerning the exact quantity of dark matter and the value of the cosmological constant. Let us thus be wary of these claimed ←245 proofs of a flat Universe. On the contrary, space has every chance of having strictly negative or strictly positive curvature.

[3]This is especially so since a number of these declarations must be contradicted by what follows, as was the case recently for the claimed discoveries of an axis of rotation for the Universe, of the first image of an extra-solar planet, and of a fossil bacterium in a Martian meteorite. We spend more time explaining to journalists why these announcements are fallacious than in divulging our own results!

Only two things are infinite, the universe and
human stupidity, and I'm not sure about the former.

—attributed to Albert Einstein

11

Open or Closed?

The most immediate question connected to the global shape of the Universe concerns the finite or infinite character of space. One may have expected that general relativity provides an answer. Indeed, every space with positive curvature has finite volume, whatever its *topology* may be.[1] As for the hyperbolic spaces, with negative curvature, or Euclidean, with zero curvature, at first sight these seem to be infinite, but this is only true if they possess the simplest possible topology. In reality, the finite or infinite character of space depends on its topology, and not only on its curvature. In fact, cosmologists most often neglect the former and only consider the latter.

This abusive simplification goes back to the very beginnings of relativity. At that time, topology remained essentially sequestered to the classification of surfaces—two-dimensional spaces—and little was known about the possible shapes of three-dimensional spaces. This is one of the reasons that cosmic topology was hardly discussed, with a few notable exceptions, like Schwarzschild, 280→
Friedmann, and Lemaître.

The model of the Universe developed by Albert Einstein and Willem de Sitter in 1932 was a great success for 65 years. It stipulated a perpetually expanding Euclidean space and implicitly put forth the hypothesis of an infinite space. The near total neglect of any topological aspects dates back to this period. In what fol-

[1]See the definition of this term on p. 58.

lowed, articles and textbooks—either specialized or not—that dealt with cosmology decreed that the spatial structure of the Universe was either the finite hypersphere, infinite Euclidean space, or infinite hyperbolic space, without mentioning the topological alternatives. Astrophysicists ignored the new developments that mathematicians, starting in the 1930s, had made regarding the possible shapes of abstract spaces. The extremely fertile ideas of topology have thus remained largely ignored in the general flow of cosmological work.

This simplification has very important consequences for those issues connected with the problem of spatial extent, since it reduces the finite/infinite dilemma to knowledge of the sign of the curvature of space—and even, in the case where the cosmological constant is assumed to be zero, to the value of the matter density. Therefore, little by little, confusion has crept in on the meaning of the adjectives "open" and "closed," as used to characterize the Friedmann-Lemaître models. Originally used to describe the *temporal* behavior of models (open if they are in perpetual expansion, closed if they are in expansion followed by contraction), they have ended up being applied also to their spatial behavior: in fact, if one clings to simple topologies and if one assumes moreover that the cosmological constant is zero, the models in perpetual expansion strictly coincide with the spatially infinite models, and the models in expansion-contraction with the spatially finite models.

However, to know if space is finite or infinite, it does not suffice to determine its metric or the sign of its spatial curvature, and it does not suffice to determine whether the density parameter is greater than, less than, or equal to one: some supplementary hypotheses are necessary—precisely those of the topology.

Confusion over the words "open" and "closed" therefore comes from mixing up the concepts of spatial curvature, temporal dynamics, and topology. The correct terminology should be the following:

- In relation to curvature, universes with positive spatial curvature should be called *spherical*, those with zero curvature, *Euclidean* (although "flat" is also commonly used), and those with negative curvature, *hyperbolic*. The sign of the curvature is fixed by the total density of matter and energy.

- In relation to their time evolution, universes which recontract should be called *closed*, those which grow forever with a velocity that approaches zero asymptotically should be called *critical*, and those which expand forever should be called *open*.

- In relation to their topology, universes whose spatial part is of finite or infinite volume should themselves be described as *finite* or *infinite*.

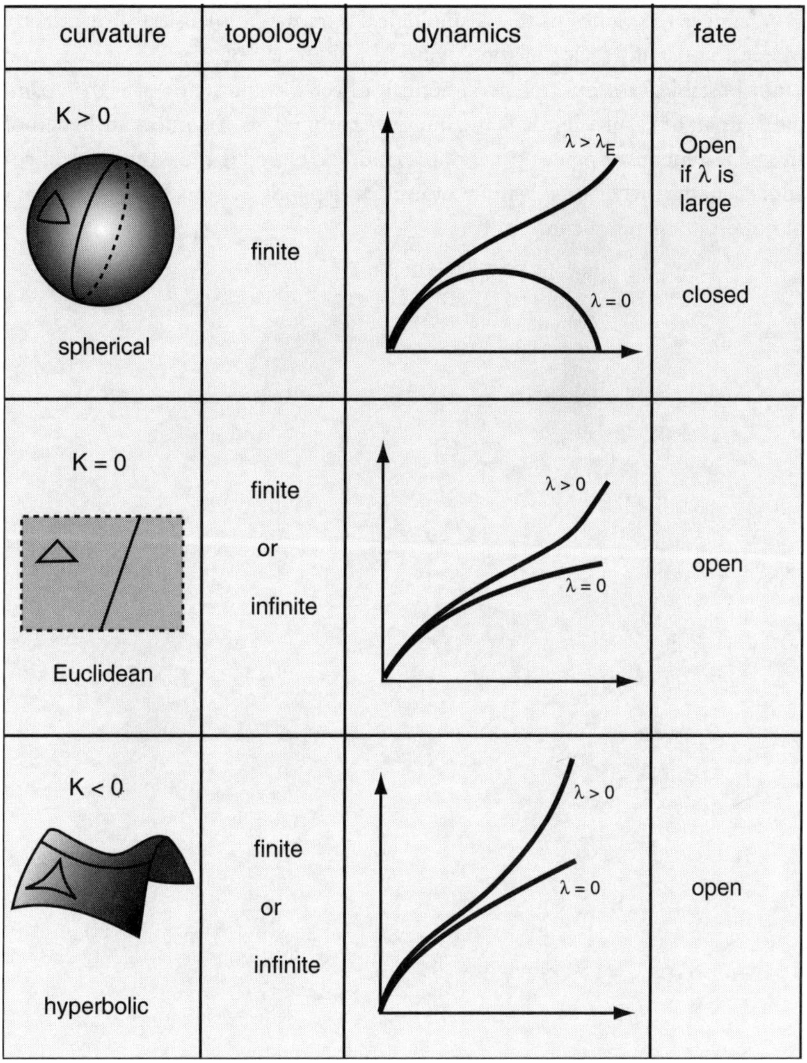
curvature
topology
dynamics
fate
K > 0
spherical
finite
λ > λE
λ = 0
Open if λ is large
closed
K = 0
Euclidean
finite or infinite
λ > 0
λ = 0
open
K < 0
hyperbolic
finite or infinite
λ > 0
λ = 0
open

Figure 11.1. Open or closed? Contrary to a currently widespread opinion, the curvature of space dictates neither the temporal evolution of the Universe (unless the cosmological constant is zero), nor the finite or infinite character of space (unless the topology is simply connected). The table summarizes the different possible cases for big-bang models. The first column shows the sign of the spatial curvature, the second the finite or infinite character of space, the third the temporal behavior of the scale factor, the fourth the open or closed character of the model.

Whatever the values of the cosmological parameters (density of matter, cosmological constant, scales of topological lengths) may be, certain combinations are not possible. For example, a spherical universe could be open if the cosmological constant is sufficiently large, but never infinite. A Euclidean or hyperbolic universe is generally open in time, but it could be closed if the cosmological constant is negative; whether it is finite or infinite depends essentially on its topology, and not on its material content.

Nine tenths of mathematics, outside of that which has been required for practical needs, comes from the solutions to riddles.

—Jean Dieudonné, *Penser les mathématiques*

12 The Knowledge of Places

The origins of topology go back to a riddle posed by the idle rich Prussians of the city of Königsberg. This port city of East Prussia (brought into the Soviet Union in 1945 under the name of Kaliningrad; today it is in Russia) is constructed around the branches of the Pregel river. The riddle consisted of deciding if, from any point in the city, it was possible to take a stroll in a closed loop while crossing once, and once only, each of the seven bridges that span the branches of the Pregel. The riddle was solved by the famous mathematician Leonhard Euler, who at that time worked not far from there, in St. Petersburg. In 1736, he published an article giving the necessary conditions that would allow such a route and, since the configuration of the bridges did not satisfy these rules, he proved that it was impossible to cross all seven bridges in a single trip.

Euler remarked that, for the first time in the history of mathematics, one was dealing with a geometric problem that had nothing to do with the metric—the mathematical object allowing one to measure distances. The only important factors were the relative positions of the bridges. Indeed, if we trace the map of the city on a rubber sheet, and if we stretch and squeeze it in any direction without puncturing, cutting, or tearing it, the nature of the problem is absolutely unchanged.

The solution to the puzzle of the Seven Bridges of Königsberg marked the birth of two new branches of mathematics which would prove to have a rich

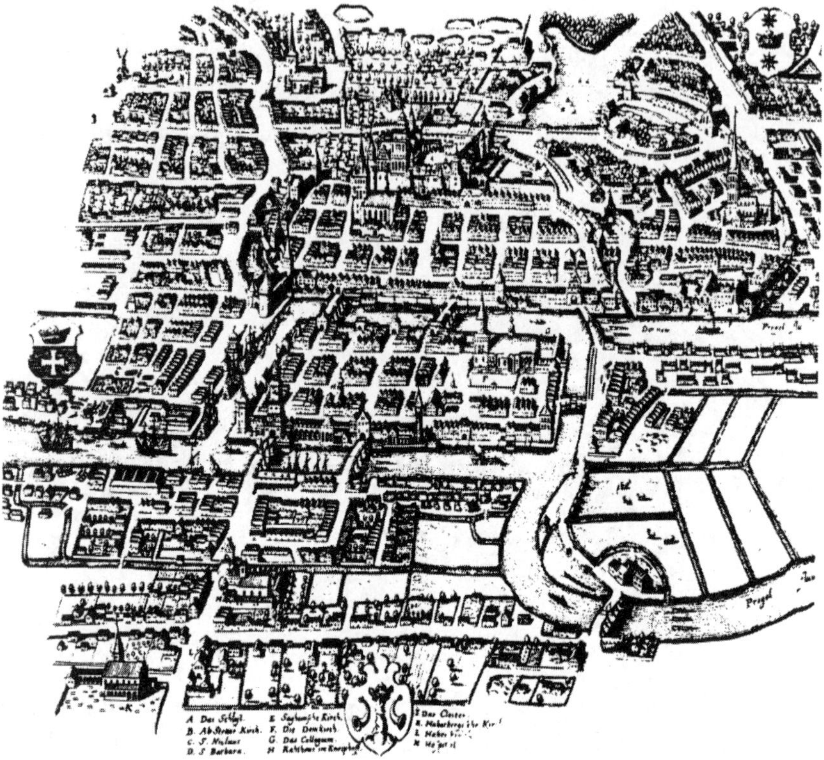

Figure 12.1. Map of Königsberg. This map of the city of Königsberg, taken from a seventeenth-century atlas, illustrates the problem posed by trying to pass over all seven bridges in a single trip.

future: topology and graph theory. Topology abstracted the distances away and only looked at the relative positions of objects in space. Etymologically, *topology* signifies the "knowledge of places." This term was only proposed in 1847, by Johann Listing, and adopted even later.[1] Graph theory, which is capable of addressing problems that may be difficult to formalize but which intuition, supported by small sketches, can quite effectively address, was primarily developed in the second half of the twentieth century, see [Biggs et al. 86].

[1] Listing's work, *Vorstudien zur Topologie*, is a German book. Euler, Gauss, and Poincaré, three big names in topology, called this discipline *analysis situs*, meaning the calculus of positions—a term already introduced by Leibniz in 1679. The French *topologie* (attested for the first time in 1876) was first a term of religious rhetoric ("the study of commonplaces or platitudes"). It took its mathematical meaning, under the influence of the English *topology*, at the beginning of the twentieth century.

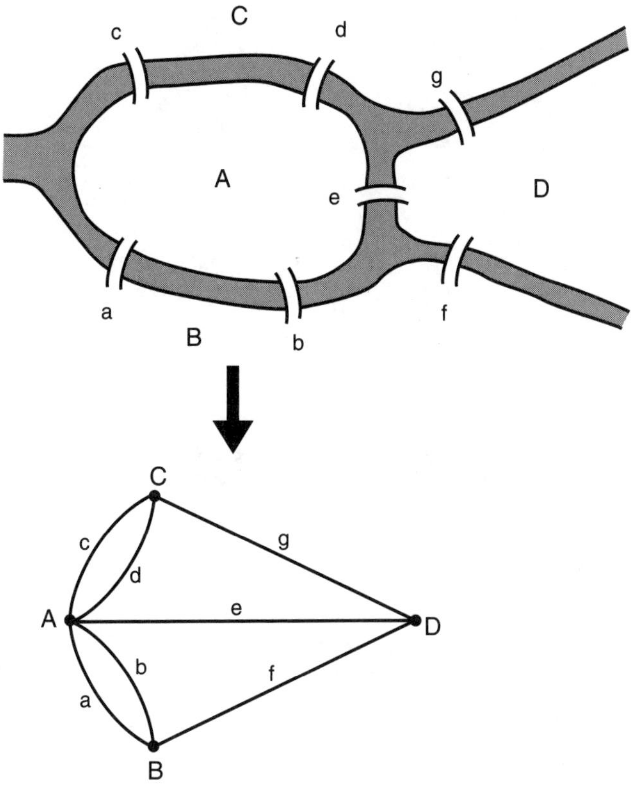

Figure 12.2. Graph of the problem of the seven bridges. The map of Königsberg can be schematized by this graph, with four vertices A, B, C, and D (corresponding to the four zones of the city that could serve as departure points) and seven arcs (crossing each of the bridges). If one changes the proportions without modifying the relative positions, the nature of the graph is unchanged.

How can the riddle of the Seven Bridges of Königsberg be connected to a question as serious as the shape of the cosmos? It so happens that the solution given by Euler perfectly illustrates the two complementary aspects of geometry as the science of space: the metric part deals with the properties of distance, while the topological part studies the global properties, without introducing any measurements. The topological properties are those that remain insensitive to deformations, provided that these are continuous: with the condition of not cutting, piercing, or gluing space, one can stretch it, crush it, or knead it in any way, and one will not change its topology, e.g., the fact that it is finite or infinite, the fact

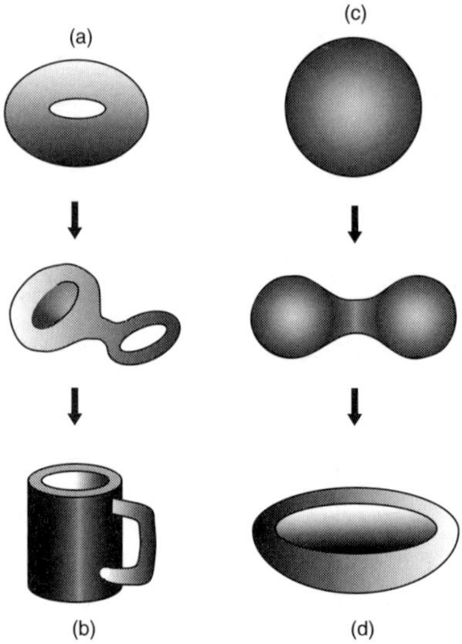

Figure 12.3. Continuous deformations. Let us imagine that the surface of a ring (a) is made of rubber, and that one can easily stretch it without tearing it. We deform part of the surface to form a bump, then transform this bump into a sort of sac which is larger than the original ring. The ring has become a coffee cup with a handle (b)! It is impossible, though, with any continuous deformation whatsoever, to obtain a bowl (d). The surface of a sphere (c), however, can do the job.

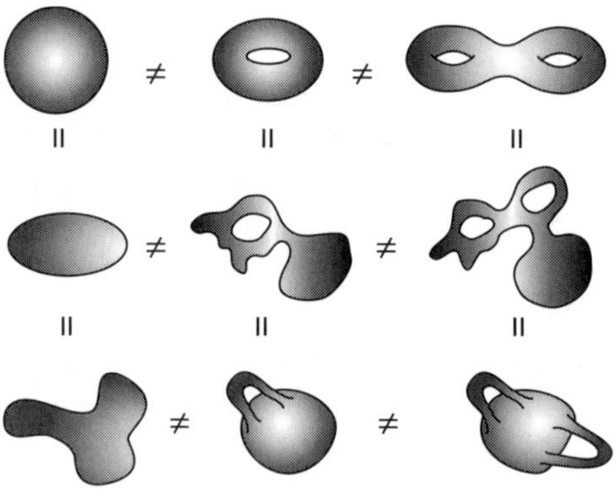

Figure 12.4. Sphere and tori. In the case of two-dimensional spaces, that is to say surfaces, the sphere has the same topology as any closed surface without a hole. Tori with one or more holes have different topology; they can be continuously deformed into spheres equipped with handles.

that it has holes or not, the number of holes if it has them, and so on. It is easy to see that although continuous deformations may move the holes in a surface, they can neither create nor destroy them.

For a topologist, there is no difference between a football and a soccer ball. Even odder, a ring and a coffee cup are one and the same object, characterized by a hole through which one can pass one's finger (although it is better not to pour coffee into a ring). On the other hand, a mug and a bowl, which may both serve for drinks, are radically different on the level of topology, since a bowl does not have a handle.

The shortest path from one point to another is a straight line, as long as both points are really facing each other.

—Pierre Dac, *L'Os à moëlle*

13
From the Cylinder to the Pretzel

Topology holds quite a few surprises. Let us take the Euclidean plane: it is an infinite two-dimensional page, that one visualizes most often within a three-dimensional space, although it has no need for this embedding to be perfectly well defined in an intrinsic fashion, without a reference space exterior to it. The local geometry of the plane is determined by its metric, that is to say, by the way that lengths are measured. Here, it is sufficient simply to apply the Pythagorean Theorem for a system of two rectilinear coordinates covering the plane: the square of the hypotenuse is equal to the sum of the squares of the two other sides. This is a local measurement that says nothing about the finite or infinite character of space.

Let us change the topology. To do so, we take the plane and cut a strip of infinite length in one direction and finite width in the other. We then glue the two sides of the strip; we obtain a cylinder, a tube of infinite length. In this operation, the metric has not changed; the Pythagorean Theorem continues to hold for the surface of the cylinder. The *intrinsic curvature* of the cylinder is therefore zero. This may appear surprising, since one has the impression that there is a non-zero curvature "somewhere," whose radius would be the radius of the cylinder.

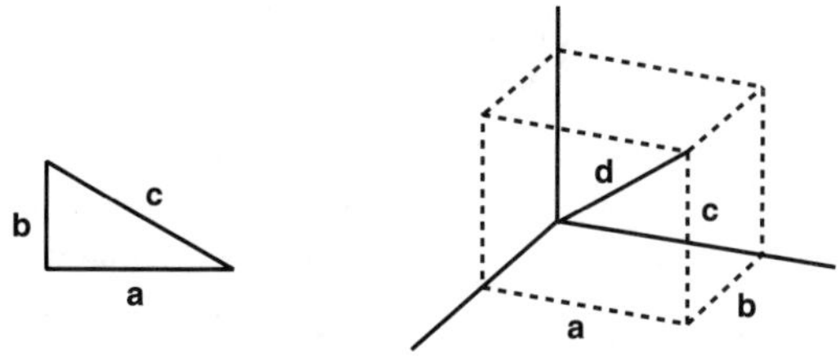

Figure 13.1. The Pythagorean Theorem. In the plane, the length of the hypotenuse of a right triangle is given by the Pythagorean Theorem: $c^2 = a^2 + b^2$. This formula generalizes to Euclidean spaces of higher dimension. For example, the diagonal of a parallelepiped is given by $d^2 = a^2 + b^2 + c^2$.

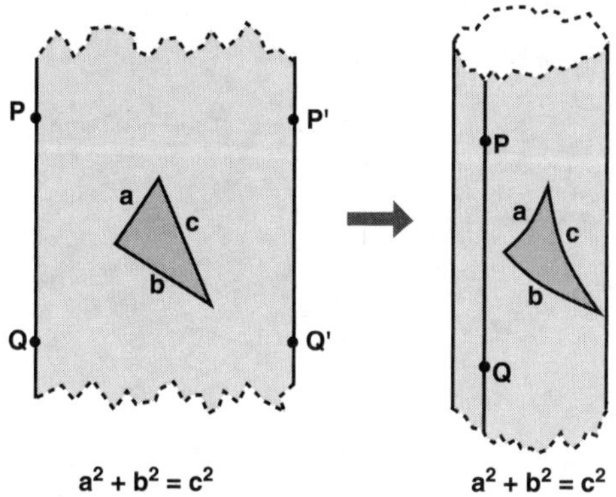

Figure 13.2. Constructing the cylinder. After having cut a strip of paper from the plane, we glue the opposite edges, which is the same as identifying the points P with P' and Q with Q'. The resulting surface is a cylinder. To us this seems to have curvature. In reality, the cylinder is flat: if a square triangle is traced, the Pythagorean Theorem remains perfectly valid.

However, this "somewhere" calls into play a space exterior to the cylinder: the one in which we visualize it. In this sense, the cylinder has a so-called *extrinsic* curvature. Nevertheless, a flat being, some sort of geometric paramecium living on the surface, would have access neither to this exterior space of higher dimension, nor to the extrinsic curvature of the cylinder. Tied to its two-dimensional space, it could make all of the necessary verifications (measuring the sum of the angles in a triangle or the ratio of the circumference of a circle to its radius), and it would detect no difference with respect to the Euclidean metric of the plane. The cylinder is said to be *locally* Euclidean.

Nevertheless, the cylinder differs from the plane in many respects. Certainly, its area is infinite, just like the plane, but it possesses a finite circumference in the direction perpendicular to its axis of rotation. In other words, the cylinder is anisotropic—not all directions are equivalent; following the length of a straight line parallel to the axis, one moves off toward infinity, while if one moves in the perpendicular direction, one returns to the departure point. In the operation of constructing a cylinder from a section of the plane, some of the global properties have changed; the cylinder thus has a different topology than that of the plane, while nevertheless having the same metric. Its most remarkable characteristic is the existence of an infinite number of "straight lines" that join two arbitrary distinct points on the cylinder: those that make 0, 1, 2... turns around the cylinder. Viewed in three dimensions, these straight lines are helices with constant spacing.

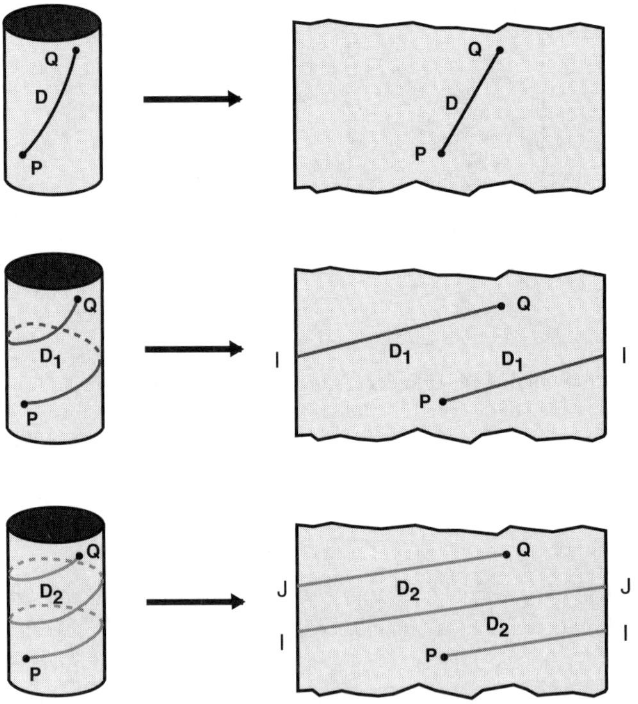

Figure 13.3. Straight lines on the cylinder. Straight lines on a cylinder are helices with constant spacing. Two arbitrary points P and Q are connected by an infinite number of helices D making 0, 1, 2, ... turns around the cylinder. If we unroll the cylinder in the plane, all of these helices indeed roll out in the shape of straight lines.

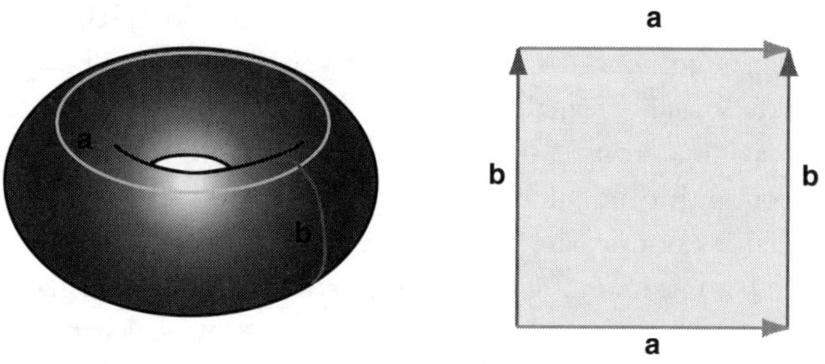

Figure 13.4. Constructing the flat torus. From the topological point of view, a flat torus is obtained by gluing the opposite sides of a rectangle. Inversely, if we cut the torus (left) along the two lines a and b, we can unroll it in a rectangle (right).

Let us continue with our cutting and gluing game. Take a tube of stretchable rubber, of finite length, and glue its two ends edge-to-edge. This is strictly equivalent to starting with a rectangle and gluing its opposite edges two by two. We obtain a torus, a surface having the shape of a ring or an inner tube.

Here, a new difficulty lies in wait. A real inner tube can be materialized in ordinary three-dimensional space; it therefore has an extrinsic curvature. However, in contrast to the flat torus, the inner tube also has a non-zero intrinsic curvature, which varies in different regions: sometimes positive, sometimes negative (Figure 13.5).

Nevertheless, the toric surface obtained by identifying the opposite sides of a rectangle has an intrinsic curvature that is everywhere zero. This flat torus, a surface whose global properties are identical to those of an a ring but whose curvature is everywhere zero, cannot be viewed within our usual three-dimensional space. One can, nevertheless, describe all of its properties without exception; its

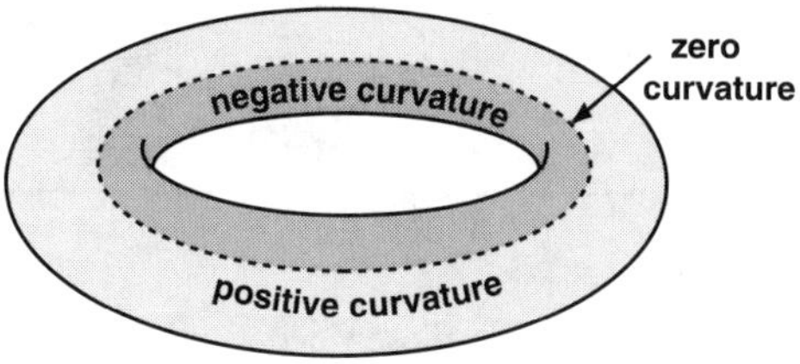

Figure 13.5. Curvature on the torus. The surface of an inner tube is not a flat torus. It has a curvature, which varies from place to place.

area is finite in the sense that it is impossible to move infinitely far away from one's departure point, and it is not isotropic, since two of its directions, named the principal directions—a small circle and a large circle—are privileged.

Let us imagine a creature living on a torus, moving straight ahead along a principal direction; it communicates via light rays with its point of departure, in such a way that it can calculate the distance travelled; at a certain moment, this distance attains a maximum, and then begins to decrease; after having made a complete circuit, the creature has returned to its point of departure. It would conclude from this that it lives in a space of finite extent. Nevertheless, by having measured the sum of the angles in a triangle in these surroundings, it has still found 180 degrees, because of which it would also deduce that it lives in a Euclidean plane. The metric (local geometry) of the torus is still given by the Pythagorean Theorem, just like that of the plane and the cylinder.

Thus, through simple cutting and re-gluing of parts of the plane, we have defined two surfaces with different topologies than the plane, the cylinder and the torus, which, however, belong to the same family of surfaces with zero curvature: the locally Euclidean surfaces. Mathematicians have found an elegant

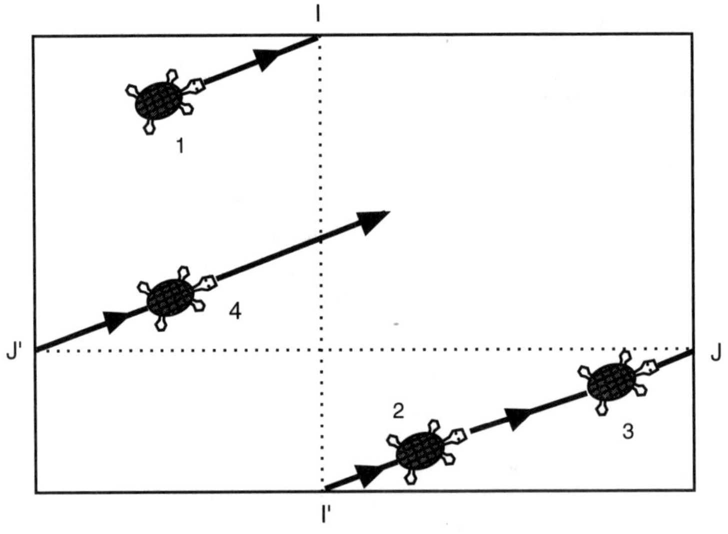

Figure 13.6. A trip on the torus. As in those video games in which the characters who leave on one side reenter on the opposite side, the turtle crosses the upper edge of the rectangle at I and reappears on the lower edge, at the equivalent point I'. Following his path in a straight line, he reaches the right edge at J, reappears at J', and so on. The torus is thus equivalent to a rectangle whose opposite edges are identified two by two.

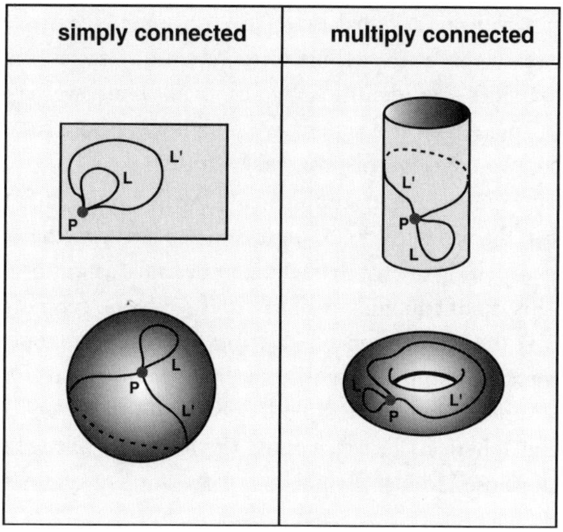

Figure 13.7. Laces and connectivity. In a plane or on a sphere, all laces can be indefinitely tightened, without obstacle. These spaces are simply connected. On a cylinder or a torus, some laces can be tightened, others not. These spaces are multiply connected.

way of characterizing these diverse topologies: the tightening of laces. A lace is a closed curve traced on a surface. On the infinite plane from which we started, we can draw an arbitrary lace, however large, from an arbitrary point; this lace can always be retightened and reduced to a point without encountering any obstacles. The topologists call such a surface *simply connected.* On the other hand, the tube and the ring do not have this property. Of course, there are laces that can be completely retightened, as in the plane; but some of them cannot be: a circle that wraps around the tube or that is traced around the ring, for example, cannot be continuously shrunk to a point. For such surfaces, where some laces cannot be indefinitely tightened because they wrap around a hole, the topology is said to be *multiply connected.* Multiple connectivity appears as soon as one performs gluings, or *identifications* of points, in a simply connected space.

The properties of simple or multiple connectedness are fundamental in topology, and the distinction holds for any type of space, whatever its curvature or dimensionality may be.

Two-dimensional spaces are, in general, easily visualized since one can mentally embed them in Euclidean space.[1] For three-dimensional spaces, we must call on a more abstract representation that I shall briefly describe, beginning, for

[1]This is true most of the time, but not always, as we have seen with the flat torus.

simplicity, with a two-dimensional example. In the nineteenth century, mathematicians discovered that it is possible to represent any surface whatsoever with a polygon whose sides one identifies, two by two. Thus, we have seen that by gluing two opposing edges of a square, we obtain a section of a cylinder, and if one connects the two edges of this, we make a torus.

The square (or the rectangle), supplemented with two identifications between its opposite sides (the top with the bottom, and the left with the right) is called a *fundamental domain*; it distinctly characterizes a certain aspect of the topology of the torus. But this is not enough; we must also specify the geometric transformations that identify the points. Indeed, starting from a square, one could identify the points diametrically opposite with respect to the center of symmetry of the square, and the surface obtained will no longer be a torus; it will no longer even be Euclidean, but spherical, a surface called the *projective plane*. The mathematical transformations used to identify points form a group of symmetries, called the *holonomy group*.

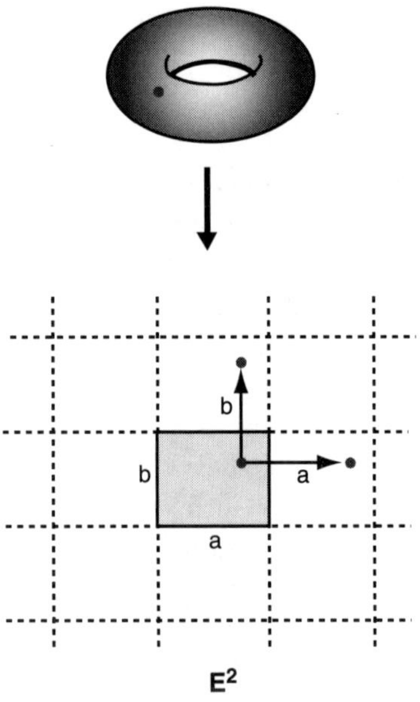

Figure 13.8. The universal covering. The fundamental domain for a torus is a rectangle, and the holonomies are translations by distances equal to the sides of the rectangle. By acting with the holonomies on each point in the fundamental domain, and by repeating the process again and again, one creates the universal covering space—here the Euclidean plane E^2.

Starting from the fundamental domain and acting with the transformations of the holonomy group on each point, one creates an endless number of replicas of the fundamental domain; we produce a sort of tiling of a larger space, called the *universal covering space.*

By construction, the universal covering space is simply connected; any lace can be shrunk to a point. The universal covering of the torus is the Euclidean plane E^2, which indeed reflects the fact that the flat torus is a locally Euclidean surface. If one cuts it along both its small circle and its great circle, and then unfolds it, it is transformed into a section of the Euclidean plane: a rectangle.

It is in their simply connected topologies that spaces take their simplest form, the unfolded form of the universal covering; the latter can be folded in many ways, each one leading to a multiply connected shape.

Let us take two tori and glue them as in Figure 13.9. As far as its topological properties are concerned, this *double torus* can be represented as an eight-sided polygon (an octagon), which can be understood intuitively by the fact that each torus was represented by a quadrilateral. But this surface, as we can see, is not capable of paving the Euclidean plane, for an obvious reason: if one tries to add

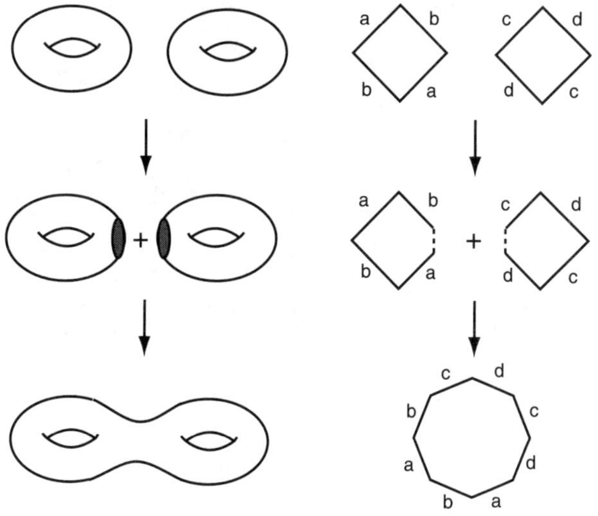

Figure 13.9. Connected sum of tori. The left column shows disks being removed from each of two tori; the edges of the holes are then glued to form a double torus. The right column shows the same operation starting from fundamental domains. The double torus can be represented by an octagon whose edges have been identified two by two in a certain way.

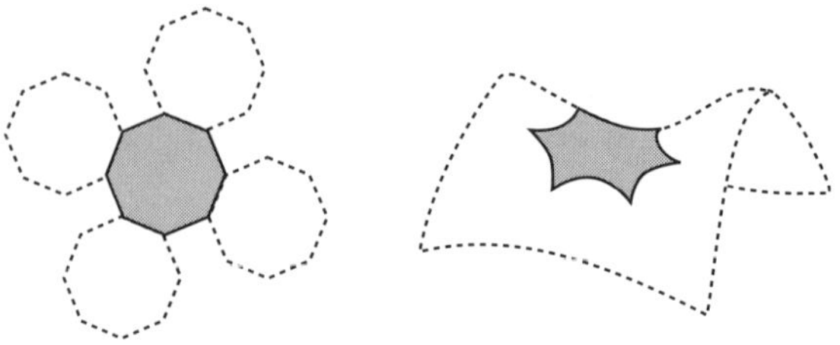

Figure 13.10. Paving the hyperbolic plane with octagons. It is impossible to pave the Euclidean plane with octagons, which implies that the double torus is not a Euclidean surface. On the other hand, the hyperbolic plane can be paved by octagons cut from the hollow of a saddle. The hyperbolic plane is thus the universal covering space for the double torus. The eight corners of the octagon must all be identified as a single point; this is the reason why one must use a negatively curved octagon with angles of 45 degrees ($8 \times 45 = 360$), in place of a flat octagon, whose angles are each 135 degrees.

Figure 13.11. Circle Limit. In this 1959 woodcutting, Escher has used the representation given by Poincaré to pave the hyperbolic plane using fish. (See Plate II. M. C. Escher's "Circle Limit III," © 2007 The M.C. Escher Company-Holland. All rights reserved. www.mcescher.com.)

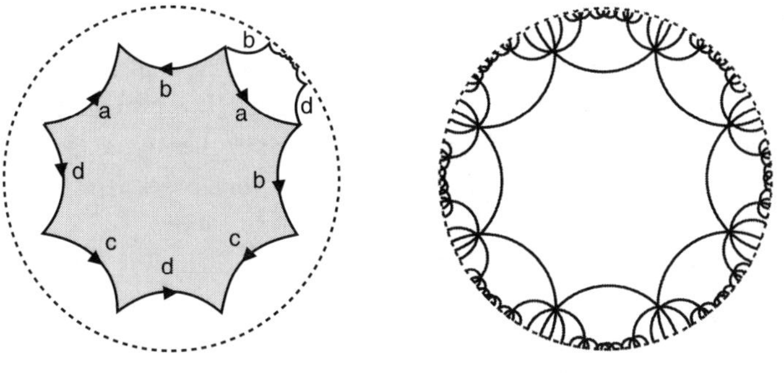

Figure 13.12. Poincaré's representation of the hyperbolic plane. By acting with the holonomies on each point of the fundamental octagon, and repeating the process again and again, one creates a paving of the hyperbolic plane by regular and identical octagons. Poincaré demonstrated that the hyperbolic plane, normally infinite, could be represented entirely within the interior of a disk, whose edge represents infinity. Poincaré's model deforms distances and shapes, which explains why the octagons seem irregular and increasingly tiny as we approach the boundary of the disk. All of the lines in the figure represent straight lines of the hyperbolic plane, and meet the boundary at a right angle.

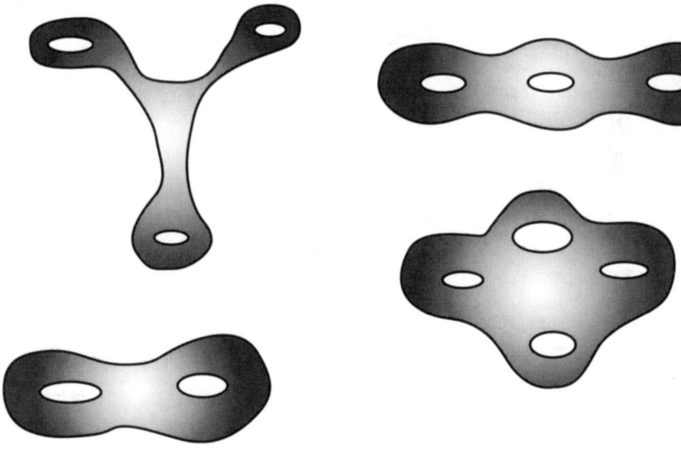

Figure 13.13. Hyperbolic pretzels. All closed surfaces having more than one hole are spaces of negative curvature, equipped with a hyperbolic geometry.

an octagon to each of its edges, the eight octagons will overlap each other (Figure 13.10).

One must therefore curve in the sides and narrow the angles, in other words, pass to a hyperbolic space; only there does one succeed in fitting eight octagons around the central octagon, and starting from each of the new octagons one can construct eight others, ad infinitum. By this process one paves an infinite space—the Lobachevsky hyperbolic plane.

A fascinating representation of a hyperbolic mosaic was given by Henri Poincaré. A certain transformation of coordinates allows us to bring infinity to a finite distance, with the result that the entire Lobachevsky space is contained in the interior of a disk. The famous Dutch graphic artist Maurits Cornelis Escher created a series of prints entitled *Circle Limit*, in which he used Poincaré's representation (see Figures 13.11 and 13.12).

More generally, a torus with n holes can be made by gluing n tori. It is topologically equivalent to a sum of n squares whose opposite sides have been identified, and this sum is itself topologically equivalent to a polygon with $4n$ edges that have been judiciously identified, two by two. The more angles there are to adjust, the more they must be sharpened to points, and the surface extended. The n-torus (with $n \geq 2$) is thus a surface of negative curvature. This type of surface is most commonly seen at bakeries, in the form of pretzels. We shall thus call them *hyperbolic pretzels.*

All of the generalized pretzels can be furnished with metrics of constant negative curvature. They all have the same local geometry, of hyperbolic type. However, they do not have the same topology, which depends on the number of holes.

What we have learned is like a handful of earth;
what remains for us to learn is like the entire world.

—Avvaiyar

14
Fascinating Shapes

Let us summarize: the shape of space is entirely specified if one is given a fundamental domain; a particular group of symmetries, the holonomies, that identify the edges of the domain two by two; and a universal covering space that is paved by fundamental domains. Classifying the possible shapes thus reduces, in part, to 250→ classifying symmetries.

Let us apply this recipe in order to list all homogeneous surfaces: two-dimensional spaces with no boundaries and no sharp points.[1] As far as the curvature is concerned, homogenous surfaces are of three types: spherical surfaces (also called elliptic surfaces), with positive curvature (like the surface of a rugby ball); Euclidean surfaces, with zero curvature (whose planar geometry is taught in high school); and hyperbolic surfaces, of negative curvature (like certain parts of a saddle or of a trumpet's horn).

Within each of these basic types, mathematicians have classified all possible topologies (here referred to as *forms*). By definition, spaces within the same class 264→ can be continuously deformed from one space to the other.

There are only two forms for spherical surfaces, both of them finite: the sphere, which can be given a wide variety of different metrical aspects depending on what sort of continuous stretching it is subjected to, and the projective plane. The sphere is simply connected; the projective plane is not. The latter surface is

[1]This eliminates, for example, disks, which have boundaries, and cones, which have sharp points.

not easily visualized; the simplest way to do so is to pass through the intermediary of its fundamental domain, a disk, whose diametrically opposite points are identified.

Euclidean surfaces can come in five possible shapes: the plane, of course, which is the simply connected prototype, but also the cylinder, the Möbius band (which is an infinitely wide Möbius strip), the torus, and the Klein bottle, all of which are multiply connected. The first three are infinite, the other two finite. These surfaces, although conceptually simple, are not all easy to visualize; thus, although the Klein bottle has no curvature, it is closed in on itself and has neither inside nor outside; it is said to be *non-orientable.*

Finally, the hyperbolic surfaces, with negative curvature, have an infinite number of topologies. Only one of them, equivalent to the Lobachevsky plane, is simply connected. All others are multiply connected, characterized by the number of holes. We have seen, for example, that the surface of a generalized pretzel is hyperbolic.

One conclusion that we can quickly draw from this classification is that, in the infinite set of homogeneous surfaces, they are all hyperbolic, up to only seven exceptions.

Such constructions can be generalized to mathematical spaces of three dimensions. The topological classification of surfaces was completed at the end of the nineteenth century, but that of higher dimensional surfaces is much more complicated, even if we restrict ourselves to the particular case of three-dimensional spaces with constant curvature—those preferentially used in cosmology. Nevertheless, throughout the twentieth century mathematicians have discovered fascinating forms, known practically to themselves alone, which can be used by physicists for the description of the real Universe.

As for surfaces, three-dimensional spaces of constant curvature can first be classified, depending on the sign of their curvature, into spherical, Euclidean, and hyperbolic families. It is then necessary to make an inventory of the topological forms within each of these families. To define and analyze their properties, the basic tools are still the fundamental domain, the group of holonomies, and the
← 272 universal covering space.

In three dimensions, the fundamental domain is no longer a polygon, but a polyhedron. This is a figure constructed from an assembly of polygonal faces, whose common edges form ridges and whose common points form vertices. Most three-dimensional spaces are thus representable in the form of convex polyhedra
← 259 whose faces have been suitably identified in pairs, along with all of the vertices.

There are 18 possible forms for three-dimensional Euclidean space. The simplest is that of ordinary space, denoted E^3. Starting from this infinite, simply connected space, it is possible, through procedures analogous to those for sur-

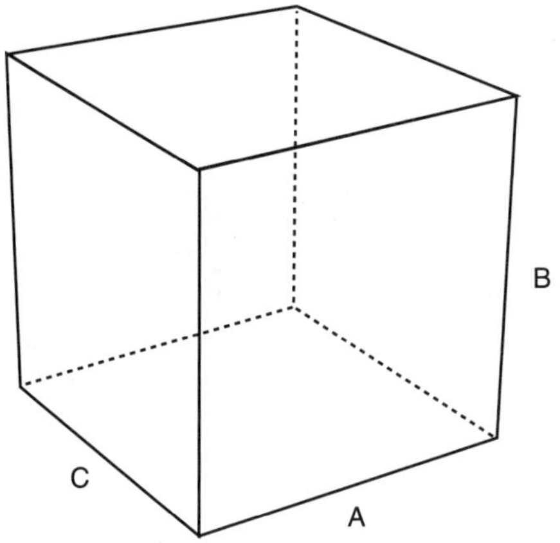

Figure 14.1. From the torus to the hypertorus. If the lengths of the edges of the fundamental parallelepiped are A, B, and C, the volume of the hypertorus obtained by identifying the opposing faces is finite, equal to $A \times B \times C$.

faces, to construct cylinders and three-dimensional tori, as well as more complicated variants. I would like to emphasize the fact that, in each of these new spaces, all of the geometric properties formulated by Euclid are seen to hold, notably that two parallel lines never meet and that the Pythagorean Theorem applies.

In the same way that we obtained the flat torus in two dimensions by gluing the opposite sides of a rectangle, in three dimensions we form a *hypertorus* by identifying the opposing faces of a parallelepiped. By traveling straight ahead within this parallelepiped, as soon as one leaves through one face, one reenters immediately through the corresponding face on the opposite side. The transformations that identify each face with its partner are the characteristic holonomies of this topology: translations by distances equal to the edge lengths of the parallelepiped.

Let us imagine ourselves moving within such a space, and let us, moreover, suppose that we give off light, like glowworms. The light we emit from our backs crosses the face of the parallelepiped which is located behind us, and reappears through the face in front, which is identified with it. By looking in front of ourselves, we therefore see our backs. In the same way, by looking to the right we can observe our left profiles, and we can see the bottom of our feet by lifting our heads. In fact, since our light is emitted in every direction, we see an infinite number of images of our bodies from every angle, laid out in a network of parallelepipeds that, extending in every direction, make up a *visually* infinite space (Figure 14.2).

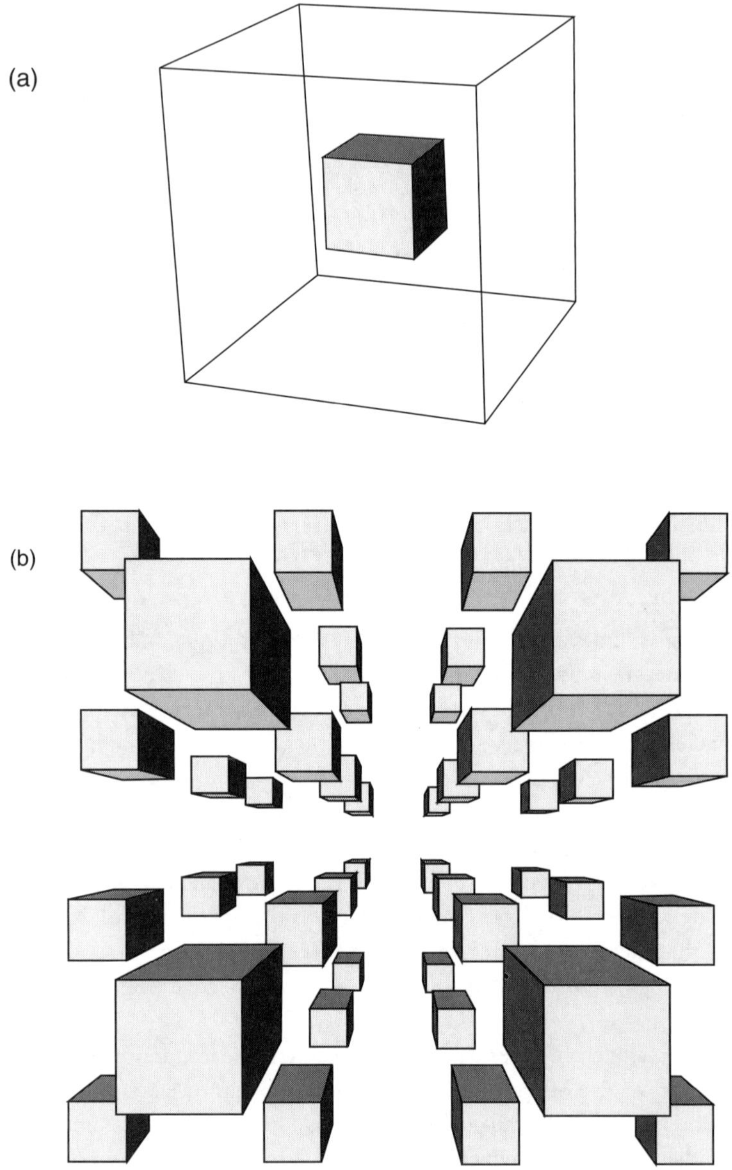

Figure 14.2. Image multiplication in a hypertorus. (a) At the center of the fundamental parallelepiped representing a hypertorus, we place a small cube with six differently shaded faces. (b) When the opposite faces of the parallelepiped are identified to form a hypertorus, the multiplication of images that results causes the small cube to be seen in countless copies, all with differing orientations.

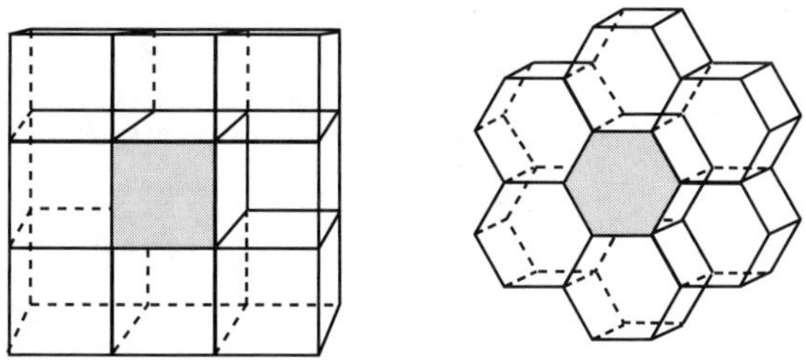

Figure 14.3. Two tilings of Euclidean space. In three dimensions, ordinary Euclidean space can be regularly "paved" by rectangular parallelepipeds or by hexagonal prisms.

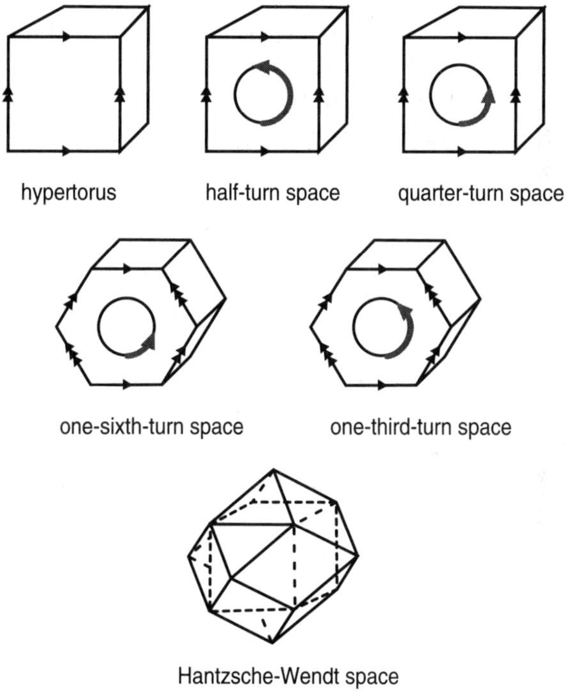

Figure 14.4. Six small Euclidean spaces. By properly identifying the opposing faces of a parallelepiped or a hexagonal prism, one obtains six finite and orientable Euclidean spaces. For the simplest of them, the hypertorus T^3, the identification is made through simple translations. For the others, one of the faces from the pair (here the front and back faces) is turned by $1/2$, $1/4$, $1/6$, or $1/3$ of a full rotation before being glued. The sixth space, discovered by Hantzsche and Wendt in 1935, has a more complicated structure based on the rhombic dodecahedron.

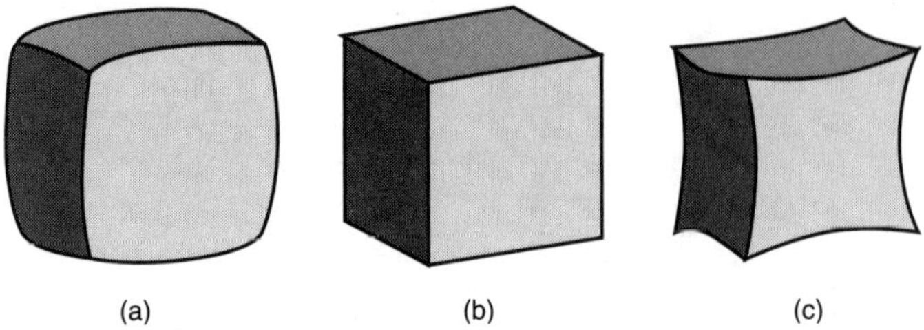

Figure 14.5. Spherical, Euclidean, and hyperbolic polyhedra. In a spherical space (a), polyhedra have larger angles than in Euclidean space (b), while in hyperbolic space (c), polyhedra have smaller angles.

Among the eighteen different forms for three-dimensional Euclidean space, eight have infinite volume and ten have finite volume. Of the latter ten, six are orientable. Their fundamental polyhedra can have the form of a parallelepiped or a hexagonal prism. In either of these cases, their repetition creates a paving of the ←272 universal covering space E^3.

Figure 14.4 clarifies the structure of these six spaces, which are of evident cosmological interest.

Indeed, even if both theory and cosmological observations indicate that the curvature of space is zero on average, a priori nothing prohibits that physical space is a hypertorus or one of its variants.

Nevertheless, there is no particular reason for the curvature of space to be strictly zero. The topological problem is posed in the same way with curved, non-Euclidean spaces. In each of these cases, the model of space that is usually considered—the simply connected one—is not the only one imaginable. One must also take into account its multiply connected variants.

The latter are representable by a more or less complicated convex polyhedron, whose faces are identified in pairs. Thus, we have just seen that the hypertorus "is" (in the sense of topological equivalence) a parallelepiped of Euclidean space whose opposite faces are homologous. However, just as for surfaces in two dimensions, Euclidean space is no longer sufficient as soon as the number of faces is greater than eight. The polyhedron must be constructed in hyperbolic space or spherical space, and be either inflated or deflated until all of the angles are exactly adjusted around a unique point (Figure 14.5).

Their volumes may be either finite or infinite. There is only one strict rule: all of the spherical topologies (spaces with positive curvature) have finite volume.

There are an infinite number of spherical spaces, but they have all been identified. The prototype (and the most voluminous of them, for a given radius of

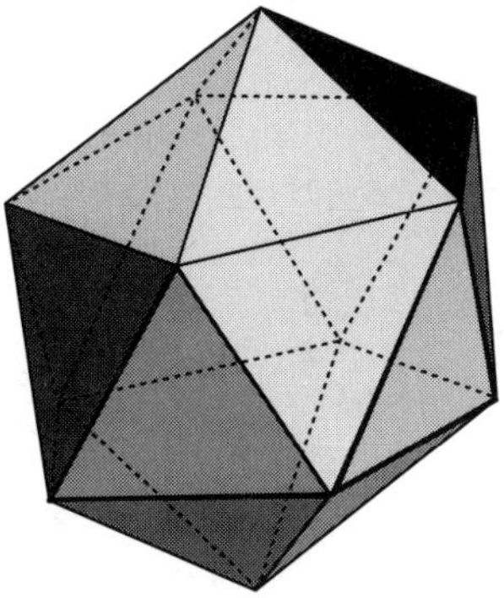

Figure 14.6. Best's closed hyperbolic space. Some finite and negatively curved three-dimensional spaces can be constructed by replacing the two-dimensional polygon of Figure 13.10 with a three-dimensional polyhedron. The shading in the figure shows which faces of the polyhedron must be glued together. If the corners of the polyhedron are correctly assembled, the space defined by the glued polyhedron must be a hyperbolic space, rather than a Euclidean space. This space, known as Best's space, has an icosahedron as its fundamental polyhedron.

curvature) is the hypersphere. This is the famous closed space with no boundary discovered by Riemann, and then used by Einstein and Friedmann in their cosmological models dating respectively from 1917 and 1922. On the other hand, Lemaître, in his 1931 model with a cosmological constant—which first introduced the concept which would lead to the big bang—preferred the topology of projective space (which he called *elliptic space*), obtained from spherical space by identifying each point with its antipode.

However, the most fascinating spaces are the hyperbolic spaces. As is already the case for surfaces, they are infinitely more numerous than the other spaces. The mathematicians have not succeeded in classifying all of them, but they know an infinite number of examples. One of the most interesting is representable by one of the five regular polyhedra, the icosahedron, in which one identifies all of the triangular faces two by two in a certain way; the interior finite, negatively curved space is called *Best's space*. Another is the Seifert-Weber space, formed starting from another regular polyhedron, the dodecahedron, with pentagonal faces.

Some of these spaces are closed (with finite volume), and others are open (with infinite volume). Only those with finite volume can be classified, even though they are far from being completely known. Many of their properties remain mysterious. For example, we do not know how to determine with certainty which of them is the smallest.

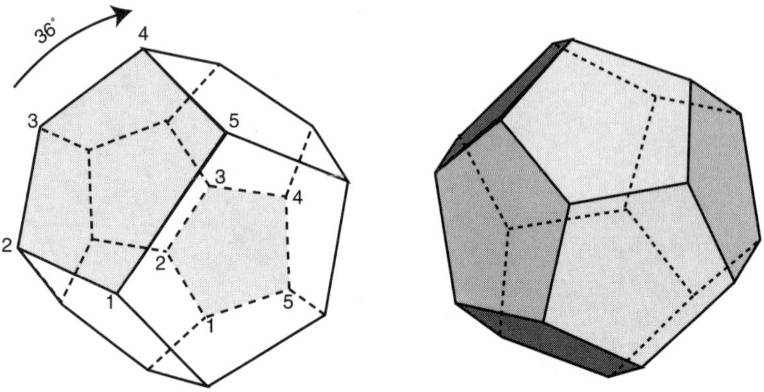

Figure 14.7. Poincaré's spherical space. If each face of a dodecahedron is glued to its opposing face after undergoing a 36° rotation (one tenth of a complete rotation), the resulting space is Poincaré's spherical space, a multiply connected variant of the hypersphere, whose volume is now 120 times smaller. However, this gluing can only be completed if one uses a positively curved dodecahedron, with edge angles of 120° rather than $\sim 117°$, as in Euclidean space.

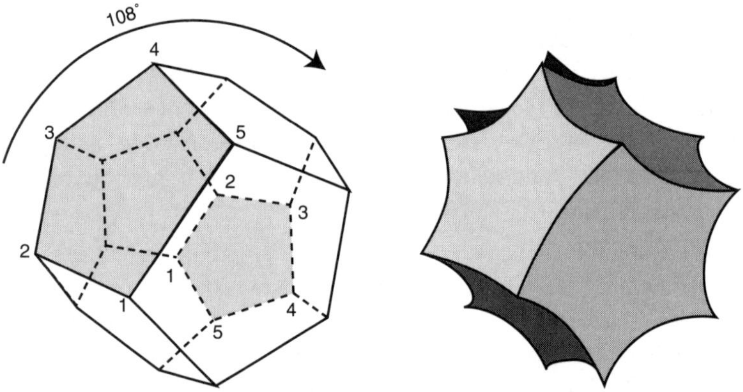

Figure 14.8. The Seifert-Weber hyperbolic space. If each face of a dodecahedron (at left) is glued to its opposite face after being rotated by 108° (three tenths of a complete rotation), the resulting space is the Seifert-Weber hyperbolic space, a multiply connected variant of the hyperboloid H^3, with finite volume. As for the octagon in Figure 13.10, the gluing can only be accomplished if one uses a negatively curved dodecahedron, with angles shrunk to 72° (on right).

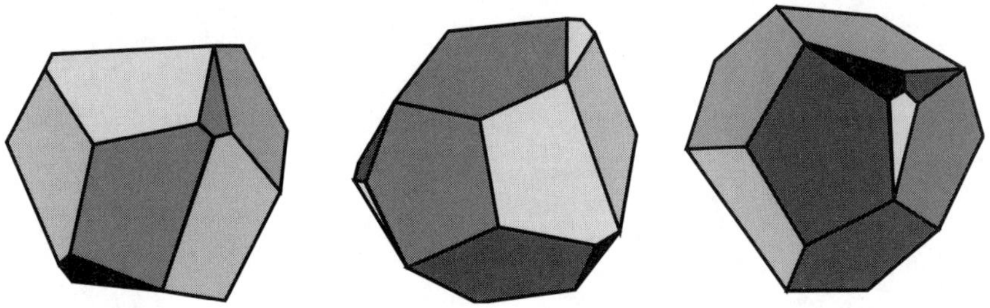

Figure 14.9. The smallest hyperbolic space. This closed hyperbolic space, discovered by J. Weeks, has for its fundamental domain an irregular polyhedron—three views of which are shown here—with 26 points and 18 faces (12 pentagons and 6 quadrilaterals), duly identified by means of complicated transformations. Of all known hyperbolic spaces, this is the one with the smallest volume.

We shall see that the general problem of the volume of space is of the greatest interest for applications to cosmology. A Euclidean space can have an arbitrary volume, no matter how small or how large. As for the volume of a spherical space, it can be no greater than $2\pi^2 R^3$, where R is the radius of curvature. This upper bound is none other than the volume of the ordinary hypersphere, which is simply connected and thus appears as the largest possible spherical space. All of the other spherical spaces, which are multiply connected and infinite in number, have a volume equal to a whole fraction $1/n$, for some n, of that of the hypersphere. Their volume can, in fact, be made as small as one pleases through the proper choice of identifications.

For closed hyperbolic spaces, each topology has a specific volume measured in units of the radius of curvature R; it is known that they cannot be made arbitrarily small. A theoretical limit was calculated in 1996, equal to $0.166R^3$, but no space has been constructed having precisely this volume. Until now, the smallest known hyperbolic space (that is to say, one whose fundamental polyhedron and holonomy group were able to be completely calculated) was discovered by an American, Jeffrey Weeks, in 1985. Its volume is equal to $0.94272R^3$.

Mathematicians work actively on these questions, but few of them are aware of the immense interest that their studies have for cosmology. Weeks has, nevertheless, posted a catalog of closed hyperbolic spaces called SnapPea on the Internet for the use of specialists. This program classifies all known spaces by increasing volume, and gives their properties: e.g., the structure of the fundamental polyhedron, the nature of the transformations in the holonomy group, the characteristic topological lengths, and so on.

The memory of the fragmented space retains a trace of time's decor.

—Maurice Couquiaud, *Trou noir*

15 Topology and Relativity

General relativity has successfully passed a certain number of experimental tests, but, like any physical theory, it is incomplete. One of the limits on its validity is well known: it does not take into account the microscopic properties of matter, described by quantum physics. Einstein was well aware of this, since, after putting the finishing touches on his gravitational theory in 1916, he passed the rest of his days attempting to unify gravity with the other physical interactions, in vain. Present day attempts at unification, including *superstrings*, *supergravity*, and *quantum cosmology*, tend to run into the same difficulties.

What is less known is that general relativity is also incomplete on the large scale: is space finite or infinite, oriented or not? What is its global shape? Gravitation does not by itself decide the overall form taken by space. The preceding examples, indeed, have shown that the curvature of space does not necessarily allow one to come to any conclusions about its finite or infinite character. In fact, in most cases finite spaces are topological manifolds admitting a hyperbolic geometry. These basic cosmological questions come from the global topology of the Universe, about which general relativity is silent. In fact, Einstein's theory only allows one to deal with the local geometric properties of the Universe. Its partial differential equations have as a solution a *metric tensor*, or, equivalently, the infinitesimal element of distance separating two events in space-time. This leads the study of the Universe, of its content, and of its physical properties to the problems of pseudo-Riemannian geometry. To a given metric element there are several, if not an infinite number of, possible topologies, and thus of possible models for the Universe. For example, the hypertorus and familiar Euclidean space are locally identical, and general relativity describes them with the same equations, even though the former is finite and the latter infinite. Likewise, the equations for a Universe of negative curvature make no distinction between a finite or an infinite space.

Models for the big bang have spaces of spherical, Euclidean, or hyperbolic type depending on whether their curvature is positive, zero, or negative. Their spatial topology is usually assumed to be the same as that of the prototypical spaces: the hypersphere, Euclidean space, or the hyperboloid, the first being finite and the other two infinite. However, there is no reason in particular for space to have a simply connected topology. In any case, general relativity says nothing on this subject; it is only the strict application of the cosmological principle, added to the theory, which encourages a generalization of locally observed properties to the totality of the Universe. Likewise, an ant in the middle of the desert would be convinced that the entire world is made of grains of sand.

Multiply connected topologies considerably enrich (as well as complicate) the field of cosmological modeling. Thanks to these, it becomes possible to consider models for the Universe where space is finite whatever its curvature may be, even if the density of matter and the cosmological constant are weak. They also lead, after detouring through highly sophisticated mathematics, to the reappearance of an idea dear to Plato and Kepler—the use of polyhedra to explain the secret 259→
architecture of the world.

At this stage, it would be wise to recall that cosmological models do not reduce to three dimensions, but in fact are four-dimensional space-times. Thus, to the problem of the topology of space is added that of the topology of time. What can be said about the topology of our space-time? An infinite spectrum of possibilities offer themselves as models. Nevertheless, some brief consideration of the physical properties of the Universe allows us to rapidly isolate a good number of inadmissible topologies. Here is why. Big-bang models for the Universe are homogeneous, meaning that their spatial part has a curvature that is everywhere uniform, and expanding. These two properties allow one to unambiguously distinguish slices of simultaneous space and the axis of cosmic time. We can therefore describe space-time as the mathematical product of a three-dimensional Riemannian space and the time axis.

This foliation considerably simplifies things. Time is represented by a space in one dimension whose points represent instants: a single number suffices to determine a particular time. Time possesses an ordered structure: on a line, one point is necessarily situated either before or after another point. (Let us note that

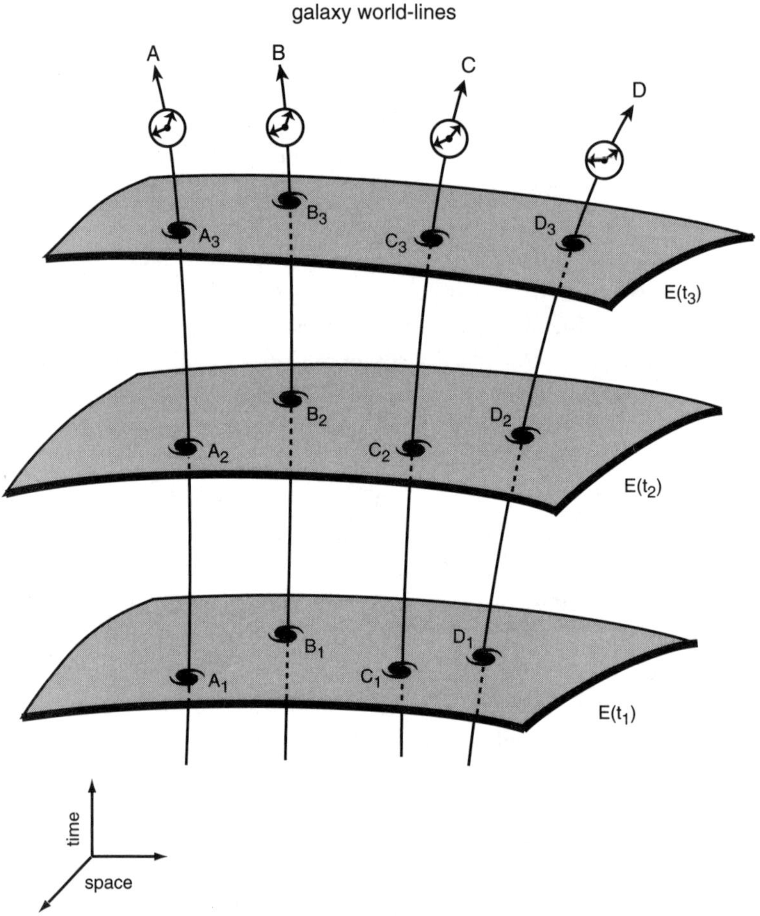

Figure 15.1. Cosmic foliation. In Friedmann-Lemaître cosmological models, space-time admits a particular foliation that allows one to unambiguously distinguish three-dimensional spatial sections $E(t)$ (which have constant curvature) and the one-dimensional time lines (known as world-lines) $A, B, C\ldots$ (which allows for the implementation of a universal time for events in cosmic history). Although it is permitted to identify points in the space $E(t)$ in order to render it multiply connected, one is forbidden from identifying points along world-lines, in order to preserve the principle of causality.

this would not be possible if time had several dimensions.) The topology of time is in the end rather poor; in contrast to that of multidimensional space, it only offers two varieties: the line and the circle. These two forms, in fact, correspond to two great philosophical conceptions, linear time and cyclic time. The latter has long prevailed in myths, such as that of the Eternal Return (see [Eliade 54]), but today it has been abandoned by physics because it violates the principle of causality, according to which cause must precede effect. As a consequence, any identification of points along the time axis is forbidden. In the framework of expanding-contracting cosmological models, one could, certainly, think to identify the big bang and the big crunch, that is to say the beginning of time with its end; but this operation is unlawful, for these points are singularities that are not even part of the Universe.

The question of cosmic topology therefore reduces principally to that of the spatial component of the Universe. For each type of possible curvature, as we have seen, there are variant big-bang models with multiply connected topologies. Most of the shapes mentioned above for three-dimensional space can be applied to the description of physical space.

Alas! For one drop of ecstasy
You would wring the universe.

—Victor Hugo, *Tout le passé et tout l'avenir*

16 Small, Strange Universes

Overwhelmed by the abstract definitions of fundamental polyhedra, universal covering spaces, and holonomy groups, the readers who are keen on cosmology have become, perhaps, impatient over the course of the preceding sections. However, their effort will now begin to bear fruit. By moving into the field of cosmic topology, we are going to discover how these mathematical notions play a fundamental role in the different ways of describing the Universe such as we observe it and the Universe such as it really is.

We live in a world of appearances, in the proper meaning of the term. Indeed, how do we perceive the shape of space, if not by observing the stars and following the paths of their light rays, which weave through the cosmic fabric? In a simply connected space, there is only, between a given observer and a given source of light, one single possible path of light—a single geodesic of space-time. From a telescope, a galaxy is seen only once in a given direction and at a given distance. There is no ambiguity. But what about in a multiply connected space? Let us imagine ourselves on the surface of a cylinder, where the straight lines (the geodesics) are the helices that connect the observer and the light source. These helical paths are infinite in number, distinguished by whether they complete one, two, three, etc., complete windings of the cylinder. We shall thus have an infinite number of light rays emitted by the source, all leaving in a straight line, but which will reach us from an infinite number of directions, giving an illusion of just as many different sources.

The same thing happens in three-dimensional space, and one may easily imagine the extraordinary optical illusion that a multiply connected space could cause if it possessed a complicated topology, folding into itself on a scale smaller than the radius of the cosmological horizon—the distance traveled by light since its emission 14 billion years ago, which is on the order of 50 billion light years.[1]

Figure 16.1 shows that if real space possesses a multiply connected topology, one would always have the illusion of living in the universal covering space, that is to say, in the unfolded version of the real space.

The models for the Universe that I call *wraparound* models are those whose spatial part is multiply connected and whose size is, in at least one direction, smaller than that of the observable universe. The French term *chiffonné* (crumpled) was suggested to me by a mathematician from Toulouse who attended one of my first lectures on the subject, in 1994. Although it has a certain poetic resonance, it crumples some of my colleagues; they remind me that, if one crumples a piece of paper, one creates angular points, ridges, and creases that change the regularity. However, there is no such behavior in a multiply connected universe: the space is rather topologically folded, without for that matter having either ridges or corners. I have in mind rather a supple fabric, that one may fold and crumple at leisure without creating false folds. Another objection one may make to the use of this term: the cylinder and the torus, the two simplest prototypes of multiply connected surfaces, hardly seem to us to be crumpled. But if we think instead of a multiply connected three-dimensional space, whose fundamental domain is a polyhedron possessing numerous faces, each face being identified to another by complex mathematical operations, we can, indeed, mentally picture the slightly misshapen interior of such a folded polyhedron as a "crumpled" universe.

A third thought experiment: let us go out on a clear night and contemplate the stars. Let us choose some: Altair in the Eagle, Deneb in the Swan, Spica in Virgo, and a dozen other less brilliant stars. Let us now imagine that these points of light are one and the same point in space, seen from different directions. Let us now begin the same mental operation by selecting a dozen other stars and bringing those back to a single entity, and so on. In total, we will have reduced the thousands of visible stars to only a hundred or so distinct points. To do this,

[1]Just so that the reader is not surprised by these figures: light does not move faster than light, but because of the rapid expansion of space, in x billion years it covers distances that are *greater* than x billion light years. See the later discussion on pp. 104 and 300.

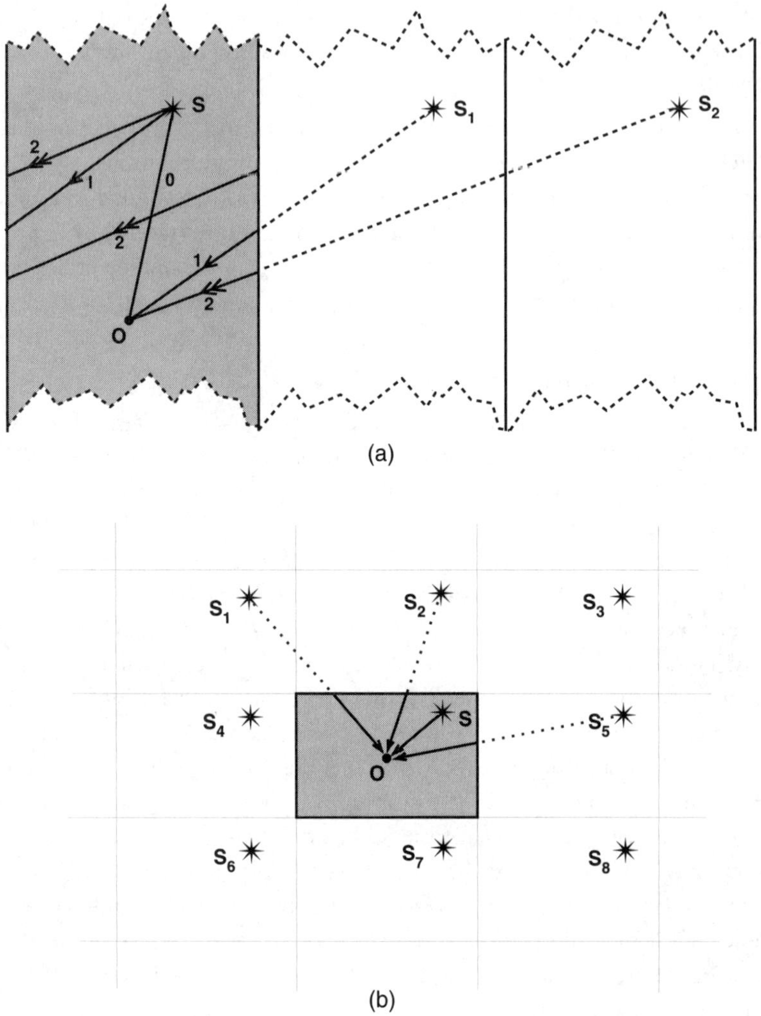

Figure 16.1. The illusion of the universal covering. (a) In the case of the cylinder, real space is the interior of the gray band, whose edges are identified. Light rays emitted by the source S arrive at the observer O along several paths, corresponding to the differing numbers of loops they have made around the cylinder. The observer O, therefore, has the illusion of seeing several distinct sources (S, S_1, S_2, etc.) aligned within a fictional space made up from the juxtaposition of endlessly repeating bands—the universal covering space of the cylinder. (b) In the case of the torus, the real space is the interior of the gray rectangle, whose opposite edges are identified. The observer O sees rays of light from the source S coming from several directions. He has the illusion of seeing distinct sources (S_1, S_2, S_3, etc.) distributed along a regular canvas that covers an infinite plane.

it is indeed necessary that the physical space that contains the stars is, in a certain way, crumpled. As we shall see, it is not a question of whether cosmic space is crumpled on such a small scale, to the extent that stars visible to the naked eye, all situated less than a thousand light years from us, show themselves in several copies. But if one replaces the stars by distant galaxies, our thought experiment fully justifies itself.

In a wraparound universe, the trajectories of light rays emitted by any light source whatsoever take a number of paths to arrive at us, each following the folds of the space-time fabric. From each star, an observer therefore perceives a multitude of *ghost images*. Thus, when we see billions of galaxies filling a space that we believe to be unfolded and extremely vast, it could just be an illusion; these billions of galaxy images could have been created by a smaller number of objects, present in a wraparound space of lesser extent. The latter creates an illusion of the infinite.

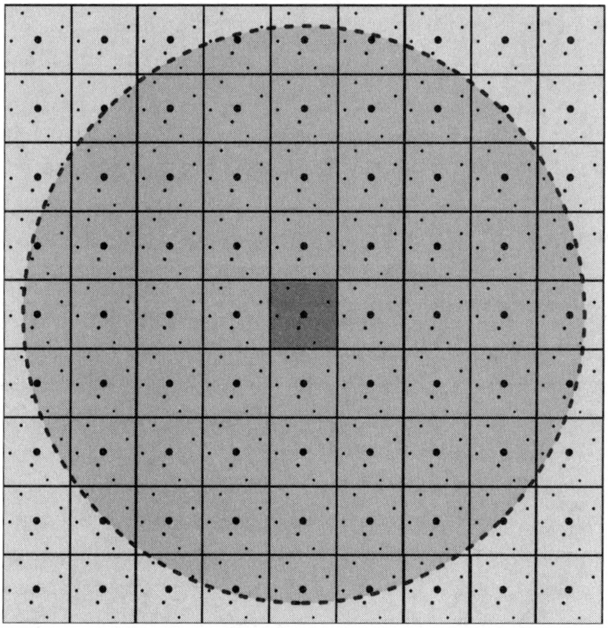

Figure 16.2. The illusion of the infinite. In practice, space-time horizons bound the view of the observer, because of which the observer only perceives a limited portion of the universal covering space. In this toric space (center square) the observable universe is found within a circular horizon (dotted line), containing only a finite number of replicas of the fundamental cell.

It is here that the metaphor of the Hall of Mirrors, developed at the beginning of this book, takes on its full meaning. God summons an astrophysicist and says to him, "I am going to place you in a room called space, and you are going to try to find its shape." "Okay," replies the scientist. Placed into a state of weightlessness so as not to be in contact with any possible surfaces, he enters a completely darkened hall. "I can do nothing, without light," he says. God, who has refrained from specifying that the walls of the room are lined with mirrors, lights a few candles suspended here and there. The astronomer sees the appearance of thousands of candles spread in every direction, some nearby, others quite distant; moreover, he notices that the paths taken by light rays are rectilinear. From this he concludes that he finds himself in an infinite Euclidean space. He cannot tell the difference between the authentic candles and their images reflected by the walls. In other words, he does not see the fundamental domain, the rectangular room covered with mirrors; rather he sees the universal covering space.

Of course, it is not a question of believing that cosmic space, if it is wrapped, is closed in by walls covered with mirrors! Instead, the fashion in which its topology comes into play, because of the identification of certain points, creates a space that is seemingly different from the real space, in an illusion similar to that of the Hall of Mirrors.[2]

Attention nevertheless! Cosmology deals with the Universe considered as a whole, which is described not only as a space, but as a space-time, shaped by matter and energy. A small wraparound Universe is thus a cosmological model belonging to the family of solutions to the equations of general relativity discovered by Friedmann and Lemaître, but whose spatial part is multiply connected, with finite volume (even though the curvature could be negative or zero), ultimately smaller than the volume of the observed universe.[3]

To construct multiply connected spaces, mathematics teaches us that one can take as a starting point one of three ordinary (simply connected) types of space with constant curvature; by identifying certain points with each other, one changes the shape of space and renders it multiply connected.

The simplest example is the case where our own space, with curvature of zero on average, would be a cubic hypertorus with a size smaller than fifty billion light years. In this case, rays of light would have had the time to make a complete tour of the Universe. This would imply that each cosmic object, each galaxy for exam-

[2]There is, nevertheless, a notable difference in the orientation of the images: in a mirror, your reflection faces you; while in a hypertorus, you would see your own back.

[3]This restriction to finite volume could be lifted. A wraparound space is closed in at least one spatial direction (like the cylinder), but not necessarily in all. The important part of our argument is that it creates ghost images. Indeed, would not an infinite tunnel covered in mirrors also multiply the images to infinity?

ple, should appear with several ghost images, observable in different regions of the sky. More generally, one could see in different regions of the sky the repetition of an identical collection of galaxies; the distant heavens of large structures, such as galaxy clusters and superclusters, seem, in fact, to be made through the juxtaposition of blocks which are nearly alike. A small wraparound model for the Universe would thus furnish a natural explanation of the fact that the observed universe, although inhomogeneous within the interior of a block containing galaxy clusters and superclusters, seems homogeneous on a larger scale—quite simply because one would see the same block repeated everywhere, by a sort of mirror effect!

If space has positive curvature, it could be wrapped while having one of many different possible topologies, such as that of projective space, or even the more subtle one of the Poincaré space, which has the dodecahedron as its fundamental domain. As we have seen, the greater the number of identification operations between the faces of the fundamental polyhedron, the smaller the volume of the corresponding wrapped space.

Finally, if space has a negative curvature, once again an infinite number of possibilities offers itself for cosmological modeling; the Seifert-Weber space, made from a dodecahedron; the Best space, constructed from an icosahedron; or even the Weeks space, whose volume is the smallest known, could be incorporated perfectly well into a relativistic big-bang model for the Universe. All of these spaces have finite volume, but it is permissible to construct wrapped Euclidean or hyperbolic models of space that have infinite volume (although all of the spherical spaces are finite). A three-dimensional Euclidean cylinder, or even a hyperbolic 272→ horn, are spaces that are closed in certain directions but infinite in others. We may easily imagine that such spaces are strongly anisotropic, because of which their application to cosmology is more delicate than that of finite wrapped spaces.

The wraparound Universe has exactly the same dynamics, the same temporal behavior, as the unwrapped Universe. It is in perpetual expansion or in expansion followed by contraction depending on its curvature and the value of the cosmological constant. However, each source appears to us in multiple copies in space-time because its radiation comes to us from different directions and different epochs in the past (Figure 16.3). The objects that seem distant to us are in fact temporally distant, and not spatially distant, since they are the images of nearby sources whose light rays, by following a very rolled up geodesic, or by turning several times around the same loop, have taken more time in coming to us. The spectral shift of light from a galaxy toward the red—which allows us to conclude that the Universe is presently expanding and which, in standard cosmology, serves to evaluate distances thanks to Hubble's law—in fact measures a *temporal remoteness.* It depends exclusively on the time of emission of the radiation that reaches us today. A "distant" ghost in a multiply connected space thus has the same redshift, and

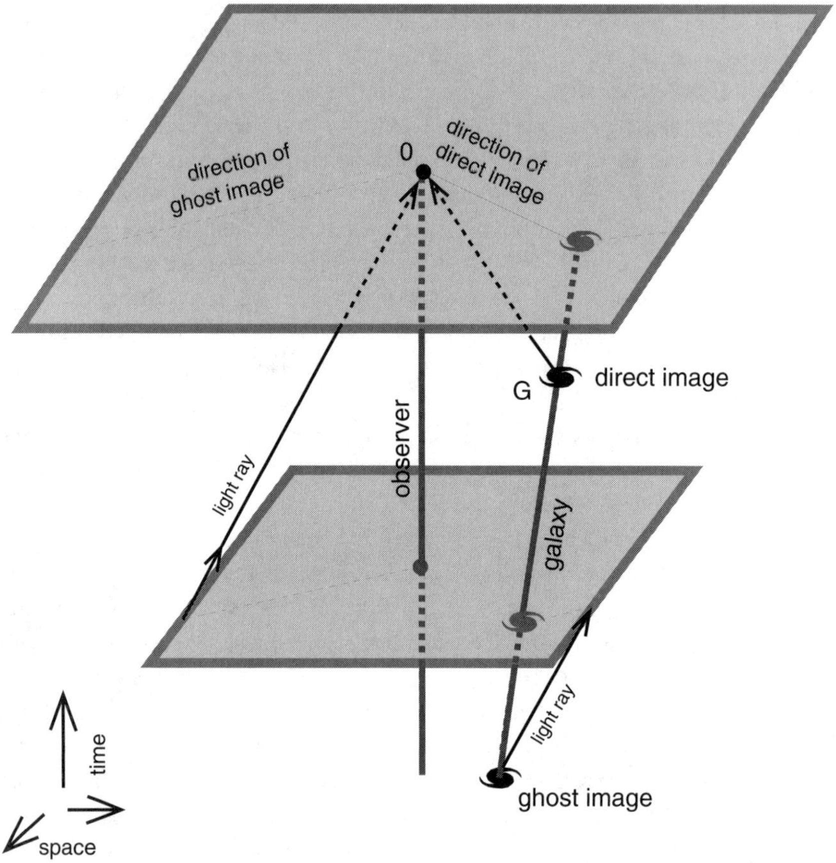

Figure 16.3. The ghosts belong to the past. Light rays come from the past. In this space-time representation of an expanding toric universe, rays of light received at the same time by an observer O come from the same galaxy G, but they have been emitted at different epochs. The "original" image of G is the most recent; the ghost images are necessarily older.

appears to us in the same position, as a "real" source occupying its place in the universal covering space.

The distinction between ghosts and real sources is not absolute; it depends on the position of the observer. Among all of the multiple images received from a single source, one can label as "real" the one whose redshift is the smallest, that is to say, the one that is closest to us temporally. This is only a convention: the other images, which can be called ghosts, are also perfectly real. However, it has the advantage that one can attempt to determine the size and shape of the fundamental polyhedron of the Universe. Observationally reconstructed, it is the ensemble of the original images that surround us and its volume extends up to the first detected ghosts. The position that one occupies in the representation of the basic cell is therefore in the center by definition. From his own point of view, every observer is at the center of the Universe, conforming to the idea that the Universe has no privileged point.

A space that is multiply connected and finite, but whose volume extends past the horizon—the frontier of the observable universe—would not, at first sight, distinguish itself from the associated simply connected model, that is to say, from its universal covering space. The determination of the volume of real space must therefore be the point of departure for every topological exploration: the smaller the space, the greater the chance for its topology to be detectable. Outside of the horizon, the wraparound model remains a purely theoretical point of view; it remains indistinguishable from the associated simply connected model, and does not affect any standard cosmological result. On the other hand, if it is much smaller, it is an occasion to considerably enrich our information. A small wraparound Universe would offer to the observer several images from the observer's past, more numerous the smaller it is, so much so that one would have under one's eyes a sort of historical photo album of celestial bodies that are born, live, and then die, including our own galaxy. We already knew that astronomers are historian-geographers since, by looking far into space, they look far into the past of the Universe. In simply connected spaces, objects seen at a given moment in the past are necessarily distinct from those seen in the more recent past; in a wraparound Universe, they can be the same objects seen at different epochs of their lives, which confers to astronomers a certain advantage over historians of societies.

The Universe (which others call the Library) is composed of an indefinite, perhaps infinite number of hexagonal galleries. In the center of each gallery is a ventilation shaft, bounded by a low railing. From any hexagon one can see the floors above and below—one after another, endlessly. The arrangement of the galleries is always the same. ... In the vestibule there is a mirror, which faithfully duplicates appearances. Men often infer from this mirror that the Library is not infinite—if it were, what need would there be for that illusory replication?

—Jorge Luis Borges, *La biblioteca de Babel*

17

Topological Mirage

A distant galaxy sends its light in every direction. One can model it with a point-like source of light, projecting its rays around it in an isotropic fashion. In a simply connected space, only one of the emitted rays reaches us, that one that takes the shortest path between the galaxy and us, and gives rise to the unique image that we see.

As I have already mentioned in Chapter 6 ("Celestial Mirages"), there may be exceptions, caused by the variation in the *local* curvature of space due to the presence of a massive body placed along the line of sight of a galaxy in the background. The light rays emitted by the latter are deflected by the intermediate body, which serves as a lens, and their multiplied trajectories give rise to a gravitational mirage, made up of several images of the same galaxy. Numerous optical illusions of this type have been detected, but the ghost images are always very close to each other, grouped together in the same direction, in the background of the lens.

Therefore, aside from these gravitational mirages, a single image from each distant source reaches us if space is simply connected. On the other hand, if it is multiply connected, light rays take a multitude of possible paths, each path producing a different image of the source. Here we have a topological mirage, where space itself plays the role of a lens. Since this type of mirage is *global*, and not local, the multiple images have no reason to be grouped in the same direction of space and the same temporal epoch. On the contrary, they are distributed in every direction in space and in every slice of the past.

For example, in a hypertoric Universe, one ray comes directly to us, while a second seems to come from nearly the opposite direction. In fact, every real image formed by rays that have followed the shortest path to reach us is accompanied by a multitude of ghost images, coming from rays that have made a certain number of tours around the Universe, in one direction or another. In principle, the potential number of ghost images is infinite. In practice, it is limited in the first place by the finite size of the observable universe, and in the second place, by the fact that the images may be highly attenuated or masked, or of a brightness too weak to be visible. The power of telescopes is therefore a limiting factor; a galaxy cluster cannot be viewed at a very great spectral shift. Also, and above all, we hit a temporal barrier: potential ghost images could correspond to an epoch of the past so distant that the object in question had not yet formed, and by consequence had not yet emitted any light! Let us not forget that, in cosmology, distant also means old, and this is even more true for ghost images.

Because of these various limitations, the number of visible images and their characteristics depend on the geometry and size of the wraparound space. The smaller this is, the closer and more numerous the images are, and thus easier to see. Inversely, the larger the space, the less its topology manifests itself. The cosmological horizon plays an exceptional role in all of observational cosmology. There are several types of horizons in cosmology: the particle horizon, the event horizon, and the photon horizon. The one that I speak of here is the photon horizon, the locus of those points in space such that the apparent speed of expansion reaches the speed of light. Beyond this, no object can be observed because the light that it emits has not had the time to catch up with us since the Universe began expanding, which is to say around 15 billion years ago. The photon horizon is therefore the frontier of observable space, a sphere whose radius is at least 15 billion light years.[1] If the size of real space is smaller than this, light rays have the time to make a complete circuit around the Universe before reaching us, and thus give rise to many phantom images. These effects become blurred when the size of real space approaches that of the horizon.

A few concrete numbers can give an idea of the gigantic cosmic mirage created by a wraparound topology. To calculate the potential number of multiple images associated with each original source present in the fundamental domain, it suffices to divide the volume occupied by the observed sample of the Universe by the volume of real space. Let us take, for example, a Euclidean space, which

[1]The precise value of the horizon radius calculated at the present epoch depends strongly on cosmological parameters: the curvature and the cosmological constant; see pp. 104 and 300.

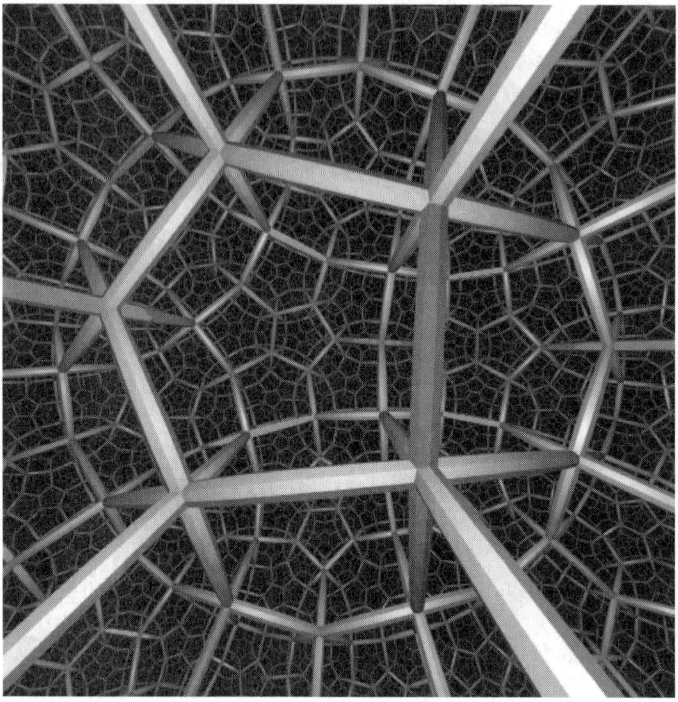

Figure 17.1. A cosmic hyperbolic crystal. A wraparound universe has a remarkable topology that allows an identification of physical space with a polyhedron, whose multiplied image makes up the world of appearances. To represent the structure of apparent space reduces to representing its "crystalline" structure, of which each cell is a reproduction of the fundamental polyhedron. Here we illustrate the apparent structure that would be offered by the closed hyperbolic Seifert-Weber space (see Figure 14.8). Viewed from the interior, one would have the impression of living in a cellular space, paved out to infinity by dodecahedra deformed by optical illusions. (See Plate III. Image from the film *Not Knot*, ©A K Peters, Ltd.)

has expanded for fifteen billion years, but whose topology is that of a cubic hypertorus whose size is equal to five billion light years. The volume of real space is therefore that of this cube, that is, 125×10^{27} cubic light years. Imagine that we have at our disposal powerful telescopes capable of detecting all objects—galaxies, quasars, clusters, etc.—up to a spectral redshift of 4 (at such distances, our present telescopes primarily find quasars, which are intrinsically the brightest objects). The observable volume would be around $5,600 \times 10^{27}$ cubic light years. There would thus potentially be 45 (5,600 divided by 125) ghost sources for each original source.

Figure 17.2. Deep space as seen by space telescope. The powerful telescopes of today capture images of millions of galaxies, of diverse shapes and colors, and seen at different ages. In this type of snapshot, nothing allows us to say if all of the galaxies are "original," or if the older galaxies are ghost images of more recent galaxies. The hypothesis of a multiply connected universe such as that of Figure 17.1 can therefore not be dismissed. (See Plate IV. Image from STScI/NASA.)

As we have mentioned previously, the proportion between ghost and original sources grows with the volume of the observed sample. If we now make use of an ideal telescope having access to the entire volume of the observable universe, and therefore capable of looking into the past up to a spectral shift of 1000, to an epoch prior to the formation of stars, galaxies, and galaxy clusters, then we would in theory be in a position to make a census of all possible cosmic objects, original and ghost. The proportion would grow to 270 potential phantom sources for each original source. In fact, one could encase 270 copies of the fundamental cube within the volume of the observable universe, and each object in the fundamental cube has a ghost image in each copy.

The case of hyperbolic wraparound universes is even more spectacular. A Universe endowed with a density equal to one third of the critical density but having the topology of the Weeks space (Figure 14.9) would be repeated 25,000 times within the observable sphere. This impressive number does not necessarily imply that the ghost images would be easy to discover. In practice, the closest ghosts would be situated at already considerable spectral shifts, comparable to those of the most distant quasars detected today.

For the cosmologist concerned with concrete observations, who is not satisfied with pure mathematical speculation, some essential questions arise: Is the topology of the Universe detectable? If we live in a small wraparound Universe, how do we differentiate between illusion and reality? Are certain galaxies that we see ghosts, and can we prove it?

Like rays coming from a powerful source, the reality of our world fades and is lost the more it moves away from us. My hand, which raises a hammer, holds the real, but my gaze, lifted to the highest places of the night, only reaches Ideas, Ghosts, an elusive surge of dreams that is going to die at the edge of that which is not.

—Jean Tardieu, *Le Ciel ou l'irréalité*

18
Hunting for Ghosts

Humankind acquired the conviction that the Earth was round well before they were able to circumnavigate it. Our rockets will never make a complete circuit through space, but we can hope to capture the light rays that have traveled to our location. Emitted in the distant past, they have perhaps had enough time to complete one or several cosmic circumnavigations. Weaving through the secret architecture of space, they bring us direct or indirect signs of the topology, and can eventually convince us that it really is wrapped. This is the prospecting work that a few dozen researchers from all over the world have begun working hard to perform.

Two large classes of observational tests have been proposed, both founded on the effect of topological mirages. The first is three-dimensional, since it applies to localized light sources like galaxies, quasars, and galaxy clusters, spread in depth in space and in time; the second is two-dimensional, since it uses the cosmic microwave background, that diffuse source throughout space emitted by the hot plasma of the primitive Universe at a unique epoch of the past, and which therefore seems to us to be situated at a unique distance.

Each of these two classes of tests can itself be subdivided into two types: a first method attempts to individually recognize multiple images of a single object: the same galaxies, the same clusters, etc., for localized sources, the same regions of the emitting plasma for diffuse radiation. The other method, which is more powerful, gives up on individual identification and focuses itself exclusively on finding evidence for statistical correlations between the positions of radiation sources. For that which comes from localized sources, *cosmic crystallography* relies on the analysis of certain regularities in the three-dimensional distribution of celestial objects; for that which comes from the diffuse cosmological background, the method of *matched circles* is founded on the angular distribution of temperature fluctuations of the fossil radiation.

Let us analyze in detail the strong and weak points, the disappointments and hopes, connected to each of these methods. In the first place, let us examine the attempts at direct recognition of individual objects. We immediately run into a difficulty: every ghost image of a single object presents different aspects of itself. The principal cause of this feature is evolution in time. The longer the trajectories taken by light rays, the more we look back in time (the amount of time taken for the light to reach us), and the more the object seems to us to be in an earlier state in its evolution. If this evolution is rapid, the object will not be recognizable!

Some sources have a relatively short time-scale of stability, which is to say that they keep the same aspect only ephemerally, because of which the time necessary for a light ray to make a circuit around the Universe and come back to us a second time suffices to render them unrecognizable, as long as the circumference of the Universe is not itself sufficiently small. Quasars, active galactic nuclei, and radio sources, as well as all of the galactic configurations of remarkable morphology, remain roughly stationary for less than one billion years, that is, the time for a photon to travel a little more than a billion light years. This length is not likely to be enough to serve as the circumference of space, since certain observational constraints stop us from supposing that the Universe is as small as that. These changing sources can be used in statistical investigations of the type I will describe farther on, but are not suitable for a ghost hunt. Individual galaxies can conserve their morphology for a longer time, but their very great number obviously makes their analysis infinitely laborious.

The most stable objects, and thus the easiest to identify, are regular galaxy clusters and giant low-activity galaxies. Regular clusters are nearly exclusively made up of elliptical and spherical galaxies, and contain a very large fraction of giant galaxies. They are themselves spherical, so much so that the angle from which we observe them has no importance. The task is thus to fix a certain number of individual characteristics capable of differentiating or identifying them: e.g., the total number of galaxies, the composition of the stellar populations at its heart, the radius of the cluster, the density distribution from the center to the exterior, the x-ray emission from the gas inside the cluster, and so on. The time-scale of stability of a rich cluster of galaxies is about five billion years, a sufficiently long time to allow photons to give us ghost images that will not be too much altered by evolution.

Beyond the fact that ghost images reveal a cosmic object at several stages of its history, there are supplementary effects that complicate the task of the ghost hunter. The longer the path taken by light, the smaller the dimensions of a source seem to us, and the weaker its brightness, since these two quantities decrease as the inverse of the square of the distance (recall that all of the ghost images occupy the same spatial position, meaning that they are at the same distance from the observer, but that their temporal distance is interpreted like a spatial distance in the apparent universe, that is to say, the universal covering space).

Finally, each ghost image shows us the original from a different orientation. One day, during a college conference, while I was presenting to schoolboys a slide showing a beautiful spiral galaxy seen in profile, a pupil asked me to show him how this galaxy looked viewed from the front. In a simply connected space, it is obviously impossible to perceive the same object in two different orientations, at least without moving oneself in space (and for a galaxy one would need to travel several million light years). In a small wraparound space, however, one can rest tranquilly in one's armchair and admire at leisure the same spiral galaxy from every side . . . as long as you have located its multiple ghost images! Our own galaxy, the Milky Way, which as seen from Earth is necessarily seen from its interior, could be viewed "from outside" if there were identifiable ghosts for it.

In conclusion, the multiple images of a given object appear to us in different directions, with different brightnesses, from different orientations and at different epochs of its history. Therefore, they do not resemble each other. This is why it is difficult to know if the Universe is wrapped through direct identification: a great part of that which we take for original galaxies could well be nothing but ghost replicas of a limited number of real galaxies, replicas that it would be impossible, in the absence of a precise stamp, to recognize as such.

Researchers have, nevertheless, made the attempt. Boudewijn Roukema and Alastair Edge have put forth the hypothesis that the strong luminosity in the domain of x-rays emitted by certain galaxy clusters could constitute a distinctive stamp. This actually comes from gas internal to the cluster, in which the galaxies bathe, which is heated to several million degrees and emits this high-energy radiation, which is detectable at a very great distance. These x-ray clusters are, moreover, dynamically stable; they may therefore prove to be more efficient than other localized sources of radiation at displaying the topology. In 1997, Roukema and Edge suggested that the Coma cluster, cluster RXJ 1347.5-1145, and cluster CL 09104 + 4109, viewed at differing spectral redshifts, but all three of which exhibit strong x-ray luminosity, could be the ghost images of a single, unique cluster in a non-orientable hypertoric topology (a sort of three-dimensional Klein bottle). Their hypothesis is difficult to refute, but the chances that this is a pure coincidence are too great for such detection of ghosts to be credible.

> The man who often contemplates the Heavens admires suns which no longer exist there, but which he perceives all the same, thanks to this phantom ray, in the illusion of the universe.
>
> Villiers de L'Isle-Adam, *L'Ève future*

19 Midnight Suns

Although the observational search for direct topological effects is not conclusive, it nevertheless sets limits on wraparound universe models, and in particular constraints on the size. In particular, present data allow us to eliminate the possibility of a very small Universe. Indeed, despite attempts at detection, we have never seen phantom images of the Sun, nor of our galaxy, nor even of nearby galaxy clusters. In 1913, in the framework of a controversy on the reality of curved space, Barrett Frankland remarked that, in a spherical space, each star would show its front and back sides in two opposite directions of the sky. Thus, one should see two Suns: the ordinary Sun, meaning the image of the star such as it was 500 seconds ago (the time taken by light to travel the 150 million kilometers that separate us from it), and the anti-Sun, situated at the antipode of the celestial sphere and showing it such as it was a long time ago (the time taken for light to travel the circumference of the Universe). From this, Frankland concluded that there should no longer be any darkness on Earth, since the anti-Sun would shine during the night! Consequently, according to him, the hypothesis was false, and Frankland rejected the existence of a spherical space.

As it happens, in a premonitory article of 1900, Karl Schwarzschild had already discussed the problem, but without resolving to exclude the possibility of a spherical space; according to him, an anti-Sun might well exist, but it was not visible because light, in traveling through space, experienced a sufficient amount of absorption to erase this secondary image. Willem de Sitter himself took up the question once again in 1916. Einstein had just published his theory of general relativity, which implied a real curvature for physical space, thus causing the concept of spherical space to pass from the realm of pure mathematical speculation to that of physically plausible hypothesis. De Sitter simply remarked that, for the anti-Sun not to be visible, it sufficed for the time taken by light to travel around the Universe to exceed the age of the Sun. The radius of physical space was at that point assumed to be constant, since the expansion of the Universe had not yet been established. In fact, this discovery furnished a new argument, since, in an expanding space, the time taken for light to complete a circumnavigation of space is lengthened. Imagine an ant placed on the surface of a balloon being blown up: however much time the ant has at its disposal, it will never finish making it all the way around the balloon if its circumference grows too quickly.[1]

The fact that we have not detected ghost images situated at the antipodes of celestial objects does not exclude the possibility of a finite space with a wraparound topology. It simply imposes constraints on the minimal size.

Let us call the smallest dimension of the wraparound space L. For example, in the case of a hypertorus, L is the smallest edge of the fundamental parallelepiped. One expects to find at least one ghost image at an apparent distance on the order of L. However, as shown by the Russians Dimitri Sokoloff and Victor Shvartsman, who pioneered this type of investigation in the 1970s, no image of a galaxy resembling ours has shown up within a distance of 25 million light years.

This lower bound on the size of a small wraparound Universe can be raised by using other distinctive objects—well known nearby galaxy clusters, such as the Coma cluster. The Coma cluster is the best-studied galaxy cluster in the northern hemisphere. We know its total luminosity, its characteristic elliptic shape, and that at its center it contains two giant galaxies, one of which is surrounded by a dozen small satellite galaxies, while the majority of other clusters, although dense in their centers like Coma, have only one dominant central galaxy. We also know that it could be dynamically stable for several billion years. In brief, Coma should be easily recognizable at great spatial and temporal distances. However, it is found nowhere in the zone explored around it, namely, a sphere whose radius reaches 400 million light years, as Richard Gott noted in 1980.

This type of test can be performed not only on individual objects, but on catalogs of objects, studied in a statistical way. In 1974, Sokoloff and Shvartsman were the first to propose measurements based on the analysis of two large surveys of clusters: the Abell survey, which catalogs 2,712 rich clusters, and that of

[1]The real reason is that the ant has a lifetime that is too short. In fact, although it may be quite surprising, it can be shown mathematically that an immortal ant would finish its tour around the expanding balloon in a finite time, whatever its rate of expansion may be.

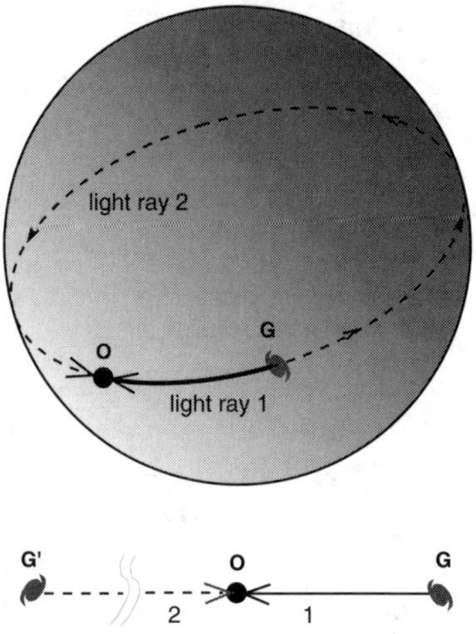

Figure 19.1. The circumnavigation of space. In a hyperspherical space, an observer O should in principle see light rays emitted by a galaxy G coming from two opposite directions. However, if the expansion of the universe is rapid enough, light ray 2 will never succeed in making it around space, because of which the observer will never see the phantom image G'.

Zwicky, which contains 9,730 clusters of all types. All have a redshift smaller than 0.2 and are therefore found at a distance of less than 2.5 billion light years. At this distance, the rich clusters are necessarily original; in fact the closest clusters that are observed are also poorer, and since the natural tendency of clusters is to enrich themselves over the course of their evolution, the poor clusters cannot represent the rich clusters in a later state.

Another observation of this type has been made by Gott around the region of Serpens-Virgo, which contains several characteristic rich clusters that are good subjects for detection: this region is found at a distance of around 1.5 billion light years and is not identifiable in any nearby neighborhood.

The constraint on the size of space imposed by these various observations is $L > 2.5$ billion light years. There is still a lot of room up to the distance of the cosmological horizon. The radius of the latter depends on the value of the cosmological parameters. For example, in a Euclidean Universe (with zero spatial curvature) without cosmological constant, it is 18 billion light years, while in a Universe with the same curvature but with a cosmological constant of 0.7, it reaches 50 billion light years. The hypothesis of a wraparound space that is smaller than observed space therefore remains perfectly viable.

There are universes of schorl, of quartz, of serpentine,
Cosmoses of graphite and fires of feldspath.

—Strada, *La Genèse universelle*

20 Cosmic Crystallography

If the hunt for ghosts seems to be a nearly impossible task, the wrapping of the Universe, inscribed in the distribution of distant galaxies, may better reveal itself in statistical terms. To prove the existence of a multiply connected topology, it would suffice to uncover repeated patterns in the arrangement of galaxies. As Karl Schwarzschild suggested as early as 1900 (Figure 20.1), if one were to find galaxies spread in a regular network, with images of the same galaxy repeating itself at equivalent points within the network, one would conclude that our Universe is a hypertorus. Patterns that are similar but more complicated would likewise reveal more subtle topologies. What type of pattern should one look for, without making an a priori hypothesis about the topology of space?

The models for a wraparound Universe have a common property: the apparent distance between two ghost images of the same source always belong to a definite collection of values, connected to the size of the fundamental polyhedron. The relations between these values are analogous to those that link the atoms of a crystal. Roland Lehoucq, Marc Lachièze-Rey, and myself have proposed a *crystallographic* analysis of the distribution of cosmic objects, aimed at eventually uncovering indications of a multiply connected Universe.

Why this reference to crystallography, a branch of the physics of solids? When one examines a crystal one is immediately struck by the existence of natural planar faces, making very precise angles between themselves, as well as an overall con-

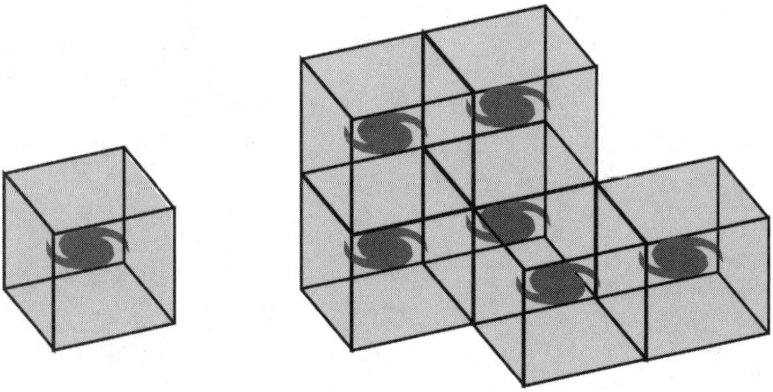

Figure 20.1. Multiple milky ways. In 1900, the astronomer Karl Schwarzschild already had imagined that our galaxy, the Milky Way, could repeat itself endlessly within a regular cubic framework, thus giving the illusion of a space that is vaster than it really is.

struction through the juxtaposition of cells placed side by side. Moreover, crystals are at the heart of most of the optical phenomena that trick our eyes. For example, the majestic sun dogs (or parhelions), which have so much affected people's imaginations, are due to the refraction of light by hexagonal ice crystals suspended in the atmosphere.

←250 In 1610, Johannes Kepler discovered the hexagonal shape of snow crystals, but his work then fell into obscurity. At the end of the eighteenth century, the naturalist René Just Haüy accidentally dropped a piece of calcite—a crystal of rhombohedric shape—on the floor of his laboratory. He noticed with astonishment that each of the shattered pieces retained a rhombohedric shape. Haüy then asked himself what he would obtain by continually breaking the calcite into smaller and smaller pieces. He conjectured that at some given moment he would reach a smallest part that still kept the shape of the calcite, a sort of fundamental rhombohedric cell (that he went on to call the "molécule intégrante," a kind of unit cell) which, when assembled in space, would make up calcite. He performed the experiment and in fact discovered the smallest rhombohedron below which the calcite decomposed into other bodies. Generalizing this principle to all crystals, he sought to determine the fundamental cell for each of them. He found six types of structures, characterized by their spatial symmetries. Then, it only remained to classify crystals. Crystallography was born.

The unit cell of the crystal bears more than a passing resemblance to the fundamental domain of multiply connected space. The cosmic crystal is observed space, made up from the repetition of the same domain. One difference from

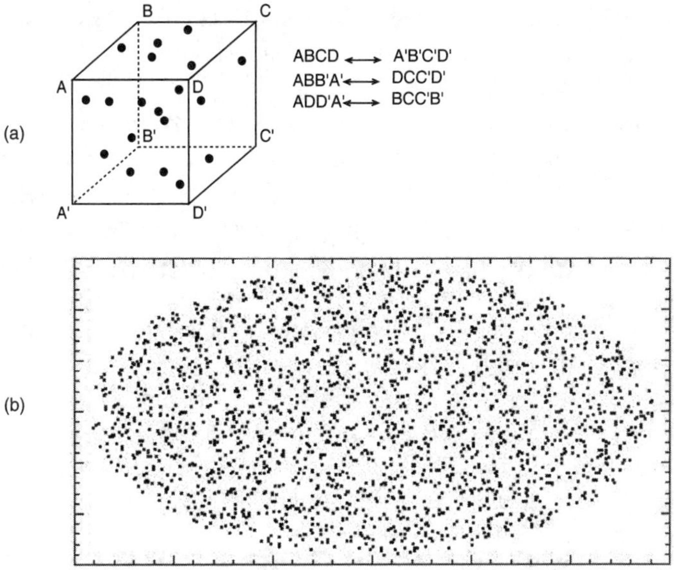

Figure 20.2. A cosmic unit cell. (a) The space here is a hypertorus, represented by the interior of a cube with edges that are five billion light-years in length, whose opposite faces are identified. Twenty galaxies are distributed randomly within the space. (b) This shows a numerical simulation of the appearance of the sky after accounting for the effect of topological mirages. Each real galaxy leads to about 50 ghost images spread throughout the celestial vault. Among the hundreds of visible images, it is impossible to separate the real images from the ghosts.

a mineral crystal is that the cosmic crystal is in space-time: the tesserae—the pieces of the cosmic mosaic—are the same piece of space viewed at different ages, which complicates the task of the astronomer who seeks to reconstruct through observation the fundamental polyhedron of the Universe. A second difference is that we find ourselves in the very center of the cosmic crystal.

It is therefore not surprising that cosmic crystallography adapts some of the methods used in the physics of solids (e.g., x-ray diffraction by lattices) to discover the "unit cells." We began by simulating universes of hypertoric shape, of a relatively small size L, and thus with a volume L^3 much smaller than the volume of the observable universe. These fictional universes contain N real sources—galaxies, for example—but seem to be populated with a multitude of objects. All of the ghost images, which are juxtaposed with the N real images, are not recognizable as such and give the impression of a very large universe populated exclusively with original sources (Figure 20.2).

In fact, the angular distribution of these sources in the sky shows no distinguishing features and resembles the apparent distribution of actually observed galaxies in every way.

Nevertheless, since in our simulated universe we have assumed the space to be multiply connected, all of the ghost images of a given source occupy particular positions in the universal network; their mutual separations must belong to a well-defined set of lengths, connected to the dimensions of the fundamental cell. By contrast, in a simply connected universe filled with randomly distributed "original" galaxies, the mutual separations between all pairs of galaxies are themselves randomly distributed. There is therefore a method that allows one to distinguish the two cases: the histogram of pairwise separations.

A histogram is a curve that jumps by steps that counts the number of pairs separated by a given distance. For example, let us take 10,000 objects, which are able to be paired up in 50 million different ways. Of these 50 million pairs, let us imagine that 36,120 are separated by 100 million light years to a precision of 1%, 45,284 are separated by 200 million light years to a precision of 1%, and so on. If all of the galaxies are distributed randomly in a simply connected universe, the histogram of mutual distances will take the characteristic form of a bell. If the space is wrapped, only the original galaxies are spread in a random way, and all of their ghost images are distributed in a regular network. The dimensions of the network will therefore be found a great number of times in the separation between two galaxies: each time that a distance comes from two ghost images of a single source. Thus, in the histogram of pairwise separations, each characteristic length of the network will accumulate a great number of pairs separated by this length. To start from a simple two-dimensional example, Figure 20.3 helps us to understand immediately why spikes appear in the histogram as soon as the space is multiply connected.

In three dimensions, and for realistic multiply connected models of space, our numerical simulations have shown the importance of this effect: in the histogram produced by a simulated toric universe characteristic spikes occur, while they are absent for a simply connected universe. When one changes the topology, the spikes remain, but with different positions and relative heights (see Figure 20.4).

The *spectroscopy* of the spikes provides a true signature of the underlying topology, albeit with the condition that it concerns catalogs of objects extending over apparent distances greater than the size of the fundamental cell. It is here that the shoe pinches. The catalogs of celestial objects that are presently available are either extended widely but not very deep, or deep but very localized in one direction of the sky. For example, the largest catalog of galaxy clusters in which one has access to three coordinates (two for the direction in the sky and one for the apparent distance, given by the redshift) contains around 1000 objects that extend

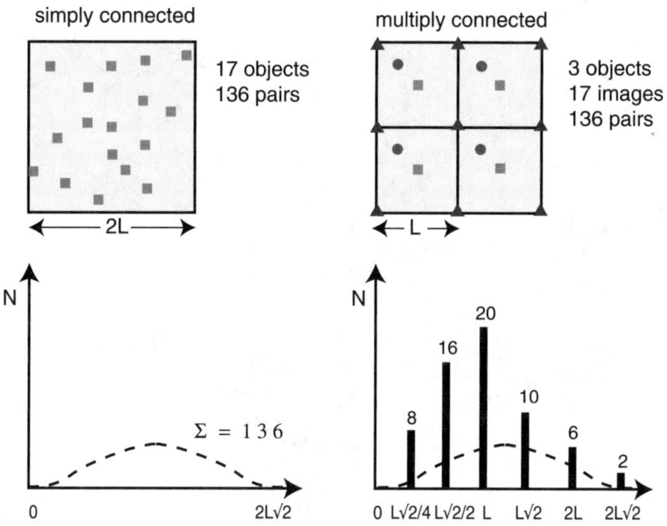

Figure 20.3. The histogram method. At left, a piece of simply connected space with surface area $4L^2$ contains 17 distinct objects, leading to 136 possible pairs. Below, the histogram of separations is a bell curve spread between zero (the minimum distance) and $2L\sqrt{2}$ (the maximum distance, equal to the diagonal of the square). The integral is equal to 136. At right, a piece of space with the same surface area of $4L^2$ is in fact a mosaic formed from four repetitions of a small toric space of length L, containing only three original objects (a square, a point, and a triangle aligned along half of a diagonal). The mosaic contains 17 topological images, leading to 136 pairs. This time, however, the distribution is no longer arbitrary, as the histogram below it shows: 8 pairs are separated by a distance of $L\sqrt{2}$, 16 pairs by $4L\sqrt{2}/2$, 20 pairs by L, etc. The accumulation of pairs having the same distance therefore gives rise to spikes.

no farther than two billion light years. The application of the histogram test to this catalog does not lead to the appearance of any peak, which confirms that the size of the Universe is greater than two billion light years, as other analyses (those of Sokoloff, Shvartsman, and Gott) have already shown.

Why confine ourselves to catalogs of galaxy clusters? Other populations of objects, such as quasars or the sources responsible for the mysterious gamma-ray bursts, are counted by the hundreds and are situated at cosmological distances. Unfortunately, although the gamma-ray bursts occupy a vast section of the sky, their distances are not known with enough precision for one to be able to apply the histogram test. As for the quasars, we have seen that they have a lifetime so short—100 million years—that their light rays would not have had the time to make it around the Universe in such a way that the ghost image of a quasar

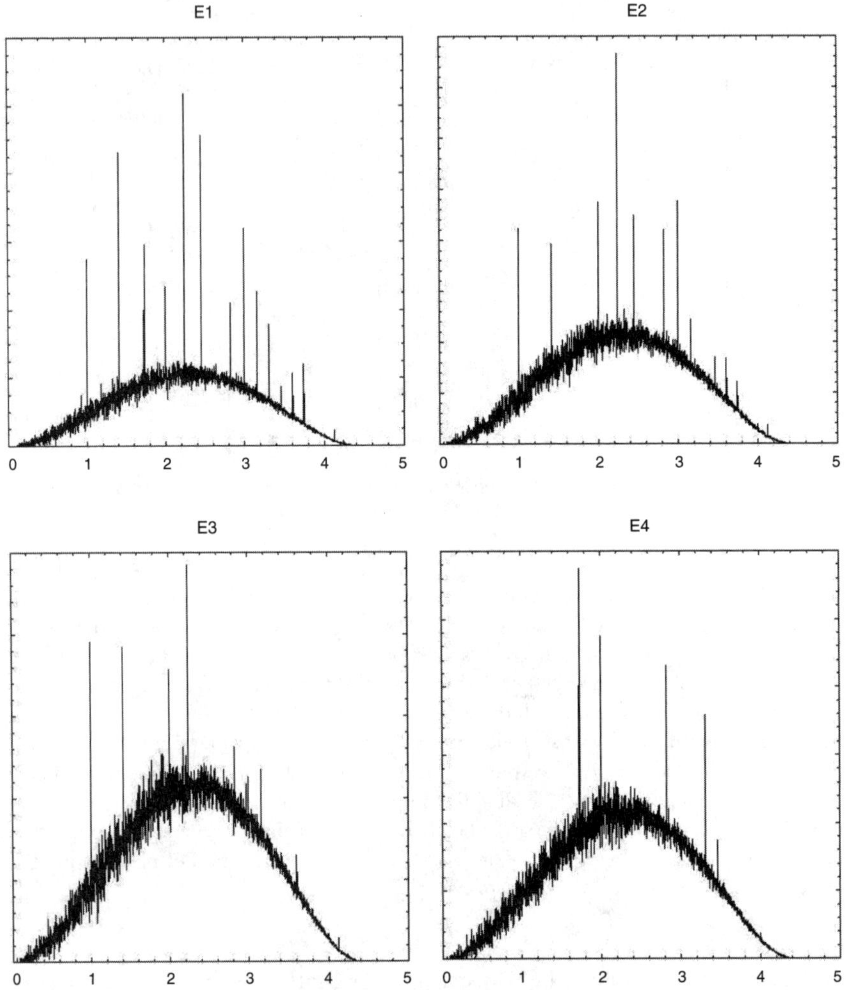

Figure 20.4. The histogram method. These numerical simulations apply the method of histograms for pairwise separations to four Euclidean wraparound spaces. The presence of peaks signals a multiply connected topology, their positions reflect the size of the fundamental polyhedron, and their relative heights characterize the holonomy group.

is also a quasar. Nevertheless, Jean-Philippe Uzan, Roland Lehoucq, and I have applied the method of crystallography to a catalog of 11,000 quasars; the absence of a peak leads us to eliminate a Euclidean space smaller than half of the horizon radius.

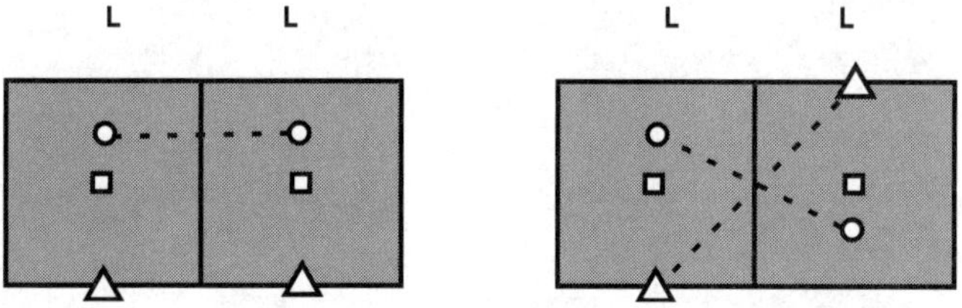

Figure 20.5. Clifford translations. At left, the topology of a torus containing three sources denoted $\square$, $\bigcirc$, and $\triangle$. Not only is the distance between sources conserved, $d[\square, \bigcirc] = d[g(\square), \mathrm{g}(\bigcirc)]$, but also the distance between the sources and their images $d[\square, g(\square)] = d[\bigcirc, \mathrm{g}(\bigcirc)] = d[\triangle, g(\triangle)] = L$ since the Euclidean translation serving as a holonomy is a Clifford translation. At right, the topology of the Möbius band. The distance between sources is conserved: $d[\square, \bigcirc] = d[g(\square), \mathrm{g}(\bigcirc)]$, but not that between sources and their images: $d[\square, g(\square)] = L; d[\bigcirc, \mathrm{g}(\bigcirc)] = L\sqrt{5}/2; d[\triangle, g(\triangle)] = L\sqrt{2}$. In fact, one of the holonomies of the Möbius band combines a Euclidean translation with a reflection, which makes it lose the property of being a Clifford translation.

We are still only at the beginning of cosmic cartography. In the same way that the Magellans and Vasco de Gamas of the sixteenth century progressively extended the world map by including countries that were more and more distant, the astronomers of today, equipped with spectrographs that allow them to evaluate the apparent distance of galaxies, are slowly constructing the three-dimensional map of the cosmos. Their exploration remains limited in depth, meaning that it is limited to fairly low spectral shifts, but the extension of the map to an appreciable volume of the Universe is only a question of time and patience. Systematic surveys of objects at large redshift are under way. When the three-dimensional catalogs of galaxy clusters are sufficiently filled out both in coverage of the sky and in depth, in around 2010, the crystallographic test could prove itself decisive in unveiling the spatial topology.

Meanwhile, we must seek to improve our techniques of crystallographic analysis, all the more so since the histogram method is not applicable to every type of multiply connected space. Figure 20.5 shows, as an example, the difference between the topology of a torus and that of a Möbius strip.

We have seen that the holonomies are always isometries, that is to say transformations that conserve distances; thus, two points of the fundamental domain separated by a certain distance have images in a replica of the fundamental domain separated by the same distance. But certain holonomies have an additional property: in these cases the distance between a given point and its image is identical to 143→ that which separates any other point and its image. These holonomies are known as Clifford translations, named after an English mathematician. We now see that

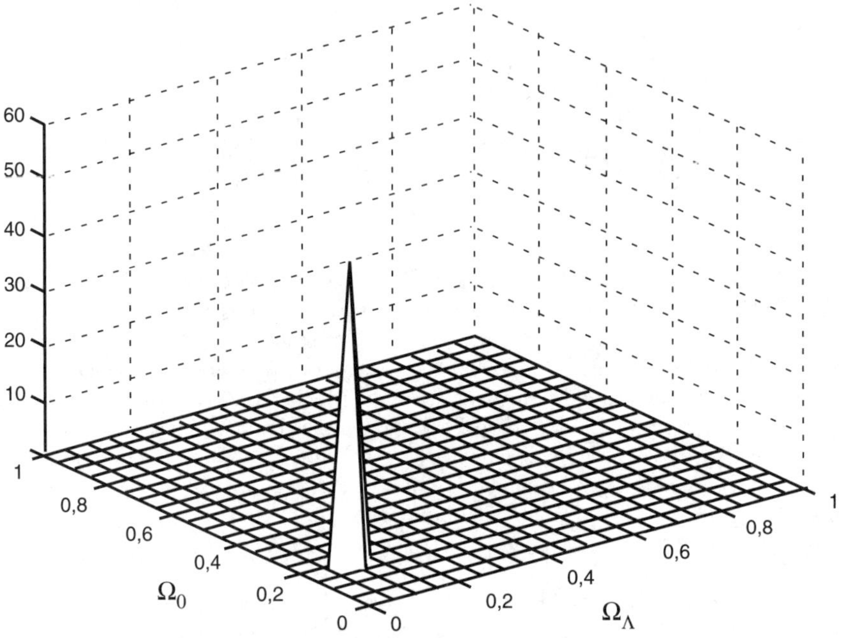

Figure 20.6. The CCP method. Starting from a catalog of galaxies, we collect into a single index all topologically correlated pairs of sources. If we obtain a peak similar to that in this figure, in addition to having proof that the space is multiply connected, we can deduce the values of the density parameter Ω_0 and the cosmological constant Ω_Λ.

there are two different contributions to the height of a spike in the histogram; that of the ordinary holonomies, which contribute little since they correspond to the number of copies of the fundamental domain within the observer horizon, and those due to Clifford translations, which carry great weight, since they correspond to the product of the number of objects in the fundamental domain by the number of copies of the fundamental domain within the observer horizon. It is only when Clifford translations exist that well-defined spikes emerge from the histogram, and clearly signal a multiply connected topology. The good news is that all of the Euclidean and spherical spaces have one or more holonomies that are also Clifford translations, which means that the histogram method applies well to these spaces. The bad news is that no hyperbolic space (of negative curvature) has a Clifford translation. We have therefore developed a new method, called collecting correlated pairs (CCP), which is intended to extract the maximum topological signal from any type of wraparound space (Figure 20.6).

The unexpected benefit of this method is the following: the extraction of the signal necessitates knowledge of the values of the two curvature parameters—the density of matter and the cosmological constant. To obtain a topological signal (if there is one!) starting from a catalog of real data, that is to say, so that the simulation can "see" any possible spikes, one must choose the right values of these parameters with a precision of a few percent. This property follows from a so-called mathematical rigidity theorem, by virtue of which the topology of 272→ a hyperbolic space uniquely fixes its curvature properties. Here is an interesting reversal of the epistemological situation. While for decades observational cosmology pushed itself to determine the curvature parameters directly while neglecting the topological possibilities, it could be that in the twenty-first century, it would be primarily the experimental determination of the topology of the Universe that would allow us to fix, with great precision, the curvature and the cosmological constant, those traditional parameters on which the ultimate fate of the cosmos depends: perpetual expansion or eventual contraction.

The statistical methods that analyze the apparent spatial positions of distant (large spectral shift) astrophysical objects are idealized and must be adapted in practice to account for the fact that we see these objects in the past and with observational biases. These biases are numerous, and too complex for me to describe here. I will only make note of one example: the proper motion of sources.

In the ideal case, the sources are fixed in a frame of coordinates that follows the expansion of the Universe (called the *comoving frame*). As a consequence, if the sources exhibit certain correlations because of a particular topology, they should be conserved over the course of cosmic evolution and thus in the cells that make up observable space. However, celestial bodies move; galaxies, for example, for the most part grouped in clusters, are endowed with proper velocities of a few hundreds of kilometers per second around the center of gravity of the cluster, which is added to the apparent motion of expansion, and which has no reason to coincide with the radial direction of the expansion. In turn, the galaxy clusters can be carried by local motions toward the attractive center of a supercluster. Thus, the galaxy supercluster to which we belong, Virgo, is in free fall at a speed of around 600 km/s toward an immense concentration of matter corresponding to a supercluster called the *Great Attractor*, which has recently been discovered.

Reality is therefore different. In the various cells of the crystallographic Universe, the positional relations between the "atoms" change slightly: the ghost images of a galaxy are shifted with respect to the expected regular positions.

Let us once again take up the comparison to candles in a room covered with mirrors. Let us suppose that, instead of being fixed, these candles are carried by a slow procession of invisible angels. All around, the mirrors reflect not only the light rays, but also the movements with which they are animated. In the eyes of

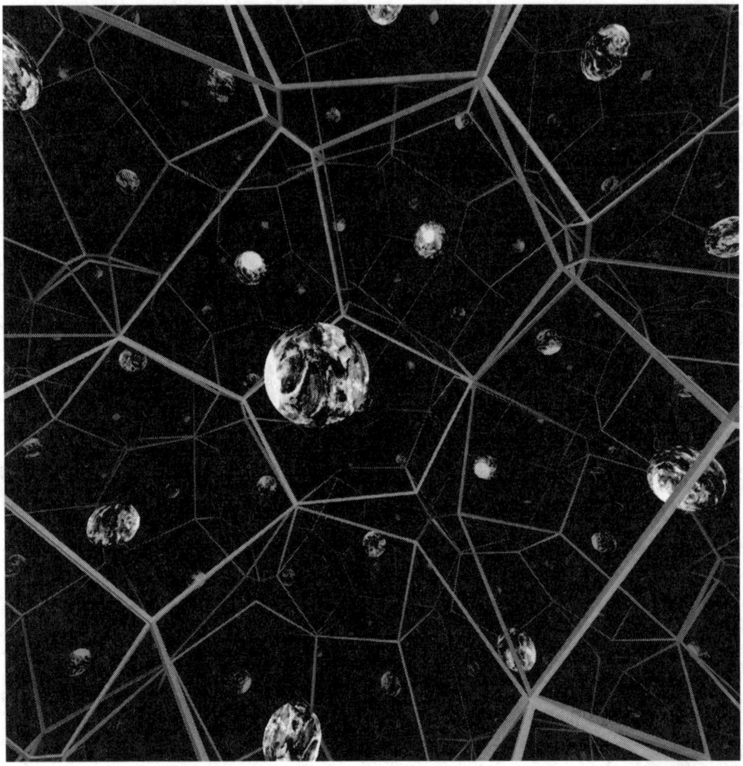

Figure 20.7. Ghosts in the Weeks space. This numerical simulation calculates the closest ghost images of the Earth that would be seen in the closed hyperbolic Weeks space, whose fundamental polyhedron is represented in Figure 14.9. (See Plate V. Image courtesy of Jeffrey Weeks.)

the astronomer, it is not a procession but rather incoherent displacements, because in the end the astronomer sees the real movement (for the closest images) and the ghost images whose direction of displacement is modified, sometimes even reversed, by successive reflections. Only an attentive observation can allow the astronomer to distinguish a procession in the foreground separating itself from a disordered background. Let us return to the real Universe: suppose that it measures only five billion light-years and that the space observed at apparently greater distances is only filled with phantom reflections. The astronomer should notice a discontinuity starting at the limit of the real Universe. In other words, the galaxy clusters that regularly recede from each other in the real part, should move in every direction as soon as we observe the ghost images.

As nothing of the sort has been observed, one might think that here is a serious objection to the hypothesis of a small wraparound Universe. In fact, this objection would hold if the Universe was static and if the proper speeds of cosmic objects were very high. However, space is expanding and this phenomenon is manifested in all of the cells of a crystallographic Universe, at a recession velocity proportional (according to Hubble's law) to the *apparent* distance of the galaxies; consequently, the expansion in distant cells seems more rapid than in nearby cells. Now, the recession velocity of galaxies at a distance of five billion light-years, 100,000 km/s, far surpasses their peculiar velocities (for example, the average velocities within the clusters, which do not exceed 1,000 km/s). The peculiar motions of ghost images would thus be imperceptible with respect to their recession velocity. It is as if the observation of moving candles was done in a rapidly expanding room; the mirrors on the walls would move away so quickly that none of the ghost reflections would seem to be approaching.

The proper motions of galaxies induce, despite everything, subtle corrections, which must be accounted for in the methods of crystallographic reconnaissance of ghost objects.

The reader should understand that the search for a clear and precise topological signature is just as difficult as the experimental determination of the curvature parameters (the density and the cosmological constant) and of Hubble's constant. For these latter numbers, it took more than a half-century of effort before we were able to begin to obtain convergent results. How much time will be needed to fix reasonable error bars on the topological parameters? Thanks to exponential growth in instrumental performance, it could be that a decade or less will suffice.

Since ancient millennia, the light of the stars has wound its way towards us and finally reached us from the bottomless depths.

—Jean Paul Richter, *Der Komet*

21 Circles in the Sky

Since the topology is a global property of the Universe, it seems judicious in testing it to use sources of radiation that occupy the greatest possible volume of space, in order to have the maximum chance of detecting phantoms. Now, today we have access to all of the observable universe thanks to the diffuse cosmological background. This is the fossil radiation from the primitive Universe, emitted only 350,000 years after what is conventionally called the Big Bang. If one compares this time interval to the 14 billion years that have since passed, it is as if a centenarian watched a video of his first cries as a baby. In its early youth, the Universe was apparently filled with a hot and dense plasma in which the photons—particles of light—remained trapped in the electric fields of charged particles like electrons and protons. Made prisoners to matter, the photons could not circulate. After 350,000 years of cosmic expansion, the temperature decreased, the electrons suddenly combined with the protons in a process called *recombination*, and the electric fields were neutralized. The photons were therefore liberated and were ←231 able to begin voyaging without obstacle in space and in time.

Since the Universe has continued to dilate and cool during the voyage, the light has cooled at the same time with it. The fossil radiation can in fact be characterized by a temperature, since it is of a particular type, known as *blackbody* radiation. When it was emitted, its temperature was 3,000 degrees Celsius; it has today fallen to 2.728 Kelvin above absolute zero (which is to say minus 270 degrees Celsius, 200 degrees less than a very cold day in Antarctica). In terms of wavelength, the original light has passed from the visible range to the range of microwaves.

The cosmic microwave background is remarkably uniform: its temperature and intensity are the same in all parts of the sky with a precision better than one in ten thousand. As a corollary, the curvature of space is also constant to within one in a hundred thousand. When Friedmann and Lemaître chose to construct their models of the Universe with a constant spatial curvature, they were guided by the pragmatic fact that they were the only calculable solutions, but also by simplicity and elegance, namely, the Copernican principle. Apparently, nature used the same criterion.

Since the discovery of the fossil radiation, cosmologists have realized that the temperature of this radiation should depend slightly on the direction in which one looks. This is for a good reason: we see heterogeneities in our material universe, wherever the density is greater or less than average. You yourself, reader, are a heterogeneity on a very small scale (according to cosmological criteria). The solar system, our galaxy, our local group of galaxies, and our local supercluster, Virgo, are other examples of heterogeneities of increasing size. We expect that all of these would have their origin in the heterogeneities of the very early Universe, even before recombination. Indeed, if at this epoch the Universe had been a perfectly homogeneous soup, without any lumps, it would remain so eternally, and we would not be here to speak of it!

The original heterogeneities were associated with temperature differences between one region and another of the primordial plasma. These fluctuations must therefore be seen as clear features in the fossil radiation. Moreover, even if the emitting regions had all had the same initial temperature, the fossil radiation should have crossed different intermediate galaxies and galaxy clusters, namely, those placed along the line of sight between us and these regions, and these paths should have caused infinitesimal differences in the temperature of the cosmic microwave background from point to point.

The detailed analysis of these fluctuations—rendered possible only since 1992, thanks to the observations of the COBE satellite, and improved since then following intense observational effort—is therefore a precious mine of information on the geometry and material content of the Universe. However, the topology of space should also be encoded in a certain way in the fossil radiation. In fact, this microwave radiation that reaches us from every direction of the sky can be considered as the image of a region of the Universe whose size was around ten million light-years across (the size of the cosmological horizon at the epoch when this radiation was emitted). If the multiply connected space had a dimension L smaller than this value, the fossil radiation would present itself today in the form of a complex mosaic, composed of juxtaposed images of the early Universe.

Several researchers have tried to show how the data collected from the COBE satellite imposed size constraints for a small wraparound Universe, through the simple fact that no particular mosaic manifested itself in an obvious way. In 1995, George Smoot, one of the principal investigators of the COBE program,[1] and his collaborators calculated the temperature fluctuations that would be produced in a universe with the topology of a torus, and noticed that these fluctuations were absent from the experimental data. From this they concluded, a little rapidly, that the fundamental cubic cell of space could not be much smaller than the horizon, which would have obviously limited the interest in multiply connected models.

Nevertheless, their analysis was only completed for the simplest Euclidean topologies (hypertori, uniquely) and did not apply at all to the spherical and hyperbolic universes. Even in the Euclidean case, the preliminary hypotheses that they imposed on the origin of fluctuations, the absence of a cosmological constant, and the particular way in which they conducted their statistical analysis limit the impact of their result. In any case, the only really credible experimental constraint affecting the minimal radius of a wraparound space, independent of any hypothesis, goes no farther than a tenth of the horizon radius. Complete freedom remained to speculate on models of small wraparound Universes, either Euclidean, spherical, or hyperbolic, and to pass them along to the test of the fossil radiation observed with better resolution.

One interesting approach, which began to be developed in 1996 by David Spergel and Neil Cornish (Princeton University), Glenn Starkman (Case Western Reserve University), and the mathematician Jeffrey Weeks, consisted in determining the topology of space by looking for very particular correlations in the temperature fluctuations of the fossil radiation.

The photons in this radiation have travelled in every possible direction in space, and thus are distributed uniformly. Let us focus our attention on the photons arriving here and now. They began voyaging at approximately the same time, at recombination, when the primitive opaque plasma condensed into a transparent gas. Since the speed of light in vacuum is constant, they all left from nearly the same space-time distance from Earth (which, at the time, was still in limbo). Thus the place of departure for the photons of the fossil radiation forms a sphere called the "surface of last scattering," centered on the observer, as it happens, Earth (Figure 21.1).

[1]For the discovery of the blackbody spectrum and anisotropy of the cosmic microwave background, Smoot won the 2006 Nobel Prize in Physics, together with John Mather.

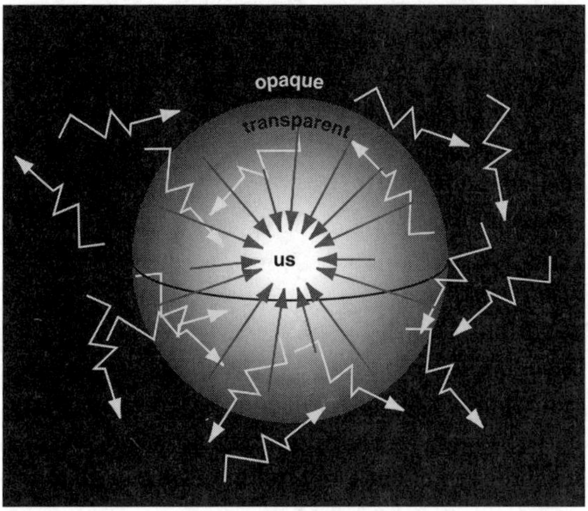

Figure 21.1. The surface of last scattering. The fossil radiation fills space-time, but the photons whose energy we can measure are those that arrive here and now. They all began their voyage at the same instant of cosmic history, around 14 billion years ago, and at the same distance from us: at the surface of last scattering, which marks the transition between an opaque universe and a transparent universe.

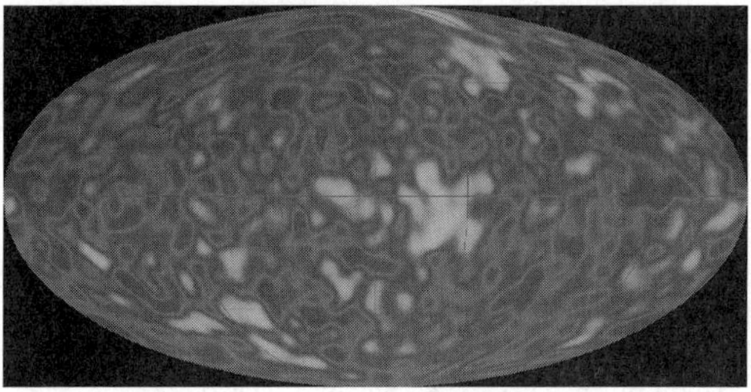

Figure 21.2. Fluctuations of the fossil radiation. Emitted at a temperature of 3,000 degrees Celsius, the radiation of the cosmological background reaches us at a temperature one thousand times smaller, in the microwave scale. This microwave map of the sky, taken in 1992 by the astronomical satellite COBE, represents the state of the Universe 14 billion years ago, at the epoch when there was only a homogeneous plasma, with the exception of minuscule variations in the temperature, encoded here by different colors. The maximal difference between the warmest points (dark red) and the coolest points (light blue) is no more than one hundred thousandth of a degree. (See Plate VI. Image from NASA COBE.)

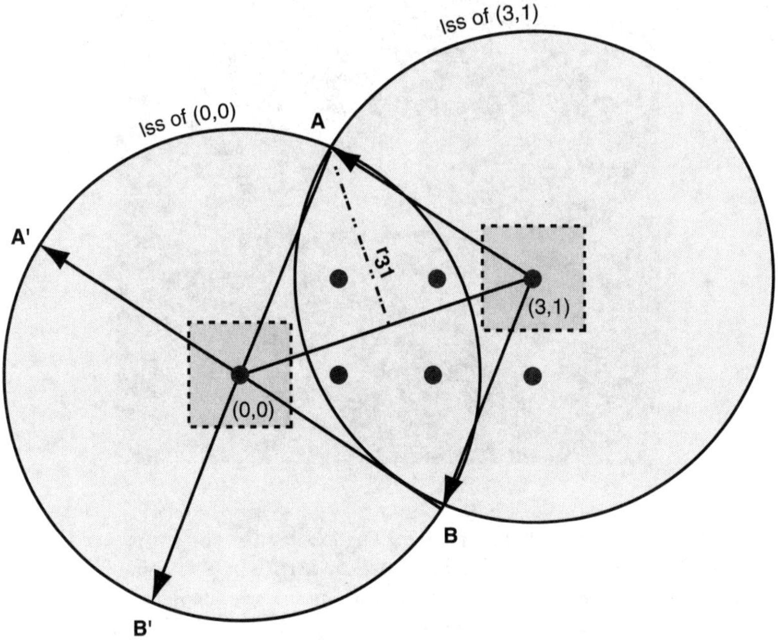

Figure 21.3. Principle of the method of matched circles. The method of matched circles is illustrated here in a wraparound space of two dimensions (a torus). The fundamental polyhedron is a square (with a dotted outline); all of the dark gray points are copies of the same observer. The two large circles (which are normally spheres in a three-dimensional space) represent the last scattering surfaces (lss) centered on two copies of the same observer. One is in position (0, 0); its copy is in position (3,1) in the universal covering space. The intersection of the circles is made up of the two points A and B (in three dimensions, this intersection is a circle). The observers (0,0) and (3, 1), who see the two points (A, B) from two opposite directions, are equivalent to a unique observer at (0, 0) who sees two identical pairs (A, B) and (A', B') in different directions. In three dimensions, the pairs of points (A, B) and (A', B') become a pair of identical circles, whose radius r_{31} depends on the size of the fundamental polyhedron and the topology.

Let us consider two different views of the Universe: one taken from Earth and the other taken from a distant galaxy. Every extraterrestrial astronomer living in this distant galaxy sees a surface of last scattering different from ours (Figure 21.3).

The map of the diffuse cosmological background that he develops will have, of course, an average temperature of 2.728 K (a reflection of the temperature of the entire Universe), but the pattern of fluctuations around this average will be

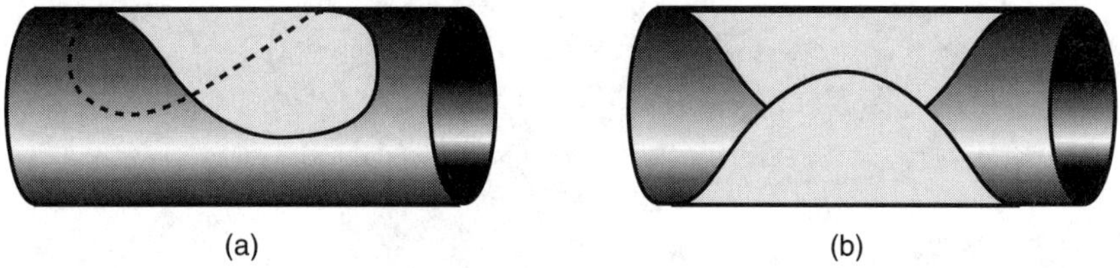

Figure 21.4. Self-intersection of the last scattering surface. In order for an observer to see matched pairs of circles in the fossil radiation, the last scattering surface must be sufficiently large with respect to the size of the fundamental polyhedron. In this example of a two-dimensional cylindrical universe, the border of the disk of paper represents the last scattering surface. In (a), the perimeter of the cylinder is larger than the disk, because of which the disk does not intersect itself: the observer will not see any pair of matched circles. In (b), the disk completely wraps the cylinder and overlaps itself: the observer will see the last scattering surface intersect itself in the form of circle pairs.

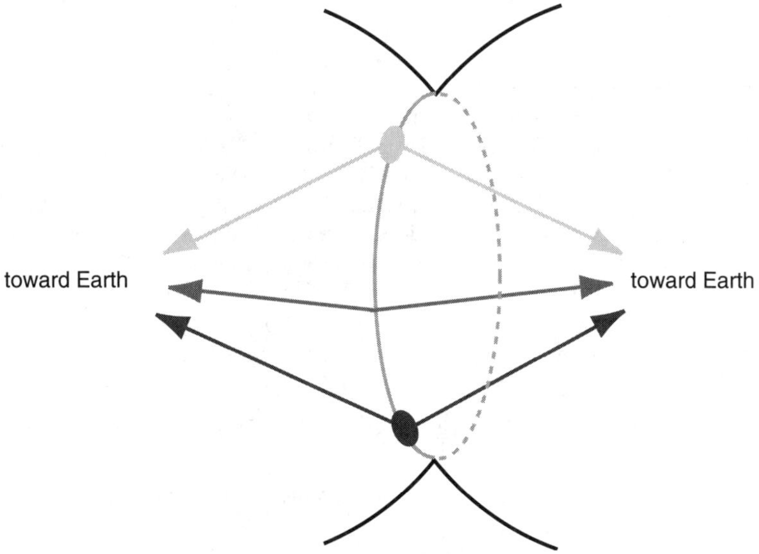

Figure 21.5. Homologous circles. If the last scattering surface wraps around the Universe, it intersects itself. The circle of intersection is visible from the Earth in two different directions, in the form of a pair of matched circles. The temperature varies slightly along the circle of intersection, sometimes cooler (the dark zone), sometimes warmer (the light zone). The corresponding photons are slightly less energetic (dark rays) or more energetic (light rays) with respect to their average energy (medium rays).

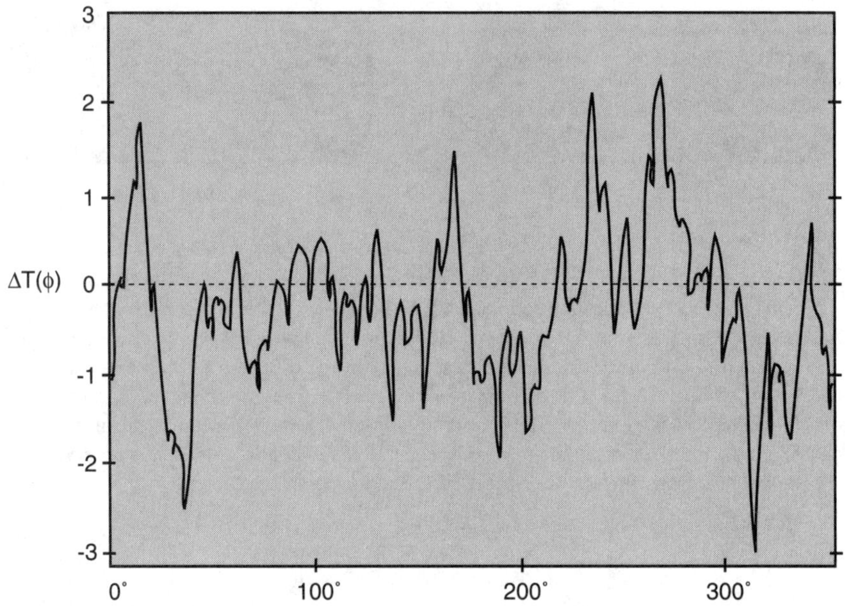

Figure 21.6. Temperature curve of a cosmic circle. Along an arbitrary circle traced on the last scattering surface (represented on the x-axis by an angle varying from $0°$ to $360°$), the temperature (the y-axis) varies by a few millionths of a degree around an average value. Although the average temperature is the same for all circles (this is the cosmological black-body temperature), the temperature curve varies in a specific way along each circle. If two circles having the same temperature curve are viewed in two different directions of the sky, they indicate a pair of matched circles within a wraparound space.

different, that is to say, the warm spots and the cool spots will not be placed at the same points as the map developed on Earth. In other words, its detailed map of temperature fluctuations of the fossil radiation will be different from ours.

Nevertheless, if the galaxy is not too distant, the two spherical surfaces intersect, along a circle. It is only along the circumference of the circle of intersection that the two observers will see the pattern of fluctuations coincide: since the corresponding photons come from exactly the same place in space-time, the astronomer here and the one over there will obtain an identical profile of temperatures (meaning the variations around the average value, the amplitude being no larger than one hundred-thousandth). Alas, unless we can exchange messages with this extraterrestrial, this concordance of measurements along a modest common circle is of no interest to us.

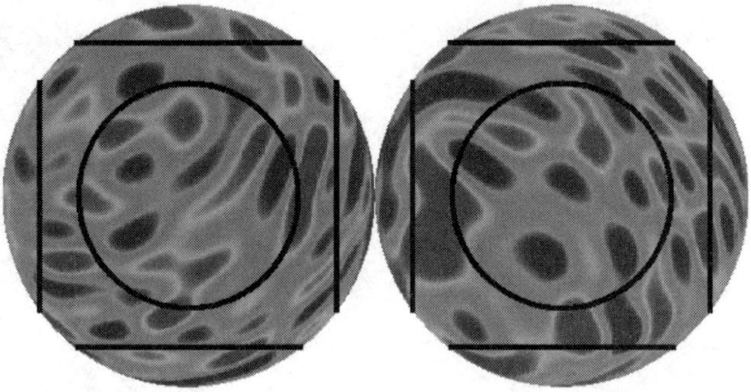

Figure 21.7. Identification of circles by their temperature curves. The WMAP and Planck satellites will furnish data on the microwave background in the form of a very precise map of the temperatures in the sky's background—represented here in the form of a western hemisphere and an eastern hemisphere. The warmest regions are in red, the coolest in blue (the maximum distance between these two extremes is no greater than 0.00001 degrees). If space is a hypertorus, and if the last scattering surface is large enough to completely wrap it in every direction, one will observe pairs of matched circles (in black). The matched circles are characterized by equal temperatures at corresponding points. Follow with your left index finger the central circle of the western hemisphere in a clockwise direction, starting from twelve o'clock. Simultaneously, follow with your right index finger the central circle of the eastern hemisphere, now turning in a counter-clockwise direction. You will pass through the same temperatures at corresponding points. The search for matched circles in the real Universe will be complicated by various sources of noise, which have been omitted in this simple illustration. (See Plate VII. Image courtesy of Jeffrey Weeks.)

However, what if this distant galaxy was nothing more than a phantom of our own galaxy, within a wraparound space smaller than the surface of last scattering? In this case, what was a single circle viewed by two different observers is equivalent to two circles seen from Earth, in different directions, by a single observer. Our last scattering surface repeats itself in the crystalline network of the Universe, and thus intersects itself several times! The circles of self-intersection should appear in pairs, in two different directions of the sky. The temperature fluctuations measured along a circle traced on the map of the fossil radiation will be identical to those seen along the homologous circle, in another region of the map. This is the method of pairs of matched circles.

The pattern of matched circular pairs varies according to the topology. The angular diameter of each pair of circles is fixed by the distance between the images. Images that are separated from us by more than two times the radius of the circle of last scattering will not produce circle pairs (Figure 21.8).

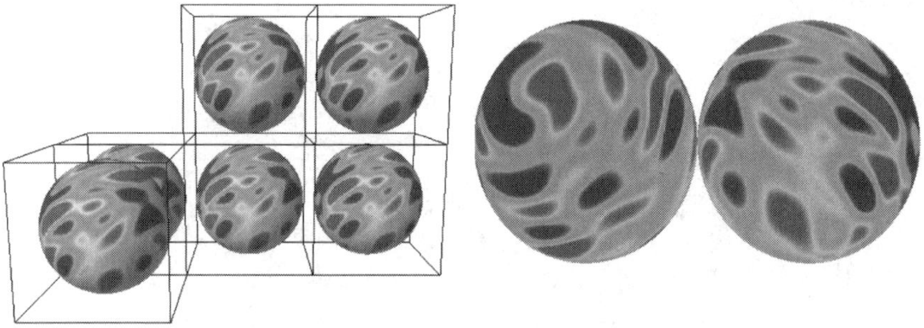

Figure 21.8. A large hypertoric universe. The space is a hypertorus, but its fundamental polyhedron is larger than the last scattering surface (network at left). The last scattering surface does not intersect itself, and no circle pairs appear in maps of the fossil radiation (shown at right in the form of two hemispheres). (See Plate VIII. Image courtesy of Jeffrey Weeks.)

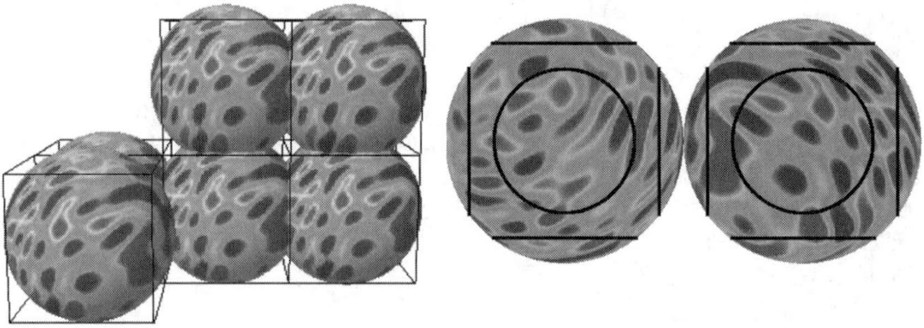

Figure 21.9. An averagely small hypertoric universe. The space is a hypertorus, but its fundamental polyhedron is slightly smaller than the last scattering surface (network at left). This is just large enough to cover space in every direction and intersect itself: one will observe three pairs of matched circles (in black in the maps on the right). Each circle corresponds with the one directly opposite to it. (See Plate IX. Image courtesy of Jeffrey Weeks.)

If the surface of last scattering is just large enough to overlap itself, it will intersect only its closest ghost images (Figure 21.9). If it is a little bit bigger, it will intersect itself more often (Figure 21.10). In fact, if it were large enough, one could observe hundreds, or even thousands, of matched circles.

Moreover, the pattern formed by the circles in the sky reveals the topology of the Universe. For example, if we observe circles like those of Figures 21.9 and

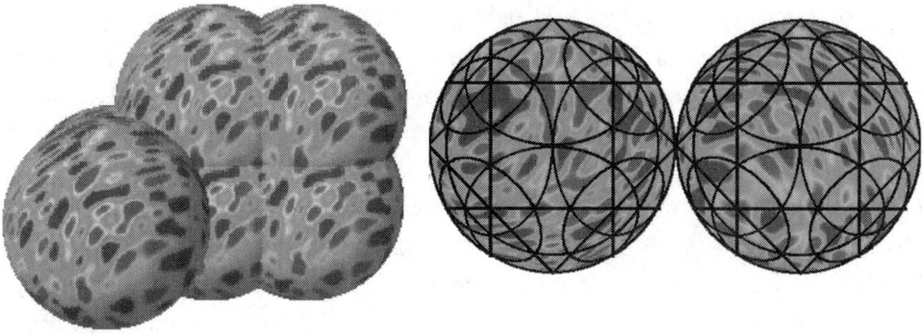

Figure 21.10. A very small hypertoric universe. The space is still a hypertorus, but the fundamental polyhedron is much smaller than the last scattering surface (network at left). This overlaps itself numerous times, and one will observe a great number of pairs of matched circles (in black in the maps on the right). (See Plate X. Image courtesy of Jeffrey Weeks.)

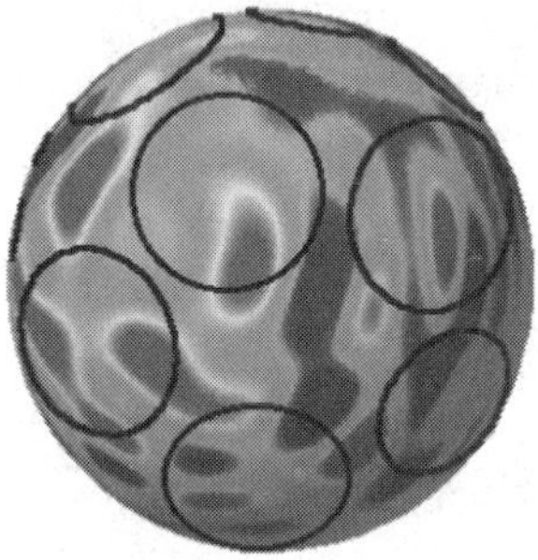

Figure 21.11. Matched circles in an icosahedral universe. If the matched circles form an icosahedral pattern, the Universe itself must be represented by an icosahedral space like that of Figure 14.6 (the Best space). For a more complicated topology, the circular patterns will be more complicated as well. In any case, the relative disposition of pairs of circles will reveal the topology of the Universe. (See Plate XI. Image courtesy of Jeffrey Weeks.)

21.10, one could conclude that the Universe is a hypertorus; if the circles resemble those in Figure 21.11, the Universe must be the icosahedral space of Figure 14.6, and so on. In practice, the pattern would in all likelihood be more complicated than in the illustrations, but the same principles apply: the relative positions of the matched circles are a signature of the topology of the Universe.

Thus, it would suffice to find pairs of matched circles in the surface of last scattering to have proof that the Universe is finite and wrapped! For the moment, unfortunately, the map of the diffuse cosmological background, constructed from data furnished by COBE in 1992, is not yet precise enough to test this method of circular pairs. Its angular resolution is ten degrees, that is, the data are incapable of describing the variation in the temperature between two points angularly closer than twenty times the apparent diameter of the Moon. On the other hand, 30% of what we see corresponds to instrumental noise and not to a cosmological signal. The measurements obtained in the Spring of 2000 by instruments on board balloons (such as the Boomerang and Maxima experiments) have reasonably better angular resolution, and have in fact allowed a decrease in the range of possible values for the curvature parameters. But they only cover a very small portion of the celestial sphere and therefore cannot be used to test the global topology of space. In order to hope to find pairs of matched circles, we will have to sweep attentively over the maps of the microwave radiation that will be furnished over the course of this decade by the WMAP[2] and Planck satellites.

If the future data of the WMAP and Planck satellites truly reveal matched circles, how could we be certain that the results are correct? There might be a defect in the cabling of the satellite, the computer programs used for analyzing the data could contain serious bugs, and so on. By chance, the information collected will be highly redundant. The first large circles would suffice to completely determine the topology of space, and would predict the position and orientation of all the other pairs! The reason for this is that, in geometric terms, the placement of the closest images of our galaxy define the structure of the entire cosmic crystal.

How do we convert the list of pairs of matched circles to an explicit description of the topology of the Universe? The American mathematician Jeffrey Weeks has found the answer.

The circle pairs directly indicate the placement of pairs of homologous faces of the fundamental domain. Let us recall, in fact, that the topology is defined by the data of the fundamental polyhedron and the group of transformations that identifies the faces. Let us suppose, therefore, that we are in a small wraparound universe. Imagine an immense spherical balloon whose center remains fixed on our galaxy and whose radius grows continuously. The balloon ends up containing the entire space and intersects itself. Let it inflate nevertheless, and let us watch it press against itself exactly like real balloons would, forming flatter and flatter intersections. When it has entirely filled the Universe, it will have taken the form of a polyhedron. The faces of this polyhedron will be identified in pairs to give the real space.

[2]See the afterword in the present edition for the WMAP results delivered in 2003.

This is how one may reconstruct the fundamental domain starting from the list of pairs of matched circles. Each face of the fundamental domain is found exactly midway between its center (our galaxy) and a ghost image of its center. We have seen that each circle in the sky is also found exactly mid-way between the center of the surface of last scattering (our galaxy) and an arbitrary ghost image of this center. Thus, the planes containing the circles coincide with the faces of the fundamental polyhedron. Once we have obtained the fundamental polyhedron, the relative dispositions of the pairs of matched circles will allow us to find the holonomies that identify them.

As with cosmic crystallography, the method of circle pairs is capable of leading to an interesting reversal of the situation in the history of cosmology: once determined, the topology then helps to fix the curvature and other cosmological parameters. Indeed, the rigidity theorem implies that a given topology has a certain volume, measured in units of the radius of curvature; therefore if we know the topology and the angular size of matched circles, we obtain a determination of the radius of the last scattering surface in units of the radius of curvature. This in turn provides an independent measurement of the density parameter, a measurement that does not rely on any assumption about the nature of the primordial fluctuations.

After having determined the topology of the Universe, we can proceed in the opposite direction and reconstruct the temperature of the photons of the fossil radiation, with their temperature fluctuations integrated and compared to those of the locally observed universe. We should thus be able to directly determine which spot in the cosmic microwave background collapsed to become our own supercluster Virgo, or other familiar structures like the Great Wall. In an unwrapped space, cosmologists have the habit of thinking that by observing the distant universe, they measure the prehistory of other regions of the Universe. If the Universe is wrapped, we shall also see our *own* origin.

The study of the fossil radiation in hyperbolic and spherical spaces is much richer, but much more difficult to treat, than in Euclidean topologies. While there are only ten distinct topologies with "flat" geometry and finite volume (of which only six are orientable), there is an infinity of spaces of positive or negative curvature, and mathematicians have already given thousands of examples that are small enough to be entirely contained within the surface of last scattering.

Many compact hyperbolic spaces can be studied mathematically thanks to a brilliant computer program named SnapPea, developed by Jeffrey Weeks. This original mathematician retired rather quickly from the academic world in order to pursue highly abstract research at home, surrounded by his family. To make ends meet, he pursues a more concrete activity, which is nevertheless playful and which does not take him very far from his preferred subjects: he invents game software which functions in multiply connected spaces. If you one day have the opportunity to play tic-tac-toe, chess, or crosswords on a Möbius strip or a torus, you owe your headache to Jeffrey Weeks! The program SnapPea is available for free on the Internet.[3] What does this strange name mean? According to its author himself, he was out of inspiration when baptizing his program; he then turned toward his wife, asking if she had an idea. Since she was at the time shelling peas, she simply responded, "Just call it Snap Pea!"

[3] SnapPea is available at geometrygames.org. However, beware! This program is only understandable by the specialists!

What I am writing at this moment in a dungeon, I have written and I will write for an eternity, on a table, with a quill, clothed, in circumstances just alike. Thus it is for everyone.

—Louis Auguste Blanqui, *L'Éternité par les astres*

22 Inflation, the Infinite, and the Folds

A scientific hypothesis does not succeed without the support and constraints of objective reality. We have seen that the model of a wraparound Universe is compatible with the data, no more. To this day, no observational proof has revealed itself to be decisive. Are there arguments of a different nature that are able to reinforce this hypothesis, or on the contrary render it less plausible? According to Epicurus, every conception of the world has its origin in philosophical astonishment. We shall therefore weigh the pros and cons by using an essentially philosophical argument.

With a multiply connected topology, the shape of space only appears in connection with the question of gravity. The fact that the quantity of matter only determines the curvature of space-time, and not its shape or its nature, causes us to discover the classical distinction in philosophy between quantity and shape, dear to Leibniz.

The diversity of multiply connected models is much larger than that of simply connected models. This theoretical richness is not a guarantee of a higher likelihood for the description of the physical world, but the multiply connected topologies have the advantage of offering models for the Universe that are finite whatever their curvature may be. Now, although physics does not forbid us from envisaging notions that imply infinity, for convenience, it does not allow them real existence; it considers that the only entities that are actually measured and the only processes that are actually executable are finite. Physicists thus in general seek to eliminate the infinities that follow from their theories.

As far as cosmology is concerned, let us for example assume the commonly accepted hypothesis, namely, that the topology of the Universe is simply connected and that space is hyperbolic or, as a limit, "flat." The physical Universe, which in this case is identical to its universal covering space, is therefore infinite. In these conditions, the fraction of the Universe that is observable is zero, since the observable universe has finite volume, and any number divided by infinity gives zero! The standard cosmological hypothesis also implies that the mass of the Universe is infinite: by the principle of homogeneity, the density of matter must be constant in space and, as infinitesimal as it may be in each cubic centimeter, integrated over an infinite volume, it leads to an infinite universal mass.

These two consequences are embarrassing enough, from many different points of view. First, on the epistemological level, a serious problem of credibility arises for cosmology considered as a scientific discipline: how does one justify the extrapolation from an infinitesimal, or even null, fraction of the Universe—the fraction that we observe—to the description of the physical Universe as a whole? It would make hardly any sense to extrapolate the curvature parameters of the Universe to areas that are infinitely large with respect to the regions in which they are actually measured.

In the second place, on the physical level, all objects that we know of, e.g., particles, photons, trees, creatures, planets, and galaxies, have finite masses. If the Universe is a physical object, why should it have an infinite mass?

On the philosophical level, cosmic infinity poses a paradox whose origin goes back to the atomist doctrine of 25 centuries ago: the plurality of worlds and beings. It was, in fact, a political agitator, Louis Auguste Blanqui, who formulated this paradox in the most striking way. He had spent more than 30 years locked up in various prisons and in 1871, during one of his incarcerations, he wrote a brochure entitled *Eternity by the Stars: Astronomical Hypotheses.* In this work he described the philosophical meditations that led him to hypothesize the infinite size of the universe. He began by asserting limitations on differentiated combinations of matter. Namely, there are only about a hundred simple elements, the atoms, starting from which every material system is constructed. Despite the incalculable number of their combinations, the result is necessarily finite, like that of the elements themselves. From there, "in order to fill up the expanse," nature must repeat out to infinity each of the original combinations. From this, Blanqui draws the most relentless of logical conclusions:

> Celestial bodies are thus classified according to originals and copies. The originals are the collection of globes which each form a special type. The copies are the repetitions: examples or prints of this type. The number of original types is bounded, while that of copies or repetitions is infinite. Each type has behind it an army of doppelgangers, whose number is without limit. ... It follows that each earth, containing one of these particular human communities, itself the result of incessant modification, must repeat itself billions of times, to meet up with the demands of the infinite. From this there are billions of earths, absolutely identical in population and material, where not one wisp varies, either in time, or in space, neither by a thousandth of a second nor the thread of a spider. ... Thus, through the grace of his planet, each man possesses within the expanse an endless number of doubles who live his life, in absolutely the same way as he lives it himself. He is infinite and eternal in the person of other selves, not only in his present age, but for every one of his ages. He simultaneously has, at each present second, doppelgangers by the billions who are born, others who die, others whose ages cover the range, from second to second, from his birth to his death.

Blanqui therefore invokes the infinity of space and the eternity of time implied by Newtonian cosmology to argue that eternity imperturbably replays, within the infinite, the same configurations. These monotonous reproductions of thousands of similar earths, and the vain character of any apparent novelty, seem all the more strange in that they were formulated by a man who wanted to change history!

What is the worth of this argument a century later, in light of relativistic cosmology, genetics, and the theory of evolution? It is true that if space is infinite and homogeneous, there are an infinite number of stars and galaxies in the Universe. Moreover, still within the relativistic framework, the models of the universe with infinite space are necessarily in perpetual expansion; they evolve, therefore, within an infinite temporal perspective. As for the simple constitutive elements of matter, the elementary particles (e.g., quarks and leptons) have replaced the atoms, but the number of distinct species is even more reduced than the number of elements. If we think of genetics, the organizing scheme for life on earth is governed by DNA molecules, which have a finite maximum size, and for which there exists a finite, although very large, number of possible configurations. All of the conditions invoked by Blanqui therefore seem to be present, in order to lead to the same conclusions. Since the planets are infinite in number in space and in time, one must expect that, on one of them at least, there exists a being with a strictly identical genetic structure to that of the author of this book, in other words another Jean-Pierre Luminet, with all his neurons in exactly the same state as mine, with all my memories, all my thoughts, and all my passing gestures. A similar lot awaits the readers of this book.

On the one hand, one may here see a philosophically unacceptable paradox: what should one think of the notions of identity, free will, and the like? On the other hand, one may remain skeptical about the reasoning. One may retort, *primo*, that just because a collection of objects is infinite, does not mean that it contains *all* objects. One may underline, *secundo*, that the hypothesis of the duplication of all beings would perhaps be acceptable in an eternal and *stationary* Universe, where the average physical properties do not evolve over the course of time, but that the stationary theory has been rejected: the Universe evolves rapidly, as attested by the observation of the fossil radiation. The physical conditions of the big bang are radically different than those which will hold in a trillion years. The various configurations of physical systems are therefore not equally probable in time. On the genetic level, the theory of evolution also seems to close off any recourse to probabilistic considerations: events are not all equally probable because they have a history.

Yes, but ... spatial infiniteness and the homogeneity hypothesis are sufficient for Blanqui's argument: at each given instant of cosmic history, the configurations that repeat within infinite space have had the same historical antecedents; each must therefore be reproduced an immeasurable number of times in various regions of space. One could still object that what counts is the interior of the observable universe, delimited by a sphere of some dozen billion light-years centered on us, and that this sphere is so "small" and the number of configurations leading to the constitution of a given individual so large that no doubling is possible.

Nevertheless ... the simplest way to settle the question is to suppose that the volume of space is finite! Now, it is only the multiply connected relativistic cosmologies that offer finite models no matter what the curvature of space may be. With the ghost images inherent to the wraparound models of the Universe, we find once again, of course, Blanqui's doppelganger earths, up to one fundamental difference: these doubles are one single and identical Earth, viewed at different instants of its history, and are finite in number since past history is bounded by the big bang. There is no longer a philosophical paradox.

Two objections can be formulated when encountering a multiply connected space—at any rate they are systematically brought up to me at every one of my lectures! The first is of a physical nature, namely, a possible incompatibility with the model of inflation. The second is of a philosophical nature, connected to the strict application of the principle of simplicity. Let us examine these two objections in turn.

The first invokes the theory of inflation, which is favored by numerous cosmologists as an explanation of the formation of large structures. This theory implies that space, a fraction of a second after its birth, would have expanded incredibly quickly for a very short time. Even if space was curved right at the beginning of its history, this excessive inflation would inevitably have rendered it Euclidean. Of course, there is no incompatibility of principle between the

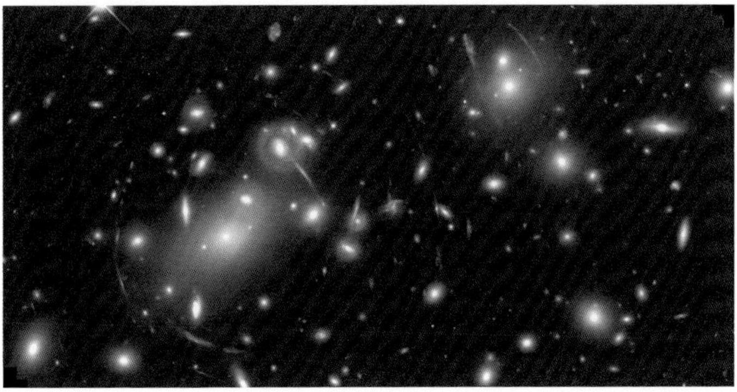

Plate I. Gravitational arcs. (See Figure 6.3. Image from STScI/NASA.)

Plate II. Circle Limit. (See Figure 13.11. M. C. Escher's "Circle Limit III," © 2007 The M.C. Escher Company-Holland. All rights reserved. www.mcescher.com.)

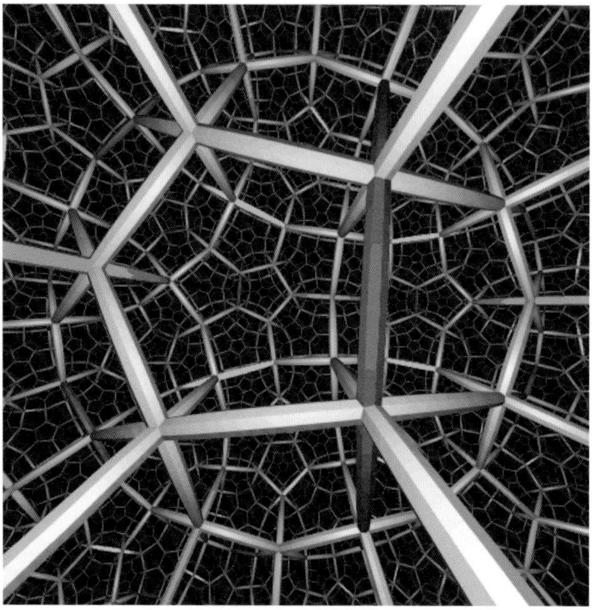

Plate III. A cosmic hyperbolic crystal. (See Figure 17.1. Image from the film *Not Knot*, © A K Peters, Ltd.)

Plate IV. Deep space as seen by space telescope. (See Figure 17.2. Image from STScI/NASA.)

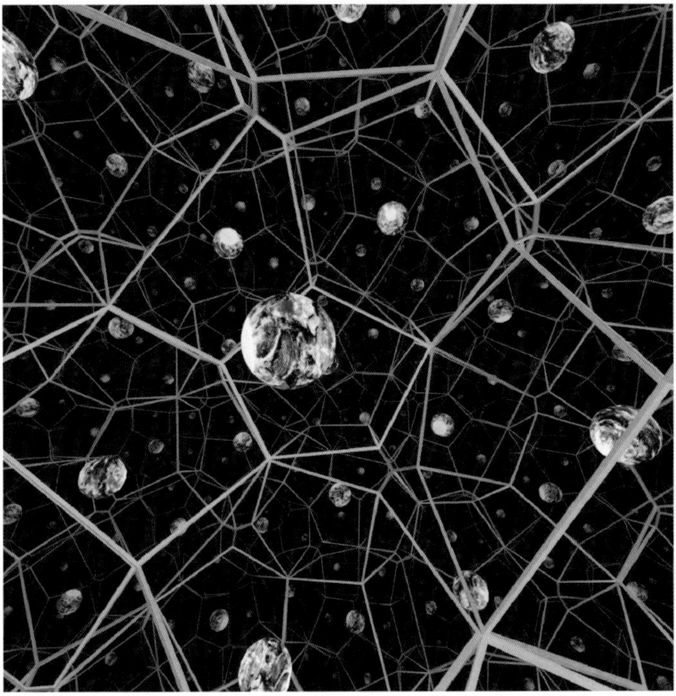

Plate V. Ghosts in the Weeks space. (See Figure 20.7. Image courtesy of Jeffrey Weeks.)

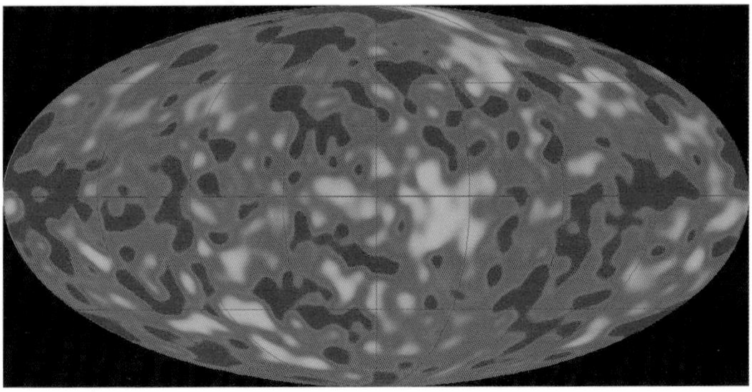

Plate VI. Fluctuations of the fossil radiation. (See Figure 21.2. Image from NASA COBE.)

Plate VII. Identification of circles by their temperature curves. (See Figure 21.7. Image courtesy of Jeffrey Weeks.)

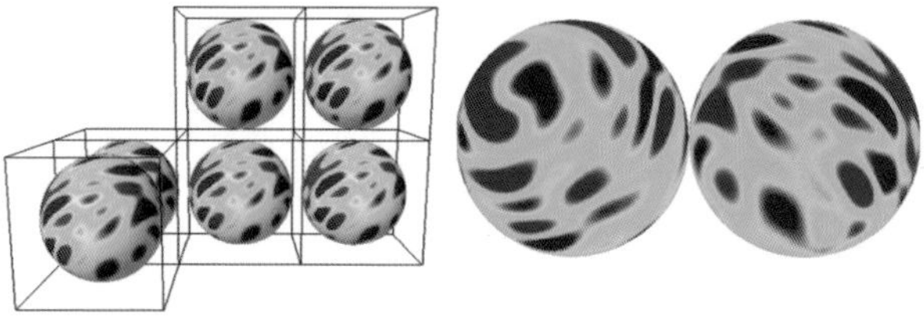

Plate VIII. A large hypertoric universe. (See Figure 21.8. Image courtesy of Jeffrey Weeks.)

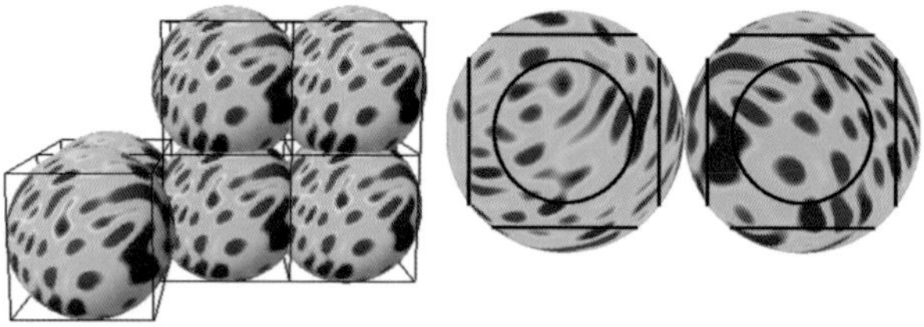

Plate IX. An averagely small hypertoric universe. (See Figure 21.9. Image courtesy of Jeffrey Weeks.)

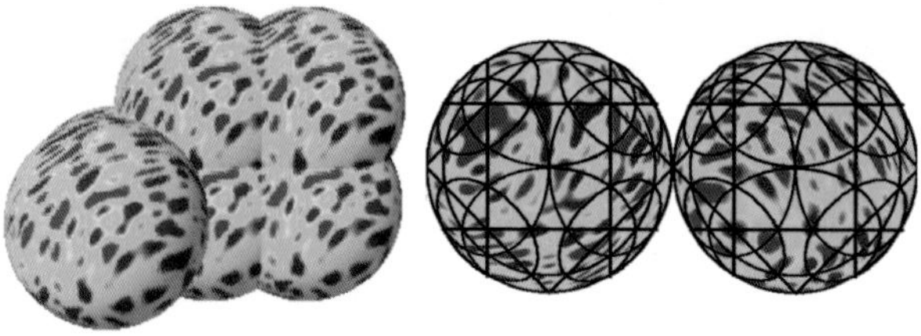

Plate X. A very small hypertoric universe. (See Figure 21.10. Image courtesy of Jeffrey Weeks.)

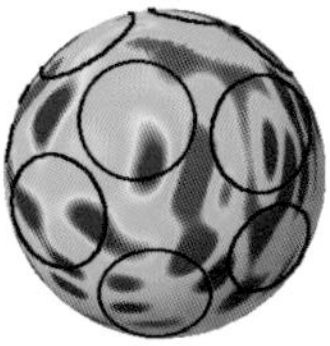

Plate XI. Matched circles in an icosahedral universe. (See Figure 21.11. Image courtesy of Jeffrey Weeks.)

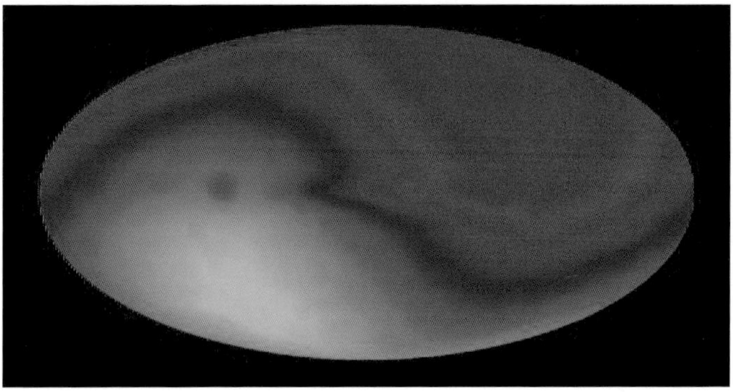

Plate XII. Dipole anisotropy. (See Figure 38.2. Image from NASA COBE.)

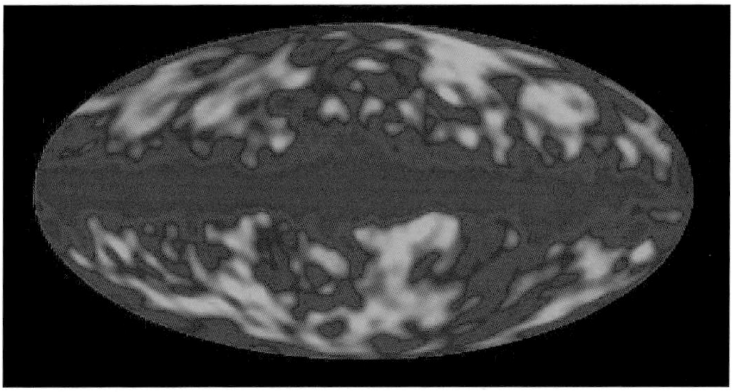

Plate XIII. Microwave emission from the Milky Way. (See Figure 38.3. Image from NASA COBE.)

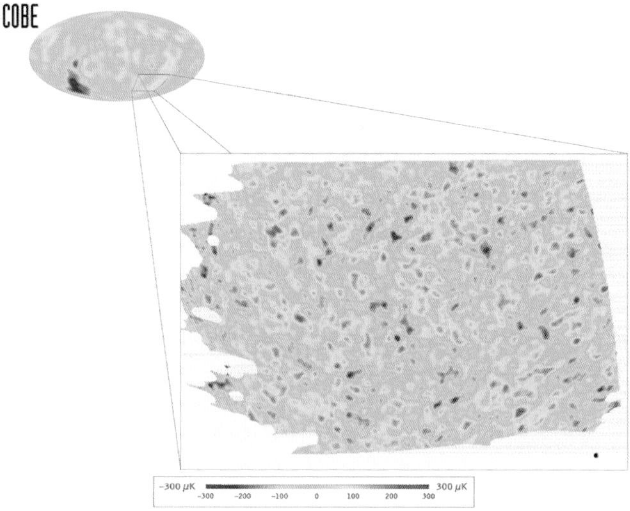

Plate XIV. Fluctuation map measured by Boomerang. (See Figure 38.4. Image © Boomerang Collaboration.)

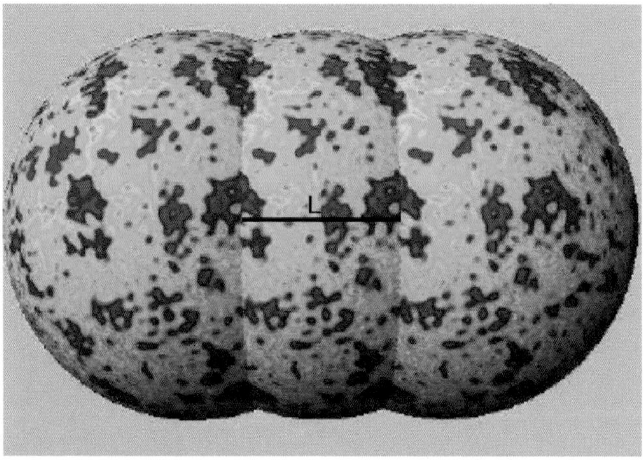

Plate XV. Simulated map of the cosmic microwave background in a small hypertorus. (See Figure 45.1. Image courtesy of Alain Riazuelo.)

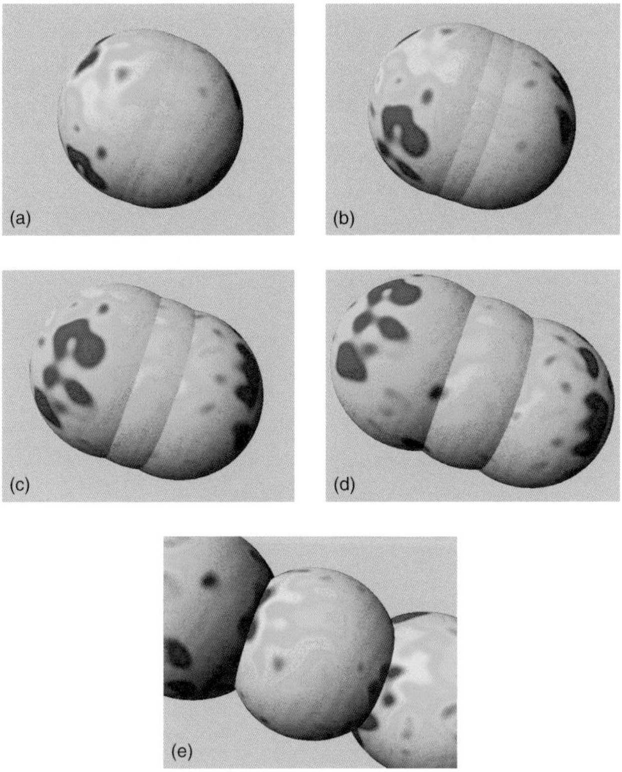

Plate XVI. Simulated map of the cosmic microwave background in a small lens space. (See Figure 45.2. Image courtesy of Alain Riazuelo.)

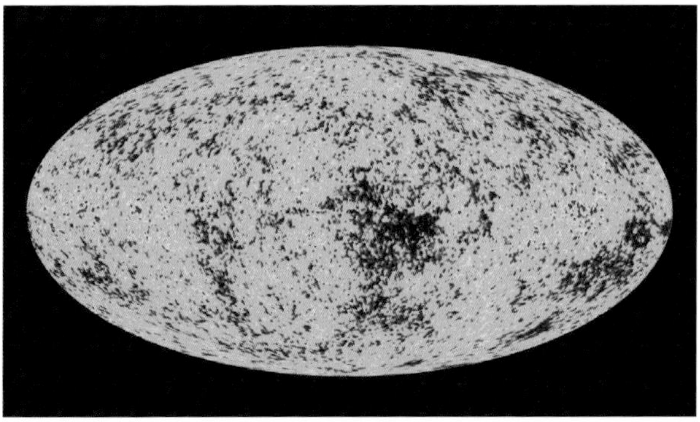

Plate XVII. WMAP map of temperature fluctuations of the fossil radiation. (See Figure 45.3. Image from NASA/WMAP Science Team.)

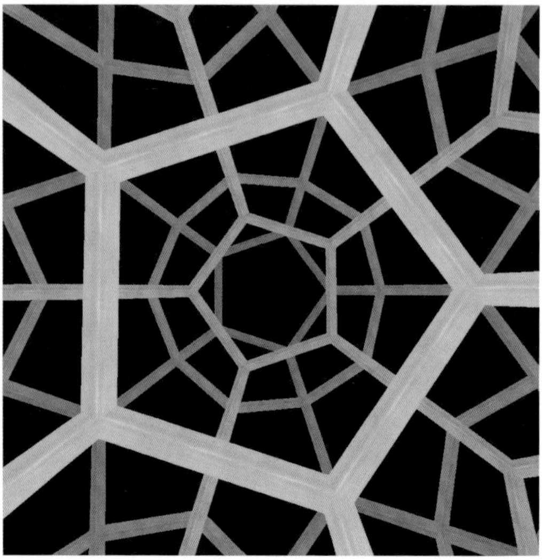

Plate XVIII. Topological mirages in the Poincaré space. (See Figure 45.8. Image courtesy of Jeffrey Weeks.)

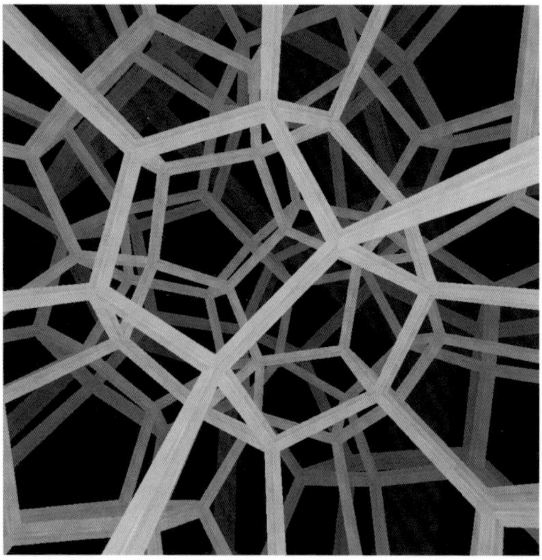

Plate XIX. Universal covering of the Poincaré space. (See Figure 45.11. Image courtesy of Jeffrey Weeks.)

inflationary models and the hypothesis of a wraparound universe: space could be both "flat" and wrapped at the same time since, as we have seen, there are 17 possible forms of multiply connected Euclidean space. On the other hand, inflation definitively forbids the eventual observation of a wraparound universe: if this process of frenetic expansion really has taken place, it may not have changed the topology of the Universe, but it has enormously dilated the volume of the observable universe. Therefore, even if space were wrapped, it would be on a scale much too large for one to be able to detect it.[1]

The credo of the physicist is that one should prefer theories that make extremely solid predictions over those that could never be excluded. Indeed, if the predictions are not verified by experiment, we have the means of demonstrating that the theory is false. Some researchers feel that the theory of a wraparound universe is among those that can never be totally excluded, since it does not make precise predictions. In contrast, they cite the theory of inflation, which predicts that space is flat, as the exemplary model of a refutable theory. This is a specious argument for two reasons. The first is that if the wrapping of space takes place on a scale smaller than the horizon, the theory is perfectly testable by the methods of cosmic crystallography or that of correlated circle pairs in the fossil radiation; the greater part of present day research in cosmic topology is concerned precisely with perfecting observational tests. The second reason is that the theory of inflation makes such a precise prediction that it can *never* be tested. In fact, one could never say that space is exactly flat, since to do so one would need results of infinite 245→
precision, which is impossible.

Let us move on to the second objection. The principle of simplicity stipulates that, for a range of models that explain some given facts, preference should be given to that one which invokes the minimal number of hypotheses. Now, the topological parameters (the nature of the fundamental polyhedron, the composition of the holonomy group) introduce extra factors in the standard cosmological models. At first sight, it may therefore seem wise to opt for the simply connected topology, all the more so since the corresponding models are easier to deal with. But to suppose because of this that the most easily handled theories have more of a chance to be true than others is to make a metaphysical hypothesis on the simplicity of nature, without having objectively specified what is meant by the degree of simplicity.

Indeed, application of the principle of simplicity is delicate, since every as yet unknown phenomenon could be a source of "natural" supplementary parameters.

---

[1]Due to a number of free and adjustable parameters in inflationary models, it turns out that if space is not flat, the possibility of a wraparound universe is not in contradiction with inflation. Recently, some researchers have built models of *low-scale inflation*, where the inflationary phase is short and leads to a detectable spatial curvature.

The notion of simplicity is therefore relative to the state of knowledge at a given moment. Should one, for example, prefer models of the Universe without a cosmological constant to those that have one? Introduced by Einstein for a reason soon after revealed to be erroneous (namely, to force relativistic models of the universe to remain static), this cosmological constant subsequently seemed superfluous, to the point that the solutions that went without it have drawn the almost exclusive attention of researchers for decades. However, following recent developments, both theoretical and observational, the situation has brusquely reversed itself. Today, the cosmological constant appears as a fundamental necessity of quantum theory and seems to have been confirmed by observational data. ← 207 Necessity and opportunism make the law: if we need a cosmological constant to render our models of the universe coherent, is it not more "simple" to assume it exists? This is an example where the strict application of the principle of simplicity has proven to be unfruitful, and has even hampered the development of the field.

In the same way, if the topological parameters should be demanded by a physical theory that is deeper than general relativity, the argument of simplicity in favor of a simply connected topology would lose all value.

It has, in fact, been found that new approaches to space-time suggest that the finite character of the volume of space could be an indispensable condition for constructing coherent models of the universe. Quantum cosmology, an attempt at describing the dynamics of the Universe at the quantum scale, is one of these approaches. Proposed in the 1960s by John Wheeler and Bryce DeWitt, it could only be made to work in extremely simplified situations, explored notably ← 34 by Jim Hartle, Stephen Hawking, Andrei Linde, and Alexander Vilenkin. In the scenario of the multiverse (see Chapter 8), where multiple universe-bubbles spontaneously spring from fluctuations of the quantum vacuum, a calculation of the probability for the appearance of a universe having such or such characteristic is possible, after making some simplifying hypotheses. These calculations favor spaces of smaller volume. Now, a Euclidean or hyperbolic space only has a finite volume if its topology is multiply connected.

In this new interpretation, the argument of simplicity therefore returns in favor of the hypothesis of the wraparound universe! In fact, multiply connected solutions become "natural," in the sense that they follow logically from the model, as opposed to being an additional hypothesis. Is not the natural character of a model the best gauge of simplicity? We have long believed that natural space was three-dimensional Euclidean simply-connected space; general relativity taught us that space-time is naturally curved by gravity. Quantum cosmology might well reveal to us that space should "naturally" be wrapped and of small volume!

A reservation should nevertheless be made to this reasoning: it is only worth as much as the models of quantum cosmology that it is founded on. We have no choice but to recognize that we still do not have at our disposal a theory capable of reliably calculating the probability that the Universe possesses such or such a configuration.[2]

[2]The same reserve should be brought to encounters with cosmological discussions that assert that the probability for the appearance of life in the Universe is zero with respect to the ensemble of all possible initial conditions at the moment of the big bang. A vision of *intelligent design* is often the unacknowledged motivation for such reasoning; in fact, one proclaims, since life well and truly exists in the Universe (or at least here, on Earth), while being perfectly improbable on the physical level, it must necessarily have been programmed by a higher Will, which has precisely adjusted the initial conditions of the Universe in such a way as to allow the dawning of the human species. This metaphysical interpretation of relativistic cosmology is called the *strong anthropic principle*.

No, I cannot believe that the stars are so distant: billions of light-years, that doesn't mean anything. One of these days it will be discovered that they are much nearer, and all that has seemed to us infinite and immense will become small and near.

—Nina Berberova, *The Black Spot*

23 Maya

Hardly any cosmology lectures for the general public pass without a listener finally asking, during the question session, "And where is God in all of this?" In general, the public expects the scientist to respond, or at least is curious to know if the speaker is a believer or not. Most often, the speaker gets out of it and cites the joke that the marquis Pierre Simon de Laplace threw at Napoleon: "Sire, I have not needed that hypothesis."

This is not a declaration of atheism, but rather a simple recognition that this type of question lies outside of the field of science. Cosmology contents itself with reconstructing the present and past events of cosmic history, starting from observations, laboratory experiments, and the theoretical models believed to best represent the Universe. Within the framework of big-bang models, it tries, as closely as possible, to approach the conditions that might have presided at the "appearance" of space, time, and matter during an event extrapolated into the past, the big bang. The mathematical nature of the big bang—that of a singularity where the curvature of space-time is infinite—implies in fact that the big bang does not belong to the space-time geometry. The big bang is therefore not even an event. It has not taken place and has no location. Similarly, it necessarily lies outside the domain of our theories!

It is not necessary, however, to see in this mysterious and inaccessible big bang a metaphor of creation or the mind of God.[1] Physics does not serve to reveal the attributes and intentions of the creator, rather it provides a means for understanding and controlling nature, with the view of dominating the environment according to our wishes.

There is nevertheless an authentic cosmological question here: that of knowing if the Universe or matter has a temporal origin; but this problem is often transformed in a fallacious way into a pseudo-problem of creation. This shift rests on the idea that creation necessarily requires an exterior agent, a cause external to the physical world. The big-bang model and the hypothesis of a beginning for the Universe that takes place at a finite time in the past suggest, at first sight, the operation of a creator. This confusion serves as a basis for the reactions of adversaries of the big bang model, for the metaphysical drifting of its partisans, and for attempts at recovery by theologians.

Historically, controversy over the big bang very quickly escaped the strict scientific framework and was diverted to its supposed philosophical and religious aspects. Alexander Friedmann, who was the first to formulate in relativistic terms the concepts of an expanding Universe and a cosmic singularity, could not keep himself from seeing metaphysical consequences, speaking in particular of creation of the world from nothing. In fact, in his general bibliography, one notes the existence of a lost manuscript, entitled *Creation* (*Mirozdanie*). Nobody knows what the contents of this might have been, but it is possible that Friedmann, a fervent believer, there developed a theological point of view. On the contrary, Georges Lemaître, the second inventor of the big bang, held to a radical distinction between science and religion, detailing that one could never reduce the Supreme Being to the rank of a scientific hypothesis. Lemaître was nevertheless also a man of faith, since he was a priest, which caused him to be unjustly suspected by Einstein of not being objective on the question. Around 1950, in England, the cosmologists Edward Milne and Edmund Whittaker drew hazardous theological consequences from relativistic cosmology. On November 22, 1951, Pope Pius XII used the theory of the big bang as evidence, before the pontifical Academy, in the affirmation of a creator: "In fact, it seems that present day science, with one sweeping step back across millions of centuries, has succeeded in bearing witness to that primordial *Fiat Lux* uttered at the moment when, along with matter, there burst forth from nothing a sea of light and radiation, while the particles of the chemical elements split and formed into millions of galaxies .... Hence, creation took place in time, therefore, there is a Creator, therefore, God exists!" A fierce

[1]The famous British cosmologist Stephen Hawking has declared that the day when humankind will have completed a "theory of everything," they shall know the mind of God.

adversary of any such amalgam, Lemaître maintained, at the Solvay Council in Brussels in 1958, that his theory remained entirely outside of any metaphysical or religious question. As one might guess, this connection between Christianity and the big bang had a disastrous influence in the Soviet Union. In the Stalinist era, the official doctrine stated that the Universe was infinite in space and time, and big-bang cosmology was pointed out as the reflection of a reactionary bourgeois idealism. Russian scientists therefore disdained to engage themselves in studies of cosmology.

Even today, the vocabulary used by certain big names in international cosmology is sufficiently blurry to lead to confusion. Because of this, someone as influential within the scientific culture as John Maddox, who has for a long time edited the journal *Nature*, could say that the cosmology of the big bang was philosophically unacceptable because it justified the views of creationists.

How is it that such errors of interpretation are perpetuated? The majority of the scientific and philosophical literature on cosmology suffers from this confusion between two completely different questions. The first: does the physical Universe have a temporal origin and, if so, what can cosmological physics tell us about it? The second: what was the external cause of the big bang at the beginning of time, and what can science tell us about it?

The first question is scientific (even if the responses are not); the second is not. The question of creation is distinct from that of the origin. Just because the origin is scientifically inaccessible does not imply that it is equivalent to a divine creation. The inaccessibility of the origin simply translates to the fact that, just as there is a horizon to our observational knowledge, there is a horizon for our theoretical knowledge. In 1830, the count Pierre Daru showed proof of a singular perspicacity when under the influence of Laplace, his colleague at the french Academy, people had acquired the certainty that Newtonian science was capable of unveiling the past, present, and future of the Universe. Daru, with a century's worth of foresight, glimpsed the fundamental limits of science, and the four verses below should be meditated upon by many of the physicists of today who have persuaded themselves that they are close to resolving the universal enigma:

> Man cannot attain the primal cause:
> Time, weight, space, matter
> Which the mind believes to understand but cannot define,
> Will hide their essence for ages to come.
>
> (Pierre Daru, *L'Astronomie*, Canto I)

It can be tempting to mix scientific research and spiritual quest, physics and tao, cosmology and Buddhism, and so forth. For those who dedicate themselves to this particular mix of genres in books for a general audience, success is nearly assured, since "what men really want is not knowledge, but certainty" (Bertrand Russell). The purpose of my book is not to deliver metaphysical certainties, but to pose a physical question: what can one say about the shape of the cosmos?

The cosmos is synonymous with order and beauty. Beyond this simple etymological observation, though, I am completely ignorant of whether the Universe is really beautiful and harmonious, and I have not formed any conviction on this subject. Is it not simply the human spirit that wants to see beauty and symmetry, where in fact there may not be any? Is it not within our brains, formed by cosmic evolution, that we select from the Universe those things that our brains are able to select, namely, the regularities alone? In the seventeenth century, the philosopher Francis Bacon wrote: "The human understanding is of its own nature prone to suppose the existence of more order and regularity in the world than it finds. And though there may be many things in nature which are singular and unmatched, yet it devises for them parallels and conjugates and relatives which do not exist."

It would certainly not displease me if future observations taught us, for example, that the structure of space is that of a dodecahedron smaller than the volume 73 → of the observable universe (this would be Poincaré's model of spherical space).[2]

For my part, I pursue the career of a researcher because knowledge is a game and a pleasure that intensifies life. One of my pleasures in fact consists in meditating on the obscure kernels that render the Universe impenetrable. I am haunted by limits: black holes, quantum uncertainty, singularities, and cosmic mirages. Modern physics teaches us that there is some blurring that reflects the very physical nature of the world, independent of the filter of our senses. In the subatomic domain, as described by quantum physics, the uncertainty principle exemplifies this type of blurring. A quantum object, for example, an electron, only has reality if one performs a measurement on it and, moreover, the measurement itself modifies its properties. There is thus an intrinsic limit to knowledge of the quantum world: the real is veiled by the very nature of the world.

On the macroscopic level, astrophysicists have discovered blurring on the scale of the Universe itself, with gravitational mirages. Light rays emitted by very distant stars encounter intermediate masses along their trajectories toward us. These masses, by curving the space in their neighborhoods, perturb the light's trajectory and lead to optical illusions that deform, amplify, and multiply the images of the background sources.

With wraparound models of the universe, the cosmic blurring is perhaps global, and not just localized in certain directions of observation. The shape

[2]These lines were written in 2000, before the observations of WMAP in 2003 and the article that we published in *Nature* where, in fact, we propose Poincaré's dodecahedral space as the best model of space capable of reproducing the data on the power spectrum of the fossil radiation; see the afterword.

of space could be subtle enough to multiply, nearly to infinity, the possible light trajectories between a distant source and ourselves, because of which we may find ourselves in a Universe whose appearance is extremely different from what it is in reality. Each real galaxy would have dozens of ghost images spread in all directions and distances, even though it may be difficult to recognize them as such. The Universe would seem to us vast and unfolded, containing billions of galaxies, while in reality it would be smaller and wrapped, containing many fewer authentic objects. Where would the illusion be, and where reality?

To conclude, the word *maya* came to mind. In Hindu thought, this Sanskrit term designates the illusory appearance that not only disguises the truth, but also leads to error. Thirty years after my first contemplations of the night sky of Provence, when I lift my head once again toward the starry firmament, I do not see the same thing. Twenty years of questioning about the shape of space have changed my gaze. In the sky, one can only see what one is prepared to see.

II
Folds in the Universe

That which keeps quiet beyond everything, is this then simply what I name Space? ... Space! An idea! A word! A breath!

—Jean Tardieu, *Le Ciel ou l'irréalité*

24
A Brief History of Space

There is no space or time given a priori; to each moment in human history, to each degree of perfection of our physical theories of the Universe, there corresponds a conception of those fundamental categories of thought known as space, time, and matter. To each new conception, our mental image of the Universe must adapt itself, and we must accept that common sense was found lacking. For example, if space is limited by a boundary, what is there beyond it? Nothing? It is difficult to imagine that, by voyaging sufficiently far in a given direction, one could reach a point beyond which nothing more exists, not even space. It is just as troubling to think of an infinitely large Universe. What would be the meaning of any measurable, that is to say finite, thing with respect to the infinite?

These types of questions were formulated in the sixth century BCE, in ancient Greece, where they rapidly became the object of controversy. The first schools of scholars and philosophers, called *presocratic* (although they were spread over two centuries and were quite different from each other), each attempted in their way to rationally explain the world, meaning the ensemble formed by Earth and the stars, conceived as an organized system. For Anaximander, from the school of Miletus, the world where observable phenomena take place was necessarily finite. Nevertheless, it was plunged within a surrounding medium, the *apeiron*, corresponding to what we today consider as space. This term signifies both infinite (unlimited, eternal) and indefinite (undetermined). For his contemporary, Thales, the universal medium was made of water, and the world was a hemispheric bubble floating in the middle of this infinite liquid mass.

We meet up again with this intuitive conception of a finite material world bathing in an infinite receptacle space with other thinkers: Heraclitus, Empedocles, and especially the Stoics, who added the idea of a world in pulsation, passing through periodic phases of explosions and deflagrations.

Atomism, founded in the fifth century by Leucippus and Democritus, advocated a completely different version of cosmic infinity. It maintained that the Universe was constructed from two primordial elements: atoms and the void. Indivisible and elementary, (*atomos* means "that which cannot be divided"), atoms exist for all eternity, only differing in their size and shape. They are infinite in number. All bodies result from the coalescence of atoms in motion; the number of combinations being infinite, it follows that the celestial bodies are themselves infinite in number: this is the thesis of the plurality of worlds. The formation of these worlds is produced within a receptacle without bounds: the void (*kenon*). This "space" has no other property than being infinite and accordingly, matter has no influence on it: it is absolute, given a priori.

The atomist philosophy was strongly criticized by Socrates, Plato, and Aristotle. Moreover, by affirming that the universe is not governed by gods, but by elementary matter and the void, it inevitably entered into conflict with the religious authorities. In the fourth century BCE, Anaxagorus of Clazomenae was the first scholar in history to be accused of impiety; however, defended by powerful friends, he was acquitted and was able to flee far from the hostility of Athens. Thanks to its two most illustrious spokesmen, Epicurus (341–270 BCE), who founded the first school that allowed female students; and Lucretius (first century BCE), author of a magnificent cosmological poem, *On the Nature of Things*, atomism continued to flourish until the advent of Christianity. It was, however, marginalized over the course of the first centuries of the Christian era, and would not again be part of mainstream science until the seventeenth century.

Plato (428–347 BCE) considered a finite universe, enclosed by an ultimate sphere that contained the stars. To speak of space (the English term comes from the Latin *spatium*), Greek terminology used different names: e.g., *khaos*, *kosmos*, *apeiron*, *kenon*, *pan* (all), and *ouranos* (sky). In *Timaeus*, a specific term was introduced, *khora*, that designates the expanse or space as a receptacle for matter, which is defined by matter. Plato went on to play an essential role in the evolution of astronomical thought by insisting upon the fact that scientists must not content themselves with contemplation of the stars, but must also use mathematics and geometry to discover the true nature of the celestial bodies and explain their movements. All of Greek astronomy, from Aristotle and Eudoxus to Ptolemy (who would create the crowning form of these ideas five centuries later), would flower around this precept of a geometric astronomy.

Aristotle (384–322 BCE) opposed himself to Plato by not developing a theory of space, properly speaking, but a theory of place (*topos*), distinct from the expanse and independent of matter. The *place* is the limit that surrounds things. The universe is not *a* place, but *the* place, the sum total of all the places occupied by bodies. Opposing himself also to the atomists, Aristotle believed that the number of bodies was necessarily finite. On the cosmological level, his conception approached that of Plato: a fixed Earth at the center of a finite world, circumscribed by the sphere that contains all the bodies of the universe. But this exterior sphere is "nowhere," since beyond it there is nothing, neither void nor expanse.

There also exists, in Greek antiquity, three great schools of cosmological thought. One includes the Milesians, Stoics, and others, and makes a distinction between the physical world (the material universe) and space: the universe is considered as a finite island plunged within an infinite extracosmic space without any properties, which surrounds and contains it. The two other schools, atomist and Aristotelian, consider that the very existence of space follows from the existence of bodies; the physical world and space coincide—they are infinite for the atomists, finite for the Aristotelians. Even today, most people have in mind one or the other of these conceptions.

The defenders of a finite universe nevertheless come across a logical difficulty: if what is beyond the world still makes up part of the world, the world cannot be 154→
bounded without there being a contradiction. This edge paradox would not be resolved until the end of the nineteenth century.

In the same way that atomism was judged to be too materialist, the Aristotelian doctrine, which implies eternal time and a non-created universe, was rejected by the first theologians of Christianity. Until the sixth century, the cosmological models of the West returned to the archaic conceptions of the Milesians, namely, a finite cosmos bathing in the void, with the added distinction that the cosmos takes the form of a tabernacle, or of a heart.

The cosmology of Aristotle, as perfected by Ptolemy and reintroduced thanks to Arabic translations and commentaries, was adapted to satisfy the demands of the theologians. Notably, that which is situated beyond the last material sphere of the world acquired the status of, if not physical, at least ethereal or spiritual space. Baptized *Empyrean*, it was considered to be the residence of God, the angels and the saints. The medieval cosmos was not only finite, but quite small: the distance from the Earth to the sphere of the fixed stars was estimated to be 20,000 terrestrial radii, because of which Paradise, at its edge, was reasonably accessible to the souls of the deceased. The Christian naturally found his place at the center of this construction.

This model of the universe imposed itself until the seventeenth century, without, nevertheless, impeding the resurgence of atomist ideas. After the rediscovery of the manuscript of Lucretius, the German cardinal Nicholas of Cusa (1401–1464) argued in favor of an infinite Universe, of a plurality of inhabited worlds, and of an Earth in motion. However, his arguments remained primarily metaphysical: the universe is infinite because it is the work of God, who could not possibly be limited in His works.

When the Polish canon Nicolaus Copernicus (1473–1543) proposed his heliocentric system, in which the Sun is at the geometric center of the world while the Earth turns around it and around itself, he kept the idea of a closed cosmos, surrounded by the sphere of fixed stars. Even if this is two thousand times further away than in the Ptolemeian model, the universe nevertheless remained bounded.

We must wait several decades more for the first cracks to appear in the Aristotelian edifice. In 1572, a new star was observed by the Dane Tycho Brahe (1546–1601), who showed that it was situated in the sphere of fixed stars, that is to say, in the celestial region until then presumed to be immovable. In 1576, the Englishman Thomas Digges (1545–1595), a staunch Copernican, maintained that the stars were not distributed on a thin layer, at the surface of the eighth and last sphere of the world, but extended endlessly upward. Digges nevertheless was not proposing a physical conception of infinite space: for him, the sky and the stars remained Empyrean, God's realm, and in this regard did not truly belong to our world.

An epistemological rupture was triggered by two Italian philosophers [Koyré 57]. In 1587, Francesco Patrizi (1529–1597) produced *De spacio physico et mathematico* (see [Brickman 43]), where he put forth the revolutionary idea that the true object of geometry was space in itself, and not figures, as had been believed since Euclid. Patrizi inaugurated a new understanding of infinite physical space, in which it obeyed mathematical laws and was therefore accessible to understanding. But it is above all his contemporary, Giordano Bruno (1548–1600), who is attributed with the true paternity of infinite cosmology. The first book of his *De immenso* is entirely dedicated to a logical definition of infinite space. Bruno argued from physical, and no longer exclusively theological, basics. His cosmological thought was inspired by the atomism of Lucretius, the reasoning of Nicholas of Cusa, and the Copernican hypothesis. From the latter, Bruno retained heliocentrism and the ordering of the solar system, but rejected the cosmological finitism. A precursor to Kepler and Newton, he also refuted the cult of sphericity and of uniform circular motion for describing celestial motion. His bold and original writings were not understood by his contemporaries, most notably Galileo. Above all they were firmly opposed by the Church. In fact, the true philosophical subversion of the end of the sixteenth century did not reside so much in the heliocentric affirmation of Copernicus as in that of the infinite multiplicity of worlds. Camped at the front ranks of the anti-Aristotelian battle, Bruno, carried away by his passion for infinity, refused to abjure and was burned at the stake in Rome.

Johannes Kepler (1571–1630), another great artisan of the astronomical revolution, tried at first to construct a universal model founded on the use of particular 259→
geometric figures: the regular polyhedra. He failed at this attempt; the ordering of planetary orbits as calculated did not correspond to the new experimental data collected by Tycho Brahe. After discovering the elliptical nature of the planetary trajectories, Kepler overturned the Aristotelian dogma of circular and uniform motion as the ultimate explanation of celestial movements. He nevertheless refused to follow Bruno in his arguments for the infinitude of the universe. He considered this notion to be purely metaphysical and, since it was not founded on experiment, denuded of scientific meaning. In 1606, in *De stella nova*, Kepler wrote: "In truth, an infinite body cannot be understood by thought. In fact the concepts of the mind on the subject of the infinite refer themselves either to the meaning of the word 'infinite,' or indeed to something that exceeds any conceivable numeric, visual or tactile measurement; that is to say to something that is not infinite in action, seeing that an infinite measurement is not conceivable." Kepler supported his argument by expressing for the first time an astronomical paradox that seemed to be an obstacle to the concept of infinite space, and which 156→
would be extensively discussed: the *paradox of the dark night.* Just like the edge paradox, this problem would not be satisfactorily resolved until the middle of the nineteenth century, although by completely different arguments.

Starting in 1609, the telescope observations of Galileo (1564–1642) furnished the first direct indications of the universality of the laws of nature. On the question of spatial infinity, however, Galileo, like Kepler, adopted the prudent attitude of the physicist: "Don't you know that it is as yet undecided (and I believe that it will ever be so for human knowledge) whether the Universe is finite or, on the contrary, infinite?"[1]

The way was definitively open for new cosmologies, constructed on the base of infinite space. Until then, the notion of space was conceived in the cosmological and physical order of nature, and not as the background of the figures and geometric constructions of Euclid. In other terms, physical space was not mathematicized. It became so thanks to René Descartes (1596–1650), who had the idea of specifying each point by three real numbers: its coordinates. The introduction of a universal system of coordinates that entirely criss-crossed space and allowed for the measurement of distances was a reflection of the fact that, for Descartes,

[1]Letter to Ingoli, quoted in [Koyré 57, 97].

the unification and uniformization of the universe in its physical content and its geometric laws was a given. Space is a substance in the same class as material bodies, an infinite ether agitated by vortices without number, at the centers of which were held the stars and their planetary systems.

This new conception of the cosmos upset philosophical thought and led it far from the initial enthusiasm of the atomists and Bruno: "The absolute space that inspired the hexameters of Lucretius, the absolute space that had been a liberation for Bruno, was a labyrinth and an abyss for Pascal" [Borges 00, 353]. As for the savants, they did not allow themselves to be discouraged by these moods and irresistibly moved towards the infinite universe.

The tendency toward the radical geometrization of an infinite space, initiated by Descartes, was consummated by the Englishman Isaac Newton (1642–1727). Newton postulated an absolute space, encompassing not only the background space of mathematics and the physical space of astronomy, but also that of metaphysics, since space was the "sensorium of God." Physical space, finally identified with geometrical space, was necessarily Euclidean (the only one known at the epoch), without curvature, amorphous and infinite in every direction. Within this preestablished framework, Newton explained celestial mechanics in terms of the law of universal attraction, from now on considered responsible for gravitation and the large scale structure of the Universe. With Newton, cosmology took root for more than two centuries in the framework of an infinite Euclidean space and an eternal time.

All the problems are not resolved in Newtonian cosmology, far from it. On the question of the distribution of stars in space, for example, Newton believed that they must occupy a finite volume since, he argued, if they occupied an infinite space, they would be infinite in number, the force of gravitation would be infinite, and the universe would be unstable. Newton, moreover, supposed that the stars were uniformly spread within a finite mass—like a galaxy, for example. But a problem of instability remained: since each celestial body is attracted by every other one, at the least movement, at the least mechanical perturbation, all the bodies in the universe would fall toward a unique center, and the universe would collapse. Newton's universe is therefore only viable if it does not admit motion on the large scale: its space is rigid and its time immobile.

Gottfried Wilhelm von Leibniz (1646–1716) rebelled against precisely such a conception. Although he also believed that space was infinite, he was in profound disagreement with Newton on several points. For him, space had no absolute character: it was a system of pure relations between bodies, which therefore had no existence independent of the latter. Likewise, Leibniz thought that the stars were uniformly distributed within infinite space since, if this were not the case, there would be a sphere that encompassed all the stars; the physical Universe would therefore be bounded by this sphere and it would have a center, which, from the Copernican point of view (in virtue of which there must be no privileged position in the universe), is completely inadmissible.

Newton and Leibniz both attempted to prove the correctness of their point of view through reasoning by the absurd: if, by admitting a certain proposition, the conclusions that one draws are absurd, then the inverse proposition is true. The German philosopher Immanuel Kant (1724–1804), in turn, sought to establish his views of space on logic. He was a partisan (against Leibniz) of absolute, rather than substantial, space, and maintained that the intuition of space guaranteed the validity of Euclidean geometry. He also believed he had put an end to the debate on the finite or infinite character of space forever, by proving that it is impossible to construct without logical contradiction a finite universe as well as an infinite universe. He concluded from this that the question has no meaning and that it is discussed in vain!

The Kantian argument is faulty, for it relies on propositions coming from supposed common sense; for example, a finite character implies limits, infinity implies the absence of limits, and so on. But common sense, whatever it may be, cannot help but be surprised on occasion. It is, after all, only an argument of authority for those ideas shared by the greatest number. However, in science, "the testimony of many has little more value than that of a few, since the number of people who reason well in complicated matters is much smaller than that of those who reason badly" [Galilei 90].

This humble reasoning on the nature of space was performed in the middle of the nineteenth century by a handful of bold geometers; Gauss, Bolyai, 161→ Lobachevsky, and Riemann discovered non-Euclidean geometries, which came to invalidate Kantian common sense.

They showed that it was not common sense that guaranteed the validity of Euclidean geometry, but Euclid's fifth postulate: starting from a point, one can trace one and only one straight line parallel to a given straight line that does not pass through this point. What would happen if one modified this postulate? Would one come across inconsistencies if one assumed, for example, that one could make an infinite number of parallels pass through this point (Lobachevsky space), or even none (Riemann space)? Absolutely not: the geometries built on these postulates are just as coherent as Euclidean geometry. From a purely logical point of view, the latter therefore loses its privileged place and becomes a very particular system among others. Notably, in 1854 Bernhard Riemann proved that a space that has no limit is not necessarily infinite. His demonstration was analogous to the arguments that allow one to show that the surface of the earth is curved and finite, although without limits. He gave the three-dimensional 272→ example of the hypersphere.

Riemann did not content himself with reasoning on abstract spaces; he wanted to apply his discoveries to cosmology. He was the first to propose a model of the Universe that was finite but without frontier, described geometrically by a hypersphere. The edge paradox was finally resolved.

In 1870, the English mathematician William Clifford (1845–1879) wrote an article describing the possibility of a variation in the curvature of space from one point to the other, which would be responsible for gravity and for the motion of matter; he suggested that some small regions of space are in fact of an analogous nature to small hills on the surface of the Earth, which is on average flat; or in other words, that the laws of Euclidean geometry are not valid there [Clifford 73]. Clifford nevertheless failed at his attempt at the geometric modeling of gravitation, for he did not envisage the possible time variations of the curvature.

The German astrophysicist Karl Schwarzschild[2] (1873–1916) published a cosmology article in 1900 that passed unnoticed, but which with hindsight seems premonitory. His original mind made him one of the rare astronomers of his time who knew a sufficient amount of advanced mathematics to understand the subtleties of non-Euclidean geometries. Also, he naturally posed himself the question of knowing if real space could perhaps be curved. In particular, he sought, starting from observation, to find a lower bound on the radius of curvature of space (when this bound goes to infinity, space becomes Euclidean). At this time, the extragalactic nature of spiral nebula had not been demonstrated, and the currently assumed model of the general arrangement of the stars was that the number of all the visible stars was no more than 40 million, that they were all confined within a space of only a few million astronomical units of radius (the astronomical unit is the distance from the Earth to the Sun, or 150 million kilometers) and that external to this there was the void of an infinite Newtonian space. Thus, our galaxy, the Milky Way, was considered as a unique island of matter lost in an unlimited ocean of void. Schwarzschild asked himself if this island might not occupy space in its entirety, with the condition that this was finite, small, and without frontier, just as Riemannian geometry offered the possibility. Estimating the average separation of the stars around the Sun and making the hypothesis that the Universe contained one hundred million stars, he found that the radius of space occupied by matter was at least a million astronomical units. This allowed him to fix a lower bound on the radius of curvature of space. He then concluded: "One may, without coming into contradiction with experience, assume that the world is contained in a hyperbolic space with a radius of curvature greater than 4 million astronomical units, or in a finite elliptic space with a radius of curvature

[2] Schwarzschild distinguished himself in 1916 by discovering, in the trenches of the Russian front, the first exact solution to the equations of relativity, representing the space-time around a black hole.

greater than 100 million astronomical units" [Schwarzschild 98]. In a note at the end of the article, Schwarzschild pushed his reflections farther by suggesting that 280→ the topology of space could be nontrivial and could generate ghost images of the Milky Way.

Cosmology developed rapidly after the completion of general relativity by Albert Einstein, in 1915. In this theory, the Universe does not reduce to a space and a time that are absolute and separate; it is made up of the union of space and time into a four-dimensional geometry, which is curved by the presence of matter.

It is in fact the curvature of space-time as a whole that allows one to correctly model gravity, and not only the curvature of space, such as Clifford had hoped. The non-Euclidean character of the Universe appeared from then on not as a strangeness, but on the contrary, as a physical necessity for taking account of gravitational effects. The curvature is connected to the density of matter and energy. In 1917, Einstein presented the first relativistic model for the universe. Like Riemann, he wanted a closed universe (one whose volume and circumference were perfectly finite and measurable) without a boundary; he also chose the hypersphere to model the spatial part of the Universe.

At any rate, Einstein's model made the hypothesis of a static Universe, with the radius of the hypersphere remaining invariable over the course of time. In truth, the cosmological solutions of relativity allow complete freedom for one to imagine a space that expands or contracts over the course of time: this was demonstrated by the Russian theorist Alexander Friedmann, between 1922 and 1924. At the same time, the installment of the large telescope at Mount Wilson, in the United States, allowed for a radical change in the cosmic landscape. In 1924, the observations of Edwin Hubble proved that the nebula NGC 6822 was situated far beyond our galaxy. Very rapidly, Hubble and his collaborators showed that this was the case for all of the spiral nebulae, including our famous neighbor, the Andromeda nebula: these are galaxies in their own right, and the Universe is made up of the ensemble of these galaxies. The *island-universes*, already envisaged by Thomas Wright, Kant, and Johann Heinrich Lambert were legitimized by experiment, and the physical Universe seemed suddenly to be immensely enlarged, passing from a few thousand to several dozen million light-years at the minimum. Beyond this spatial enlargement, the second major discovery concerned the time evolution of the Universe. In 1925, indications accumulated that tended to lead one to believe that other galaxies were systematically moving away from ours, with speeds which were proportional to their distance. This experimental result remained totally incomprehensible until the scientific community accepted, in the early 1930s, the solution proposed in 1927 by the Belgian physicist Georges Lemaître: space as a whole dilates over the course of time; it is in expansion, and 176→ this expansion carries the ensemble of galaxies along with it.

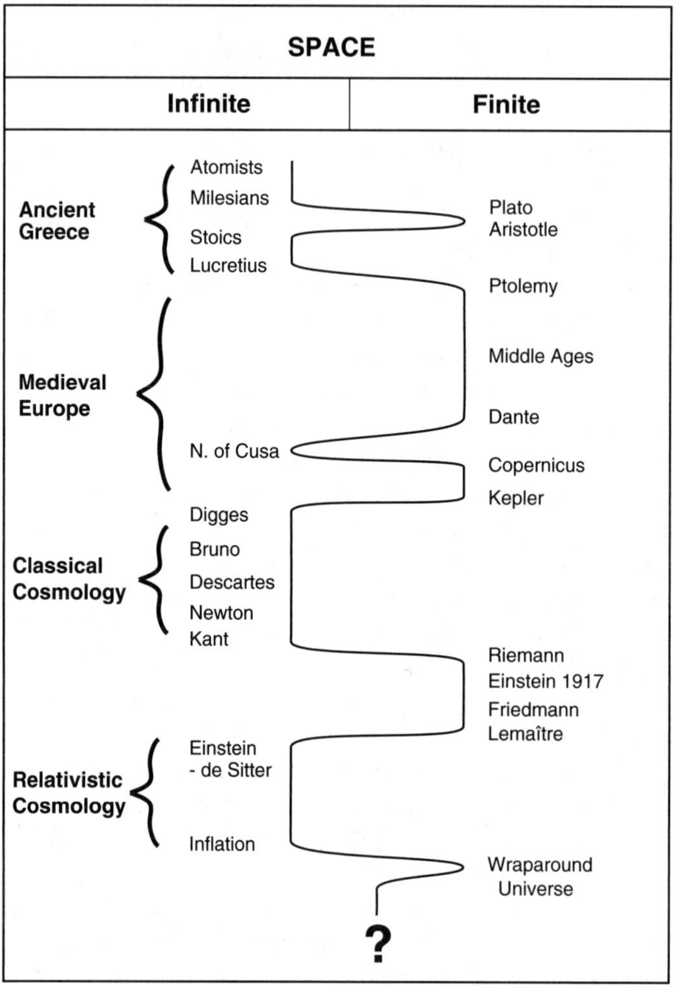

Figure 24.1. Finite or infinite? The hesitant waltz between different conceptions about the extent of space over the course of the ages ends up, naturally, in a pirouette.

The question of the finite or infinite nature of space can be perfectly well posed in the framework of the solutions of Friedmann and Lemaître. These cosmological models assume that irregularities in the distribution of matter are negligible, as a consequence of which the Universe has everywhere the same geometric properties. These properties are of only two types: the curvature, constant in space but whose sign remains to be determined, and the topology.

As far as the curvature is concerned, three families of space were considered: Euclidean space (with zero curvature), spherical space (with positive curvature), and Lobachevskian space (also called hyperbolic space, with negative curvature). Spherical space, as a rule, is of finite volume. This is the basic reason why the pioneers of relativistic cosmology—Einstein, de Sitter, Friedmann, and Lemaître—chose it as the starting point.

General relativity shows how to indirectly measure the curvature of space. Its value depends on the average density of matter and energy that it contains. If the real density passes a critical threshold, the curvature is positive and space is finite. For 60 years, observational cosmology has sought to determine the curvature of 217→ space by taking stock of all the forms of energy and matter that contribute to it.

However, for the geometries of the two other families, Euclidean and hyperbolic, the finite or infinite character of space no longer depends simply on the curvature and on the energy density: it depends on the topology. There is presently no general physical theory that allows one to predict the global topology of the Universe starting from its local characteristics.

The world has no outside, no beyond, for it contains and embraces everything.

—Guillaume d'Auvergne, *De Universo*

25 The Edge Paradox

If the Universe is finite, it seems necessary for it to have a center and a frontier. The center poses hardly any conceptual difficulty: it suffices to place the Earth there, like the geocentric systems of antiquity (appearances lead one in this direction), or the Sun, as Copernicus did in his heliocentric system. The notion of an edge of the Universe is on the other hand more problematic.

In the fifth century BCE, the Pythagorean Archytas of Tarentum described a paradox that aimed to demonstrate the absurdity of having a material edge to the Universe. His argument would have a considerable career in all future debates on space: if I were at the extremity of the sky, could I extend my hand or hold out a stick? It is absurd to think that I could not; and if I could, what lies beyond is either a material body or space. I could therefore move beyond this once again, and so on. If there is always a new space toward which I can extend my hand, this clearly implies an expanse without limits. There is, therefore, a paradox: if the Universe is finite, it has an edge, but this edge can be passed through indefinitely.

This line of reasoning was taken up by the atomists, such as Lucretius, who presented the image of a spear thrown to the edge of the Universe, and then by the partisans of an infinite Universe, such as Nicholas of Cusa and Giordano Bruno.

It is clear that, if one conceives of the Universe as a space enclosed by some kind of envelope, such as the surface of the sphere of fixed stars as imagined by Plato and Aristotle, the paradox is unsolvable. But over the course of centuries, the defenders of the finite Universe attempted to find satisfying explanations. One of them, which came from the Aristotelian doctrine as revised in the Christian Middle Ages, proposes a gradual border: the physical world, the realm of corruptible elements, changes itself progressively to the spiritual world, of incorruptible nature. This solution resolves the paradox in two ways: either the spear, made of terrestrial elements, must fall again toward its natural place, the Earth; or it will indeed pass the frontier, but will be transmuted into an ethereal element. A less convoluted solution, proposed by the Stoics, is the moving edge: the material world is finite, but it is surrounded by an infinite void. Throwing the spear beyond the edge will simply enlarge the cosmos, by pushing back the frontier.

It was not until the development of non-Euclidean geometries, in the nineteenth century, that it was possible to resolve the paradox logically. These geometries allow one to conceive of three-dimensional spaces that are finite but without 161→ edges, just like the surface of a sphere in two dimensions.

In the twentieth century, research on the global topology of space, whether Euclidean or not, also ends up with solutions of finite volume but no frontiers. Applied to cosmology, all of these new geometries allow for consideration of a finite Universe without any contradiction.

These notions are nevertheless hardly intuitive. Even today, in the minds of many people, it is rather the Stoic conception that prevails. All those who, in relation to the big-bang models proposed by modern cosmology, ask themselves into what the Universe expands, have this mental image of a cosmos-bubble with a moving edge, blowing up within an empty and infinite space. Nevertheless, this image should be abandoned. Relativistic cosmological models identify the Universe with space—or more precisely, with a more general physical and geometric entity, *space-time-matter*. Therefore, the Universe, whether it is finite or infinite, cannot inflate into anything else, since there is no space outside of itself.

Thus, in the same way that the concept of the cosmic center has been eliminated by the cosmological principle, the notion of the edge of the Universe is eliminated by the *principle of the content*: the physical Universe contains all that is physical and nothing else. This declaration may seem trivial, but it is more profound than it seems. It says, in particular, that the Universe is not a physical object like others. Every object has an edge—even if it is not distinct, as is the case for the Sun or a galaxy. However, the Universe has no edge. Space and time are not empty receptacles in which the material world can be placed, in the manner of an object. They are an integral part of the Universe. To take up an apt expression by Nicholas of Cusa: "The fabric of the world has its center everywhere and its circumference nowhere" [of Cusa 85].[1]

---

[1]The saying is generally attributed to Pascal, who repeated it in his *Pensées*.

Suddenly, one evening, one sees the night black and superb
At the hour when under the great shroud everything is quiet.

—Victor Hugo, *La Comète*

26
The Dark Night Paradox

At first sight, there is no reason to be astonished that the night is dark. The Sun sets and no longer gives light, leaving only the stars to weakly illuminate the night. Let us suppose, however, that space is infinite, and uniformly filled with sources (both stars and galaxies). In whichever direction we look, we should find a light source more or less far away along our line of sight. The sum total of their luminosities should render the night sky as bright as at noon, indeed even more so: the background of the sky should resemble a radiant vault continuously covered with stars, in the manner of a gigantic sun. Why is it not like this? If there are other stars of the same nature as ours, how is it that they do not all outdo our Sun in brightness? This is the paradox of the dark night. To firmly grasp it, let us liken the universe to a forest and the stars to trunks of identical trees, which are widely spaced. The closest trunks seem the largest, the more distant trunks seem smaller; but it is clear that, if the forest is sufficiently large, the trunks overlap to form a continuous background, seeming to surround us with a circular wall.

The problem was formulated for the first time by Kepler, in 1610 [Kepler 65]. That same year, *The Starry Messenger*, published by Galileo, had put forward the first telescopic observations, and revealed many more stars than one had been able to imagine. The paradox of the dark night was apparently an obstacle to the concept of homogeneous and infinite space. As Kepler did not share the ideas of Giordano Bruno, according to which the Sun is a world like others, among stars scattered out to infinity, he responded to the paradox by opting for a finite model of the universe, bounded by a wall or a vault. In this case, the stars are too small in number to cover the entire sky, and there is no longer any reason for the background of the sky to be bright. 154→

Although this explanation runs into the edge paradox, Kepler saw this to be the only possible exit to explain nocturnal darkness. Meanwhile, over the course of the eighteenth century, the Newtonian concept of an infinite space imposed itself, to the detriment of the closed cosmos defended by Kepler and his predecessors. It was within this new framework that discussion on the dark night was taken up again in 1721 by the famous English astronomer Edmund Halley, who concluded that "if the number of Fixt Stars were more than finite, the whole superficies of their apparent Sphere would be luminous"; then in 1743 by Jean-Philippe Loys de Chésaux, who calculated that the sky, during the day as well as the night, should be 90,000 times brighter than the Sun!

At the beginning of the nineteenth century, in the village of Bremen in Germany, a physician and amateur astronomer passed his nights scrutinizing the sky with the help of a telescope installed on his roof. Heinrich Olbers discovered the asteroid Pallas and a few comets. In 1823, he looked anew into the transparency of cosmic space: "If there really are suns in all of infinite space, . . . their number must be infinite, and thus the whole sky should appear as bright as the sun. For every line that I imagine drawn from our eyes will necessarily encounter some fixed star, and consequently, every point of the sky should send us stellar light, that is, solar light" [Olbers 83]. This time, the remark hit the bull's-eye, so much so that the paradox of the dark night is often rebaptized as *Olbers's paradox*.

How does one respond to this within the framework of Newtonian cosmology? Various solutions have been proposed. First of all, are the other stars truly analogous to the Sun? The answer is yes. Although sizes, masses, luminosities, and colors vary, each star remains on average comparable to our own personal star. In particular, their surface brightnesses are nearly identical. The solution to the paradox cannot come from there.

Another hypothesis that was put forward supposes the Universe to be filled with diffuse matter, which absorbs a large part of the stellar luminosity. We know today that the space between galaxies is filled with gas and very fine dust. But we know as well that these particles cannot absorb radiation without emitting it again in one way or the other. The paradox of the dark night cannot be thus resolved.

A third solution was proposed: could it not be that the arrangement of stars in the Universe is very particular? Newton himself imagined a single galaxy lost in an ocean of void. In this case, the number of stars would be finite and the paradox resolved. It is nevertheless difficult to accept such a privileged status for the galaxy. As Carl Charlier remarked at the end of the nineteenth century, certain hierarchical distributions[1] of the stars in the Universe, although still extending to infinity, would scale down the calculation of the sum of luminosities. However, this resolution of the paradox would demand an unlikely arrangement of celestial bodies. Today, telescopes have in fact discovered a sort of hierarchical organization of the visible matter in the Universe, in the form of stars, galaxies, clusters, and superclusters, but this arrangement does not at all correspond to the one proposed by Charlier.

In 1848, a radically new solution was found by Edgar Allan Poe [Poe 48]. In his prose poem *Eureka*, he explained that the dark of the night depends on the finite amount of time that has passed: "The only mode, therefore, in which ... we could comprehend the voids which our telescopes find in innumerable directions, would be by supposing the distance of the invisible background so immense that no ray from it has yet been able to reach us at all." The author, in fact, knew that to observe objects far off in space was equivalent to observing the past. Let us therefore suppose that the stars did not exist for more than, let us say, a billion years. Light propagates at a finite speed of 300,000 km/s. In a billion years, it can travel no farther than one billion light-years. If a star is more distant than that, its light simply has not had the time to reach us. Thus we can only receive light if the stars that have emitted it are sufficiently near. Therefore, even if space is infinite and the distribution of stars unlimited, the sky is not uniformly bright because the stars (and a fortiori the entire Universe) have existed only for a finite time. The paradox was resolved.

By understanding that nocturnal darkness is rich in lessons on the temporal finiteness of the world, Poe anticipated modern scientific cosmology by several decades. But since scientists never take inspiration from poets in developing their theories, Poe's explanation was unknown for a long time.

Relativistic cosmology, with its big-bang models, differed profoundly from that of the preceding centuries. It proposed at least three possible responses to the paradox, each of them sufficient to resolve it unequivocally: the finiteness of space, already envisaged by Kepler, the finiteness of time, suggested by Poe, and the cosmological redshift.

For the finiteness of space, relativistic cosmology allows finite models for the Universe as well as infinite ones. The difference with respect to Kepler's reasoning comes because the finite spaces envisaged in relativity have either a non-Euclidean geometry or a multiply connected topology, which allows them to evade the edge paradox. ←154

For the finiteness of past time, everything depends on believing that the Universe only existed, in a state comparable to that seen today, for a limited time,

[1] Actually, Charlier anticipated what in modern language are called *fractals*.

improperly called the age of the Universe, which is around 14 billion years. This immediately implies that no star had yet appeared 14 billion years ago. There is therefore a spatial frontier from beyond which we don't receive light. It is not a physical barrier; there is nothing in space that marks it.

We are like a sailor on the high seas, who can see nothing beyond the horizon. This is why this fictive limit on our observations is called the cosmological horizon, and why this horizon darkens the night.

Relativistic cosmology furnishes a third possible explanation, which relies on the fact that the Universe is expanding. This dilation of space modifies the laws of propagation for light in the Universe; the effect translates into a shift in frequency, 179→
the redshift seen in galaxies, and an attenuation of its energy.

In fact, we indeed receive the radiation from distant stars analogous to the Sun, situated, for example, in faraway galaxies. A given star emits visible light, but we only end up receiving from it some infrared radiation: the optical wavelengths have been shifted toward this domain of lesser energy. Each photon emitted by the star reaches us at a lower frequency, with diminished energy. In total, we receive from the star a luminous power that is less intense and less visible. After everything is calculated, there is no longer anything astonishing in the fact that the background of the sky is not bright.

Relativistic cosmology therefore resolves the paradox of the dark sky without any contradiction, by mixing in a certain way the second and third explanations, as well as perhaps the first. In doing so, it allows one to understand how an apparently banal observation, the dark of the night, opens the way to some of the most profound teachings about the spatial and temporal structure of the Universe.

To tell the truth, big-bang models have thoroughly changed the problems associated with the dark night, through an effect that neither Kepler, Olbers, nor their successors could have suspected: it seems that the background of the sky *is* bright, even at night. Of course, it is not as bright as the surface of the Sun, nor does it shine at the same wavelengths. However, according to models of the big bang, the entire Universe was so hot around 14 billion years ago that each of its points was as luminous as the surface of the Sun. Each direction leaving from our eye reaches a point of this past Universe. And, by the same reasoning as that of Olbers, even in the absence of every star, we should be surrounded by this enormous bright object, the early Universe.

This is not a new paradox. We indeed receive this radiation, but it is shifted toward long wavelengths and weakened. Fourteen billion years ago, the Universe was an immense shining object emitting visible radiation in every direction. But this radiation has experienced the same redshift as that of the galaxies. Since it is very old, the shift is very strong: redshifted by a factor greater than 1000, it has transformed the light into microwaves. This electromagnetic fossil radiation, a

←231 vestige of the primitive epoch, was detected for the first time in 1964. Today it is being exhaustively observed under the name of *cosmological background radiation.*

Further Reading

Edward Harrison. *Darkness at Night: A Riddle of the Universe.* Cambridge, MA: Harvard University Press, 1989.

Thus we may perhaps, one day, create new Figures that will allow us to put our trust in the Word, in order to traverse curved Space, non-Euclidean Space.

—Francis Ponge, *Texte sur l'électricité*

27 Non-Euclidean Geometries

In book I of his *Elements*, Euclid poses the five "requests" that, according to him, define planar geometry. These postulates would become the keystone for all of geometry, a system of absolute truths whose validity seemed irrefutable. One of the reasons for this faith is that these postulates seem obvious: the first of them stipulates that a straight line passes between two points; the second that any line segment can be indefinitely prolonged in both directions; the third that, given a point and an interval, it is always possible to trace out a circle having the point for its center and the interval as its radius; and the fourth that all right angles are equal to each other. The fifth postulate, however, is less obvious: given a straight line and a point not belonging to this line, there exists a unique straight line passing through the point that is parallel to the first.[1] Since this "parallel postulate" was more complicated than the others, the mathematicians following Euclid would try, for many centuries, to prove it from the four preceding ones, all in vain. In the nineteenth century, there occurred one of the great sudden

[1]The original formulation of this postulate is different from this more popular version, due to the Scottish mathematician John Playfair (1748–1819), who demonstrated that it was equivalent to the one given by Euclid.

revolutions in the history of mathematics (and also in human thought, as will be seen by what follows): two new geometries that do not satisfy the fifth postulate, but that are perfectly coherent, were discovered. In one of these geometries, called *spherical geometry*, no parallel line satisfying the conditions can be traced. This is the case for the surface of a sphere; the straight lines become great circles, whose planes pass through the center of the sphere, and since all great circles intersect each other at two diametrically opposed points (in the manner of the terrestrial meridians, which meet at the poles), no "straight line" can be parallel to another.[2] In the other geometry, called *hyperbolic geometry*, through any given point there passes an infinite number of lines parallel to another straight line.

Euclid thus displayed uncommon depth of view. Not only is the fifth postulate indispensable, since it cannot be derived from the others, it is moreover this postulate that uniquely characterizes planar geometry. If it is infringed upon, the geometry is fundamentally changed in nature: now non-Euclidean, it allows one to model a space endowed with curvature.

Of course, three-dimensional Euclidean geometry already allowed one to speak of curvature with respect to objects of lesser dimension. It is obvious that the curvature of a circle, a one-dimensional space, becomes greater as the radius shrinks. Likewise for the sphere, a two-dimensional surface: the larger its radius, the weaker its curvature; in the limit, as its radius tends toward infinity, its curvature tends toward zero, which is to say that the sphere approaches the Euclidean plane. At any rate, in nature, planar geometry is an idealization. Most surfaces have a curvature that varies from one point to the other, being sometimes positive, sometimes negative: e.g., a hilly terrain, the surface of a piece of clothing, or the surface of the human body. But the discovery of non-Euclidean geometries allowed, for the first time in history, one to define a curved three-dimensional space in a coherent way, and as a consequence to use it to model the physical space in which we live.

The parallel postulate allows two other equivalent formulations. One concerns the sum of angles in a triangle, the other the ratio between the circumference and radius of a circle. If the postulate is satisfied, it is strictly equivalent to dictating that the sum of the angles in a triangle is equal to 180 degrees and that the ratio of the circumference of a circle to its radius is equal to 2π—this is Euclidean geometry, with zero curvature. If no parallel can be traced, the sum of the angles in a triangle is greater than 180 degrees and the ratio of the circumference to the radius is smaller than 2π: one is in a spherical geometry, with positive curvature.

[2]The sphere also does not satisfy either the first or second postulates: for two points situated at the antipodes, there is an infinity of straight lines passing from one to the other, and every straight line is of finite length (its circumference).

If an infinity of parallels can be traced, the sum of the angles in a triangle is smaller than 180 degrees, the ratio of the circumference to the radius is greater than 2π, and one is in a hyperbolic geometry, with negative curvature. The discoverers of non-Euclidean geometries were four mathematics geniuses named Lobachevsky, Bolyai, Gauss, and Riemann.

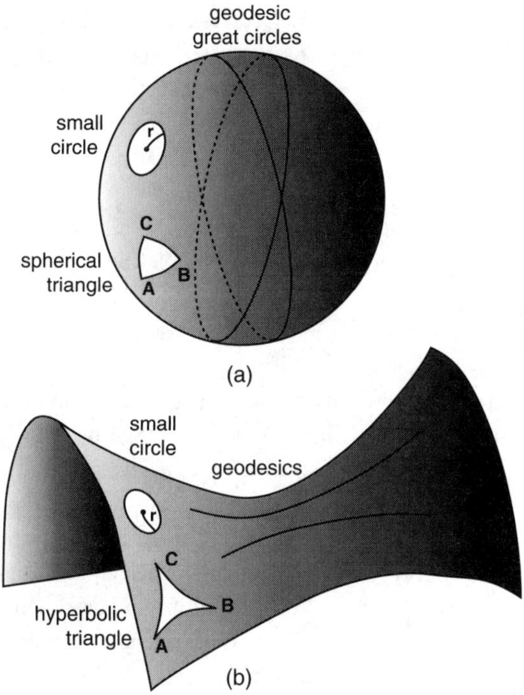

Figure 27.1. Two non-Euclidean surfaces. (a) The sphere is a finite space in two dimensions, of uniform (the same at every point) and positive curvature. The geodesics—equivalent to the straight lines of the plane—are the great circles, all centered on the origin and having the same radius as the sphere. Therefore, there are no parallel geodesics. The geometry of the sphere is not Euclidean: the sum of the angles in the spherical triangle ABC is greater than that of the angles in a planar triangle (180°), and the ratio of the circumference of a small circle to its radius is smaller than 2π. (b) The hyperboloid, of which a *saddle region* is represented here, is an infinite two-dimensional space, with uniform but negative curvature (at each point, the two radii of curvature are centered on opposite sides of the surface). There is an infinite number of geodesics that are parallel to each other. The geometry of the hyperboloid is not Euclidean: the sum of the angles in a hyperbolic triangle ABC is smaller than 180°, and the ratio of the circumference of a circle to its radius is greater than 2π.

Nikolai Lobachevsky (1792–1856) was teaching at Kazan University, in Russia, when he published his results on an "imaginary geometry" (which would later be called hyperbolic), in 1829. One often reads that non-Euclidean geometries were conceived by way of pure mathematical speculation, without any reference to the physical universe. This is completely inaccurate. Lobachevsky, for example, declared that there was no branch of mathematics, no matter how abstract, that could not one day be applied to phenomena in the natural world. He used the best astronomical data of his time to calculate the sum of the angles in a cosmic triangle with three corners formed by stars. Of course, he found no difference from 180 degrees, but he remained convinced that there was nothing particularly natural about Euclidean geometry. He would not live long enough to see the triumph of his ideas: after dismissal from his academic position in 1846, his health went into rapid decline.

Without being aware of Lobachevsky's work, the Hungarian János Bolyai (1802–1860) developed hyperbolic geometry in a 24 page text entitled *The Absolutely True Science of Space*, as an appendix to a larger treatise written by his father, who was himself a mathematician. This accomplished violinist, fencer, and dancer in the imperial army of Austria, fluent in nine languages including Chinese and Tibetan, would leave behind twenty thousand pages of unpublished mathematical manuscripts. "Out of nothing I have created a strange new universe," he wrote, to which his father replied: "For God's sake, please give it up. Fear it no less than sensual passions because it, too, may take up all your time, and deprive you of your health, peace of mind, and happiness in life" (quoted in [Davis and Hersh 81]).

Carl Friedrich Gauss (1777–1855) taught himself to read and count at three years of age. Celebrated at first for his calculations and astronomical predictions, he became a professor at the University of Göttingen, where he established himself as the greatest mathematician of his time. He was also able to use his intelligence for practical ends: in the lecture hall of Göttingen, he carefully studied the financial news; thanks to judicious market speculations, he amassed a considerable personal fortune, which he administered with a masterly hand until his death.

Gauss had made the same discoveries as Lobachevsky and Bolyai, several years before them; he wrote up some notes and sent some to friends in his letters, but did not dare to publish them officially. For him, in fact, non-Euclidean geometry could have no real existence.

However, Gauss was the first to test the curvature of physical space by measuring the sum of the angles in a large triangle. Over the course of the 1820s, he made a precise survey of the triangle formed by the mountain peaks of Brocken, Hoher Hagen, and Inselberg, in the south of Germany. The longest side measured 100 kilometers. Gauss found the sum of $179°59'59.32''$ and concluded that, up to observational errors, space in the neighborhood of Earth is Euclidean.

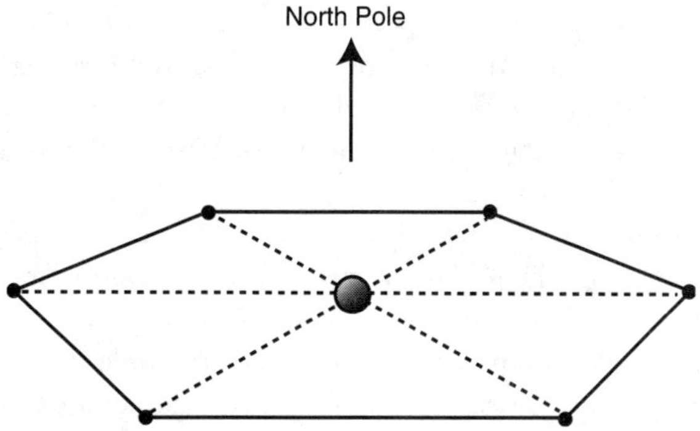

Figure 27.2. The triangle method. By measuring the distances between six rockets launched in the plane of the terrestrial equator, and by comparing their distances with respect to the Earth, it would be possible to determine the curvature of space in the neighborhood of our planet.

A modern version of Gauss's method for directly measuring the curvature of space would be to launch six rockets in the plane of the equator, each one moving radially away from the Earth, while remaining perfectly equidistant to each other (Figure 27.2).

One would thus have a series of equilateral triangles. If space is flat, the distance between each pair of rockets would grow while remaining equal to their distance from the Earth—which would also grow. If space is spherical, the distance between the rockets will grow more slowly than their distance to the Earth, while in the hyperbolic case their mutual distance would grow more quickly than their distance from us. This method combines the measurement of the sum of angles in a triangle and that of the circumference of a circle. In fact, the rockets would remain on a circle, with the effect that their mutual distances would reflect the augmentation of the circumference of the circle. Moreover, the angle formed by the straight lines that would connect each pair of adjacent rockets would be greater or less than 60 degrees (the value for flat space), according to whether the curvature of space was positive or negative.

It would probably be useless to attempt this experiment; in fact, the difference from Euclidean geometry depends on the size of the triangle used with respect to the radius of curvature of space. The larger the triangle, the more important the difference—if there is a difference. It is obvious that a bacterium living on the surface of a large balloon does not perceive the curvature and has the impression of living on a plane. In the twentieth century, general relativity has taught us that, due to the weakness of the gravitational field caused by the Earth or by the Sun, the curvature of our surrounding space is so weak that a triangle of 100 km is much too small to show it. Any other method for locally measuring the curvature would run into the same difficulties, unless carried out near a black hole or at the surface of a neutron star, which for us is—at least for the moment—totally inaccessible.

Present-day observations teach us that the average radius of curvature of the Universe is at least 10 billion light-years. Would we need to send rockets on a voyage at the speed of light for several billion years in order to form triangles that are sufficiently large to have a hope of directly revealing the curvature of the Universe? Luckily, no! Measurements of this type can be carried out, not with rockets, but with natural objects already placed at a good distance and equidistant to each other: the warm spots of the fossil radiation. This method has recently been used (the Boomerang, Maxima, and WMAP experiments) to set limits on ←245 the curvature of cosmic space.

Strangely, Gauss, Lobachevsky, and Bolyai did not think to clarify the consequences of the inverse hypothesis, in which the sum of the angles in a triangle is greater than 180 degrees. It had been known, however, since antiquity, that on a sphere, the sum of the angles for a triangle made up of three arcs of great circles (the equivalent of straight lines) is always greater than 180 degrees. This spherical trigonometry had been the subject of innumerable treatments, at first useful for astronomy, and then for geography and navigation. The celestial vault, in fact, offers a spherical concavity, and if one wants to mentally trace a straight line between two stars, one must use an arc of a celestial great circle centered on the Earth. In the same way, if one desires to travel from one point of the terrestrial surface to another one that is very distant, the only authorized paths are situated on the convex surface: there is no question of digging a direct tunnel through the interior of the planet. The shortest possible path is thus an arc of a great circle, a fact well known to navigators.

One could have thus noticed that spherical trigonometry, in two dimensions, led to a coherent geometry, although it was not Euclidean. Why the resistance against passing to higher dimension? Probably because this implies a physical space that is finite but without frontier, encircled by straight lines. Now, up to this point, geometry had not been the science of space, but the science of figures in space. As a homogeneous, indefinite expanse of three dimensions, physical space constituted a neutral reality that gave rise to no speculation. This picture had been established in the eighteenth century by the philosophy of Kant, when ←143 he made Euclidean space a category of reason.

We must wait for Bernhard Riemann (1826–1866) for the acceptance of the validity of spherical geometry in three-dimensional space. In 1854, the young mathematician gave his inaugural dissertation at the University of Göttingen entitled "On the Hypotheses which Lie at the Basis of Geometry." The story of this lecture is an astonishing anecdote that shows how the sciences, just like universal history, sometimes progress unexpectedly through small accidents and fortuitous circumstances.

At the age of 27 years, Riemann wished to teach at Göttingen; to demonstrate his abilities, each candidate had to present an oral dissertation in front of the grand assembly of the university. The rules asked that each candidate propose, some time in advance, three subjects, among which the professors chose that which seemed most promising. Traditionally, it so happened that the professors always selected one of the first two subjects on the list. Riemann had therefore carefully prepared his first two subjects and slightly neglected the last, which in fact concerned the foundations of geometry. However, he had not accounted for the presence in the jury of the great Gauss who, although a septuagenarian, remained the dominant figure on the faculty. Naturally, Gauss felt that the third subject was the most interesting. Panicked at the prospect of this seminar, on which his entire career depended, Riemann threw himself madly into the preparation of his lecture and subsequently fell ill, to the point that he missed the date of presentation and only recovered his health in April. He then prepared his lecture for seven weeks, an amount of time that Gauss had requested as an adjournment for medical reasons. Riemann finally delivered his oral dissertation on June 10, 1854.

It was in this lecture that he demonstrated the possibility of a geometry in which space is finite without having a frontier: "The unboundedness of space possesses in this way a greater empirical certainty than any external experience. But its infinite extent by no means follows from this; on the other hand if we assume independence of bodies from position, and therefore ascribe to space constant curvature, it must necessarily be finite provided this curvature has ever so small a positive value. If we prolong all the geodesics starting in a given surface-element, we should obtain an unbounded surface of constant curvature, i.e., a surface which in a flat manifoldness of three dimensions would take the form of a sphere, and consequently be finite" [Riemann 73].

Riemann also proposed to extend geometry to spaces having an arbitrary number of dimensions. For this, geometry must be understood as the study of possible *varieties*. In the most general sense, a variety is nothing more than a collection of objects from a set. Geometry then becomes the study of the rules and conditions placed on the objects from the set in order to determine the distances between elements.

For example, in a two-dimensional space, each point is defined in a unique way by an ordered pair of real numbers (x_1, x_2): its coordinates. In three-dimensional space, each point is defined by a triplet of numbers (x_1, x_2, x_3).

In an abstract n-dimensional space, each point has a unique address given by an n-tuplet $(x_1, x_2, x_3, \ldots, x_n)$, an ordered set of n real numbers. The rules for determining the distances between elements in the set are summed up by the metric, a sort of generalization of the Pythagorean Theorem whose coefficients determine the curvature of the space.

Riemann is not only the creator of the most profound lecture in the history of geometry, but in 1858 he accomplished an exploit of the same order, with a completely innovative article of eight pages, on number theory. Perhaps he would have performed a third exploit, if he had not died of tuberculosis at 39 years of age.

Riemann's inaugural dissertation was not published until after his death, in 1867. Another year passed before the Italian Eugenio Beltrami (1835–1900) crowned the edifice of non-Euclidean geometry: his model of the *pseudosphere*, describing a hyperbolic space of constant negative curvature by means of an equation analogous to that of the hypersphere, put in evidence the common nature of the hypotheses of Bolyai and Lobachevsky, and of Riemann. The way was now definitively opened for Clifford, Einstein, Weyl, and others to model certain natural phenomena like gravitation in terms of non-Euclidean geometry.

Further Reading

Boris Rosenfeld. *A History of Non-Euclidean Geometry: Evolution of the Concept of a Geometric Space*. New York: Springer-Verlag, 1988.

WHO includes diversity, and is Nature.

—Walt Whitman, *Kosmos*

28
Cosmos and Logos

Discourse about the *cosmos* has considerably changed its purpose over the course of time. Homer used the word to describe the finery of women, the ornaments of the warrior, the construction of a poem, and physical, moral, or social ideals. The notion of cosmos applied to the organization of the Universe appeared with Pythagoras and the presocratics, around the sixth century BCE, to designate the order and beauty of the world. According to them, reality was not the chaos described two centuries earlier by Hesiod in his *Theogony*; it was, on the contrary, beautiful and well arranged. Cosmos was thus opposed to chaos, the order on high contrasted with the disorder down here. In his poem "On Nature," Heraclitus affirmed: "Among the things distributed by chance, the most beautiful: the cosmos." In *Timaeus* Plato makes of the cosmos the image of an intelligible model, and in his *On the Heavens*, Aristotle uses *cosmos* to designate the ordering of the sky.

The philosophers of antiquity were therefore the first to speculate on the general organization of the Universe, on the nature of space, time, and matter, in other words, to apply the *logos*—human reason incarnated by language—to the cosmos. In brief, they were the first to do cosmology. Nevertheless, it would take two millennia before such discourse on the Universe finally carried this name.

The term *cosmology* was mentioned in English for the first time only in 1656, in Thomas Blount's dictionary *Glossographia*, where it is defined as discourse about the world, which corresponds exactly to its etymology. It was taken up again, in Latin, in the work by Christian Wolff, *Cosmologia generalis* (1731): "General cosmology is the science of the world or of the universe in general, to the extent that it is a composed and modifiable being."

Wolff sought to approach questions on the physical nature of the world by disengaging himself as much as possible from myth and religion. Despite this, one constant in all of the later history of cosmology is the reproach which would be heaped upon him for mixing physical statements with philosophical or religious reflections. Thus, the French term was used for the first time by Pierre de Maupertuis (*Essai de cosmologie*, 1750), but the author affirmed that, after having noticed that philosophers have always wanted to explain the system of the world, it is impossible to achieve a complete solution of this problem. When the word was admitted the following year into the French encyclopedia, Jean le Rond d'Alembert, author of the article, noted that the subject is hardly deserving of interest, since it reduces to the question of knowing if one can find God in nature. According to Kant (*Critique of Pure Reason*, 1755), "no observation would be able to confirm the thesis of rational cosmology, whose objective goes beyond any possible experiment." In the nineteenth century, Auguste Comte would once again say that it was only possible to study the solar system, as opposed to the Universe, which would always escape us.

It is easy to understand the reticence of positivists toward the discipline. If astronomy is there to speak of stars and their organization, what could cosmology really speak about? Of what is hidden behind this organization? One then falls on the idea of the Supreme Being and on religious questions. Of the place of us humans in this organization? If one is not once again in religion, one is at least in philosophy. So it is with other questions, such as: is the Universe born and, if yes, when and how?

In brief, if one takes interest in the Universe and does not make use of a telescope, what can one say? Apart from theological and philosophical discourse, a third category has taken the risk of entering into the subject: literature. However, in this case it is more a matter of using the image of the cosmos established at a given time to develop mystical tales, fantastical reveries, or utopian discourses. The cosmic voyages imagined by Cicero (*The Dream of Scipio*), Hildegarde of Bingen (*Scivias*), Dante (*The Divine Comedy*), or Cyrano de Bergerac (*The Other World*) were produced by exceptional people, able to relate their experience of the intangible, since to speak of the Universe as a whole requires knowing how to place oneself outside of it. Now, since the fall of Aristotle's system of a closed world, this had become unthinkable. The Universe would from now on be mixed up with Newton's infinite space. Its observational part would not stop growing and deepening, but its most distant abysses seemed forever inaccessible to human experience. Ideas about its general organization could not be supported by any properly cosmological observation. The models were inevitably stuck on metaphysical presuppositions. This was notably the case with Newtonian cosmology and its adaptations by Wright, Lambert, and other cosmologists of the eighteenth and nineteenth centuries.[1]

The situation changed completely in the twentieth century, with the advent of general relativity. The Universe took on a new meaning: it was no longer a frame containing objects; it was the ensemble of a frame plus objects. For the first time, it became legitimate to hold a rational discourse on the whole. Relativistic cosmology notably contradicted the arguments of Kant and Comte: it is possible to arrive experimentally at a determination of the fundamental cosmological parameters, and even to prove the finite character of space.[2]

Nevertheless, one should not believe that the interpenetration of physics and philosophy has ceased with relativistic cosmology. Cosmology is never innocent. However sophisticated the mathematical formalism on which it is founded, it assumes a philosophy of nature, where it is possible to attribute to the world a shape, a structure, an order. The word *Universe*, a synonym of *cosmos*, but taken from Latin, suggests the notions of unity and of diversity. The Universe is the unity of the diverse. There is a unity to the diverse if there are universal natural laws. These laws are supposed to act at every point in space and at every moment in time, and therefore to be indifferent to space-time. They reassemble that which space and time tend to disperse. For many physicists in search of unity, cosmology has as its aim to explain the abundant multiplicity visible in the material universe—from elementary particles to galaxy superclusters—by an underlying unity, that is to say a certain secret order, a certain harmony.

Einstein, inventor of the theory of general relativity, which today serves as the principal scientific basis for any cosmological study, did not himself keep clear of metaphysical presuppositions. Among the numerous possible solutions to his equations, he chose one very particular solution in 1917, describing a finite and static universe, doing so for aesthetic reasons. This choice, in fact, caused him to miss the final layer, which would confer upon relativistic cosmology its status as an experimental science: the discovery, with cosmic expansion, that *the whole* evolves, that the Universe has a history—a true history, not a simple creation followed by nothing. And this history, inscribed in each galaxy, in radiation, and in each element, could be rationally decrypted.

[1]See, for example, "The Grandeur of Space," in [Lachièze-Rey and Luminet 01].

[2]This is so on the condition that the hypothesis of homogeneity is justified.

Another example of a presupposition that today underlies all relativistic models of the universe is the cosmological principle, according to which there is no privileged position within space. Now, this is not, strictly speaking, demonstrable, since we never observe a slice of simultaneous space: every observation plunges us into the past of the Universe.

The task of modern cosmologists consists in developing a coherent description of the observable universe. It is not obvious that this is possible. In the mid-1970s, the Swedish physicist Hannes Alfvén gave vent to a fierce epistemological critique of cosmology. Is it still a myth or does it in the end constitute a science? But how does one construct a science on a unique object? How should one study an object from which, by definition, one cannot extract oneself, in order to view it from a distance?[3] The undeniable authority of this Nobel laureate in physics rattled more than one as to the validity of the cosmologists' scientific approach.

It is true that a number of cosmological assertions, as delivered by some champions of the discipline, expose themselves to epistemological panning. For example, one often reads, even in the most serious journals and under the pens of seasoned professionals, that the big bang is the violent explosion that gave birth to the Universe. Now, not only does such a definition create confusion between the beginning of the Universe (which lies outside of any language) and the beginning of the intelligibility of the Universe, but each of the terms used is improper.

Even more debatable, or even dangerous, are the comments that were made after the discovery of fluctuations in the cosmological radiation made by the COBE satellite in 1992: "If you are a believer, it is as if you are looking at the face of God," "They have found the Holy Grail of cosmology," and "The greatest discovery of the century, if not all times." These are not the exclamations of visionaries who believe to have seen the features of Jesus in the COBE map, but well and truly declarations made by eminent cosmologists. Such collusions between cosmology and belief, between science and show business, are not of a nature to clarify the ideas held by the public. They show how difficult it is for some to unburden themselves of the philosophical-mystical clutter that covers discourse on the whole, and seem to justify, despite two centuries of "progress," the criticism of d'Alembert. For my part, in these excesses I see either signs of great naivety or, for those who are not at all dupes, great media cunning!

[3]The neurosciences have momentarily run into the same type of argument: some people believe that a human brain could never attain an objective knowledge of the human brain. However, the human brain in fact has a few tricks up its sleeve.

Why? From where, the universe? Where is it going?

—Jules Laforgue, *Éclair de gouffre*

29 Cosmic Questions

Modern cosmology covers three large areas:

- *geometric* cosmology, which speculates on the local and global properties of space-time, its shape and its frontiers, including singularities and horizons;
- *physical* cosmology, which treats the material processes which take place in the universe at such or such stage of its evolution; and
- *observational* cosmology, which looks for the brute facts: the spectral shifts of galaxies, the cosmological background radiation, gravitational mirages, and the distribution of quasars, galaxy clusters, and other large structures.

Since observational cosmology is carried out by an observer localized in a small region of the Universe, who only has access to very partial information about space-time as a whole, the observer is confronted by two basic problems: does the observable universe possess properties that reflect those of the Universe in general, or do these properties rather result only from the particular position of the observer?

If space is infinite or exactly Euclidean in nature, it is a matter exclusively for geometric cosmology, since no measurement relating to the observable universe 245→ would be able to shed light on it. If space is finite in nature, although it is still an assuredly geometric question, it could nevertheless be elucidated experimentally.

As for physical cosmology, some properties that are not directly observable may play an important role, and leave measurable traces. For example, the topological defects that might have formed in the primitive Universe, such as cosmic strings, domain walls, or textures, could leave specific imprints in the temperature fluctuations of the cosmological background radiation. Likewise, if space is closed (whether by curvature or by topology) on a scale that is presently larger than the horizon, it may have nevertheless had effects in the distant past of the universe, and may have repercussions today on the large scale distribution of galaxies or other observable characteristics of the universe.

Historically, geometric cosmology immediately occupied the forefront of the scene. There are at least two reasons for this: the first is that before the use of astronomical telescopes by Galileo, and before the development of a mathematical theory of gravitation by Newton, observation and physical reasoning were absent ←169 from the cosmological question. It was founded nearly exclusively on philosophical assumptions, such as that of geometric harmony, or on religious ones.

The second reason ties in with the very meaning of geometry: etymologically, *geometry* means "measurement of the Earth," which is to say, by extension, "measurement of the world, measurement of the universe." Now, measurement has always been preeminent in scientific practice. Cosmology, the science of the Universe as an object, passes first through its measurement. And since the Universe has not only extent but also duration, this requires knowing how to measure not only space but also time.

Moreover, as soon as one speaks of measurement, one is inevitably led to speak of the boundaries of the object to be measured, and to thus broach the essential cosmological questions, which challenge human imagination, for example:

- What is space?
- What is time?
- What is the void?
- Did time have a beginning?
- Will the Universe end?
- Is space infinite?
- Did the big bang really take place?
- Was there something before the big bang?
- What is the matter in the Universe made of?

- When and how did the stars and galaxies form?
- Does space have extra hidden dimensions?
- Are there other universes?

One cannot provide a firm response to these questions, but one can at least eliminate many models and theories, and give plausible responses as provided by relativistic cosmology. Accompanied by their particular strengths and weaknesses, these solutions (which include the big-bang models) remain idealizations of the real Universe.

Space alike to itself
that it grows or denies itself
—Stéphane Mallarmé, *Quand l'ombre menaça*

30 Expansion and the Infinite

The Universe is expanding. What does this really mean? Most people imagine an original huge explosion, as the term *big bang* suggests, and the metaphor is constantly used in popular accounts. Some speakers even have the tendency to mime a gesture of expansion with their hands, as if they were holding a piece of space or an immaterial balloon in the process of inflating. The public imagines some matter ejected at prodigious speeds from some center, and tells themselves that it would be better not to be there at the moment of explosion, so as not to be riddled through with particles. None of this is accurate. At the big bang, no point in the Universe participated in any explosion. Put simply, if one considers any point whatsoever, we notice that neighboring points are moving away from it. Is this to say that these points are animated by movement, given a speed? No, they are absolutely fixed, and nevertheless they grow apart.

To unravel this paradox, it is necessary to make more precise what one exactly means when speaking of a fixed point. The position of a point is fixed by coordinates: one number for a line (the miles along a highway), two numbers for a surface (latitude and longitude), and three for space in general (length, width, and height). A point is said to be fixed if its coordinates do not change over the course of time. In an arbitrary space, curved or not, the distance between two points is given by the so-called metric formula, which depends on the coordinates and generalizes the Pythagorean Theorem. In principle, therefore, the distance between two points does not vary. In an expanding space, on the other hand, this distance grows, while the points do not move, even by a millimeter, meaning that they strictly conserve the same coordinates. These fixed coordinates are known as *comoving* coordinates. In relativistic cosmology, galaxies remain fixed at comoving positions in space. They may dance slight arabesques around these positions, under the influence of local gravitational fields, but the motion that moves them apart from each other resides in the literal expansion of the space that separates them. It is a form of motion that no human being has experienced. It is therefore not surprising that our intuition rebels against its implications, and seeks other less radical interpretations. Thus, when Friedmann in 1922 and Lemaître in 1927 discovered that the expansion of the Universe was rendered possible by the laws of general relativity, and became convinced that this model should describe physical reality, they met many detractors—first and foremost among them being Einstein himself!

If space is expanding, it would seem at first sight that it must be finite: either because one visualizes some sort of big balloon that inflates into an external space, or because one starts from the principle that an expanding thing must necessarily grow in size and that, if this thing has a size, it is necessarily finite. One might also deduce, incorrectly, that at the big bang the Universe as a whole was necessarily contracted into a point of zero volume.

However, infinite systems may perfectly well expand or contract. How can something that occupies an infinite space dilate, and starting from what? This is one of the questions asked of me most frequently. In an attempt to explain, let us begin with a very simple example, that of a one-dimensional infinite universe: mathematically, it would be a straight line, on which each galaxy would be represented by a number (see Figure 30.1).

At a certain instant (time A), the galaxies are separated by equal intervals, which means that their distribution is homogeneous. If the Universe is expanding, at the following instant (time B), the galaxy initially denoted by 1 occupies the position that galaxy 2 had, while galaxy 2 is now found where galaxy 4 initially was, and so on. We are really dealing with a uniform expansion within an infinite space containing an infinite number of galaxies. If one goes back in time to describe the big bang, all of the galaxies approach each other, and when they are all touching, space is still represented by a straight line of infinite length!

In generalizing to three dimensions, one can in the same way imagine that if space is infinite, the big bang, meaning the beginning of expansion, began at every point of a space that was infinite beforehand, as compressed as it may have

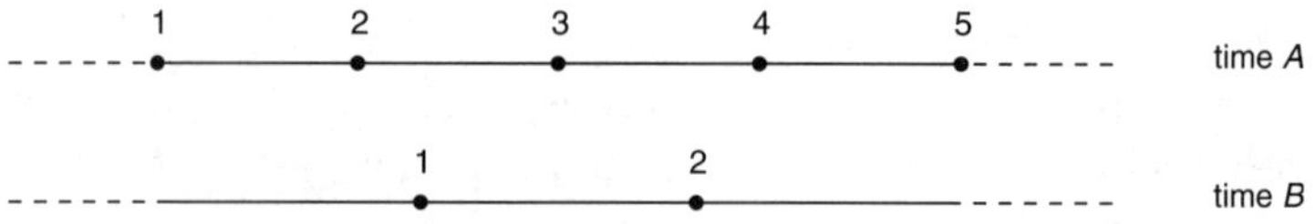

Figure 30.1. An expanding one-dimensional universe.

been. If one cuts infinite space into cubes, then each of these cubes will in the future become a larger cube, or in the past a smaller cube; but no cube is the center of the infinite space.

How could an infinite space dilate or contract and still occupy the same volume? The German mathematician David Hilbert gave a simple analogy. He imagined a hotel containing an infinite number of rooms, all of them occupied. A new customer arrives. Where do we put him? Fear not! The receptionist only has to transfer the occupant of room 1 into room 2, that of room 2 into room 3, and so on. The last to arrive can take possession of room number 1. Let us now imagine that car after car of tourists unload, an infinite number of them. Here again, the receptionist can perfectly well satisfy everybody: he moves the occupant of room number 1 to room number 2, that of room number 2 into room number 4, that of room number 3 into room number 6, and so on. In this way, one out of every two rooms will now be free (each room with an odd number)!

Just as in Hilbert's hotel, an infinite space can accept an additional amount of space without becoming more cramped.

The spawning galaxy in flight is a rainbow trout which goes back against the flow of time towards the lowest waters, towards the dark retreats of duration.

—Charles Dobzynski, *La dernière galaxie*

31
Galaxies in Flight

Since the time of Newton, we have known that white light, passing through a prism, is decomposed into a spectrum of all colors. Violet and blue correspond to the shortest wavelengths or, equivalently, to the largest frequencies; red corresponds to the largest wavelengths and to low frequencies. In 1814, the German optician Joseph von Fraunhofer discovered that the light spectrum from stars is streaked with thin dark lines, while that from candlelight has bright lines. These phenomena remained puzzling until 1859. It was then that the chemist Robert Bunsen and the physicist Gustav Kirchhoff analyzed the light created from the combustion of different chemical compounds (burned with the now famous Bunsen burner) and saw that each of them emitted light with its own characteristic spectrum.

At nearly the same time, Christian Doppler discovered in 1842 that moving the source of a sound produced shifts in the frequency of sound waves, a phenomenon experienced by anyone listening to the siren of an ambulance passing by. The French physicist Armand Fizeau noticed the same phenomenon with light waves: depending on whether a source of light was moving closer or farther away, the received frequencies are either raised or lowered with respect to the emitted frequencies. The shift becomes larger as the speed of displacement is increased. If the source is getting closer, the frequency grows, and the light becomes more "blue"; if it moves away, the frequency lowers and the wavelengths stretch out, becoming more "red," with respect to the spectrum of visible light. Since this shift affects the whole spectrum by the same amount, it is easily quantified by looking at the dark or bright lines, which are shifted together, either toward the blue or toward the red, and it provides an incomparable means of measuring the speed of approach or retreat for light sources.

Shortly after this discovery, astronomers began an ambitious program of spectroscopy, with the aim of measuring the speed of the planets and stars by using their spectral shifts.

The formula for the Doppler-Fizeau effect is quite simple: $v = cz$, where v is the speed of the source, c the speed of light (300,000 km/s), and z the relative shift in wavelength, that is to say the difference between the received wavelength and the emitted wavelength, divided by the emitted wavelength.[1]

The proper speeds of celestial bodies, whether planets, stars, or galaxies, are no more than a few hundred kilometers per second, which is small with respect to the speed of light. The Doppler-Fizeau effect is therefore weak, and requires precise spectroscopy to be measured. But let us imagine for a moment a universe where the speed of light would only be 1,000 km/s; the starry sky would offer an extraordinary spectacle to the eye: the stars would change color, those that flee would change into rubies, and those that approach into amethysts!

In the second decade of the twentieth century, some American astronomers such as Vesto Slipher and Edwin Hubble applied the program of systematic measurement of speed to the spiral galaxies—which produce spectral shifts similar to those of stars. As long as the speeds deduced from the Doppler-Fizeau formula were no more than a few hundreds of kilometers per second, nobody questioned the fact that the spectral shifts of galaxies indicated the speed of their motion with respect to the Earth, just as for stars.

However, over the course of the 1920s spiral galaxies were discovered whose spectral shifts, systematically oriented toward the red, were greater than 0.1, which implies speeds of flight of more than 30,000 km/s! In a letter addressed to de Sitter in 1931, Hubble announced his preoccupation: "We use the term 'apparent velocities' in order to emphasize the empirical features of the correlation. The interpretation, we feel, should be left to you and the very few others who are competent to discuss the matter with authority." In fact, it was Georges Lemaître who, in 1927, was the first to understand that the redshift of galaxies did not
←184 correspond to physical speeds: it is space that is dilating, dragging galaxies along with it in apparent recession.

[1]For the lovers of formulae: $z = (\lambda_o - \lambda_e)/\lambda_e$, where λ_o is the observed wavelength and λ_e the emitted wavelength. We see immediately that if the source moves away, z is positive, while if it is getting closer, z is negative.

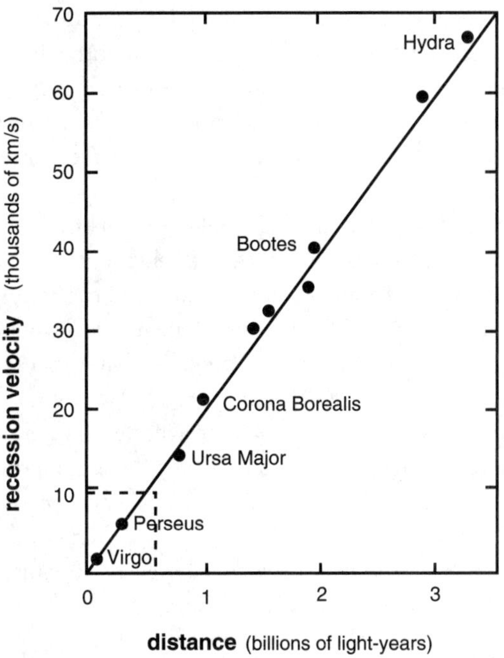

Figure 31.1. The velocity-distance relation. During the 1920s, measurements of cosmological spectral shifts were made only on relatively nearby galaxies, belonging to the Virgo and Perseus clusters. Later measurements extended to much more distant clusters, like Bootes and Hydra. The proportionality relation has been verified up to distances of 3 billion light-years. Beyond this distance, the curvature of space bends the line toward the bottom or the top.

Although the theoretical perspective found itself radically changed, it remained nevertheless tempting to interpret these redshifts through the Doppler-Fizeau effect. The terms *recession velocity* and *speed of flight* were rapidly adopted by the astronomers and popularizers. These speeds seemed to be a completely natural consequence of the expanding motion of matter from the big bang. One pictured an initial firework throwing galaxies to the four corners of the Universe, like debris from an explosion. Perched on one of the cooled embers, we would be contemplating the other embers, moving away from us at great speed. However, this image is erroneous and leads to a completely false physical interpretation of the cosmological redshift. In fact, as astronomers explored the Universe more and more deeply, it turned out that the galaxies they detected were affected with redshifts that were ever larger. In 1930, Hubble was worried by redshifts of 0.1, but what can one say today of the quasar discovered in early 2000 with $z = 5.82$? As techniques progress, the records fall, month after month. The most recent data from the Hubble space telescope and the European VLT (very large telescope) in Chile mention galaxies whose spectral shift is 10. This means that they are observed such as they were 13.2 billion years ago, when the Universe was only 500 million years old!

If one believes the formula for the Doppler-Fizeau effect, these galaxies must be receding at speeds five or six times greater than that of light, which is of course absurd. According to the theory of relativity, no object can travel faster than light. Even in a Newtonian model of the Universe, where speed would not be bounded by any limit, any *observed* galaxy would necessarily have a speed of flight smaller than that of light: if this were not so, its light rays would not have the time to reach us, since the distance to cover would grow faster than their speed.

The formula for the Doppler-Fizeau effect is therefore incapable of accounting for the cosmological redshift. A number of teachers of physics and astronomy believe themselves to be doing the right thing by appealing to the formula known as the *relativistic Doppler-Fizeau effect*, which corrects the regular formula in a way such that galaxies at large spectral shift now have speeds that are smaller than that of light.[2] Their idea is that at high speed special relativity is substituted for Newtonian mechanics, if one wants to correctly interpret the world. Indeed, by using the relativistic formula, a quasar at $z = 4$ only has a speed of flight equal to 92% of the speed of light, and everything seems to once again be in order. Nevertheless, they have done no more than to replace one incorrect explanation with another. To calculate the speed of the quasar in this way presupposes that special relativity is applicable on cosmological scales; however, this is a theory of flat space-time without gravity, while in cosmology the curvature of space-time plays a primary role.[3]

Ascribing cosmological redshifts to a Doppler-Fizeau effect, either relativistic or not, leads to an unfortunate lack of understanding of big-bang cosmology, and causes one to miss one of its most mysterious beauties.

Big-bang cosmology is founded on the theory of general relativity, which transcends both Newtonian mechanics and special relativity, and introduces concepts that have no equivalents in these older theories. These concepts are not easily accessible to the general public, since our intuition is forged by natural processes that take place in a world moving at small velocity with weak gravity.

Just as the constancy of the speed of light leads to surprising but true statements, such as the twin paradox, the curvature of space-time leads to its own

[2] Again, for the lovers of formulae: $1 + z = (1 + v/c)/\sqrt{1 - v^2/c^2}$.

[3] Even if space has zero curvature, the Universe as a space-time has a non-zero curvature as soon as it is in expansion.

menagerie of phenomena. One of them involves the slowing down of clocks in a gravitational field. This gravitational redshift is produced when the wavelength of light emitted from a massive body is shifted toward longer wavelengths during its journey toward the observer. This shift has nothing to do with any kind of Doppler-Fizeau effect, since the observer is not in relative motion with respect to the body emitting the signal.

The cosmological redshift is in fact another such phenomenon specific to general relativity, which has no counterpart in either Newtonian physics or special relativity. According to the interpretation initially given by Lemaître, it is not caused by galaxies moving through space like projectiles; it is the spatial fabric itself that expands. The cosmological redshift results from the fact that the radius of curvature of the Universe was smaller in the past, when the Universe was younger. This radius of curvature is measured by a *scale factor* a, which could be, for example, the radius of a sphere that includes the observable universe. Since space expands, a grows over the course of time; the light rays coming from the past have been stretched along the way, between the moment they were emitted by a distant galaxy and the moment we detect them. The correct formula for the cosmological redshift becomes $1 + z = a_o/a_e$, where a_o is the scale factor of the Universe today and a_e the scale factor of the Universe at the time when the light was emitted from the galaxy.

Now, there is no longer any contradiction. By exploring the more and more distant past of the Universe, big-bang cosmology predicts a scale factor r_e tending toward zero, which in turn implies redshifts tending toward infinity.

Although the maximum values observed today for galaxies are around $z = 10$, modern cosmology has discovered a much older light source, whose redshift reaches 1100: the cosmological background radiation. This implies that it was emitted during an epoch when the scale factor was 1,100 times smaller than it is 231→
today.[4]

The frontier of the observable universe—the cosmological horizon—necessarily corresponds to the limit where the redshift is infinite. This does not prohibit that, in certain models of the Universe, space can be immense and accommodate galaxies situated hundreds of billions of light years away from us. However, we cannot receive any radiation from them: in terms of a speed of expansion, these distant regions move away at a speed greater than that of light.[5]

---

[4] To deduce a corresponding time scale in the history of the Universe, as one often does in scientific popularizations, we must pass to a particular model whose curvature parameters, meaning the density and the cosmological constant, are fixed. For plausible values of these parameters, the time of emission occurred between 300,000 and 1 million years after the big bang.

[5] In general relativity, space can expand more quickly than light, because space represents neither matter nor energy.

There, where worlds seem, with slow steps,
Like an immense and well-behaved herd,
To calmly graze on the ether's flower.

—Giovanni Pascoli, *Il ciocco*

32 The Rate of Expansion

In 1927, Georges Lemaître published a revolutionary article in the *Annales de la Société scientifique de Bruxelles* entitled "A homogeneous universe of constant mass and increasing radius, accounting for the radial velocity of extragalactic nebulae." As the title suggests, Lemaître showed that a relativistic cosmological model of finite volume, in which the Universe is in perpetual expansion, naturally explains the redshifts of galaxies, which at that point were not understood. In particular, the article contained a paragraph establishing that 42 nearby galaxies, whose spectral shifts had been measured, were moving away at speeds proportional to their distances. Lemaître gave the numerical value of this proportionality factor: 575 km/s per megaparsec, which means that two galaxies separated by 1 megaparsec (or 3.26 million light-years) moved away from each other at an apparent speed of 575 km/s, and that two galaxies separated by 10 megaparsecs moved apart at a speed ten times greater.

This unit of measurement, the kilometer per second per megaparsec, shows clearly that the speed of recession depends on the scale. In 1377, in his *Book of the Heavens and the World*, the scholar Nicole Oresme had noted that, at dawn, one would not notice anything if the world and all living creatures had grown by the same proportion during the night. In Lemaître's theory, on the contrary, the separation between two points in space grows faster with greater separation, which renders it perceptible.

Lemaître's article, published in French, passed unnoticed until 1931, when it was finally read by Arthur Eddington, who published an English translation. Unfortunately, this version omits the paragraph in which Lemaître established his law of proportionality. Meanwhile, in 1929, the great American astronomer Edwin Hubble had published the experimental results he obtained with his collaborators and described a general law, according to which the recession velocity of a galaxy is proportional to its distance. This law, identical to Lemaître's, with the same proportionality factor, would from now on carry the name of *Hubble's law*. It forms the experimental basis for the theory of the expansion of the Universe, of which the big-bang models are the fruit.

By a curious historical irony, Lemaître would fall victim to two injustices. Not only does the law of expansion not carry his name, even though he was the first to discover it, but he is also not generally credited with the invention of the theory of the big bang: it is Hubble, once again, even though the latter had always disavowed this interpretation of his measurements!

Moreover, Hubble believed that the proportionality factor between recession speeds and distances was constant over the course of time. This proportionality factor was also baptized *Hubble's constant*. However, the development of cosmological models coming from general relativity, initiated by Friedmann and Lemaître, showed that this "constant" is not one: it depends on time, so much so that it is more correct to call it the *rate of expansion*. Of course, the more distant the galaxy, the greater its apparent speed of recession, but the relation between the two is a little more complicated than a simple law of proportionality. It depends on the curvature of space. The measurement of differences with respect to the strict law of proportionality is in fact one of the experimental methods that allow for evaluation of this curvature. The rate of expansion is therefore a crucial quantity in all current cosmological models, and enormous efforts have been made in order to determine its exact value.

How does one do this? According to the Hubble-Lemaître law, one must measure both galactic redshifts (by using the spectral lines) and, independently, the distances to these galaxies. The first task is done with great precision; the second, on the other hand, remains a fundamental problem.

The construction of a trustworthy scale of extragalactic distances has been 192→
an immense challenge for astronomers since the 1920s. For a long time, there were two opposing camps, one arguing in favor of a lower value for the rate of expansion, the other in favor of a higher value, with the two values differing by a factor of two!

In reality, each of the two camps evaluated the distances in different ways. Today, the situation has improved considerably thanks to a better understanding of the sources studied and new techniques of observation. Recent measurements

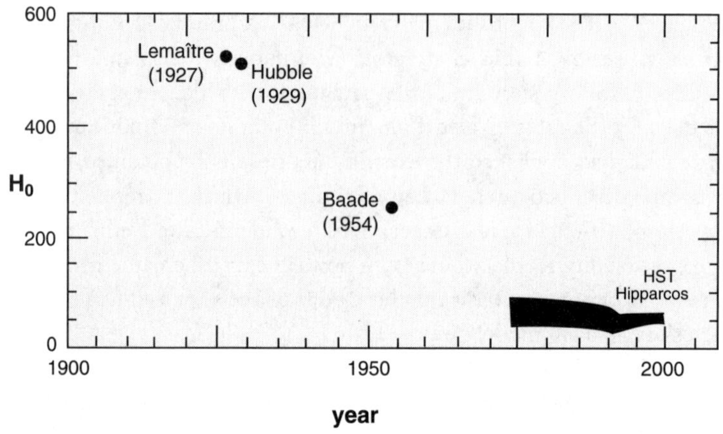

Figure 32.1. Uncertainties in the rate of expansion. This historical picture shows the various values of the rate of expansion adopted by astronomers, starting from the pioneering works of Lemaître and Hubble and continuing through the most recent data obtained by orbiting telescopes. Over seventy years of observation, the value has decreased by a factor of almost ten!

converge toward a value of 70 km/s per megaparsec (almost ten times smaller than that initially estimated by Lemaître and Hubble[1]), within an error of only a few percent.

A question often asked by the public about the expansion of the Universe is the distance scales on which it effectively acts. The solar system, for example: does it grow under the effect of cosmic expansion? The answer is no. The rate of expansion is imperceptible on such small distances. And the entire galaxy? No again. Just like any other individual galaxy, it is far too connected by its own gravity for its stars to be dispersed by the expansion. And nearby galaxies? Still no! United in the middle of groups or clusters, galaxies experience their own peculiar velocities in random directions. In fact, the spectral lines of the Andromeda galaxy, situated 2 million light-years away, are shifted toward the blue and not the red, indicating a speed of approach of 300 km/s. This is not the effect of a contraction of the spatial fabric between it and us, but simply a movement of free fall, due to gravity! Some other galaxies exhibit a shift toward the blue; but they all belong to our small local group, including about twenty members, each moving with its own peculiar velocity around the center of gravity of the group.

[1]The fact that it has decreased by a factor of ten in the space of eighty years does not reflect its real dependence as a function of cosmic time, but rather the enormous errors in human measurements!

In fact, the distance scale beyond which the expansion acts can be read directly from the value of the expansion rate: e.g., 70 km/s for 3 million light-years, or 700 km/s for 30 million light-years. As long as local speeds are greater than these values, they erase the tendency toward expansion. In all of the clusters, the galaxies have local speeds of a few hundred kilometers per second around the center of gravity of the cluster. The clusters themselves can participate in currents of matter at a large scale, under the gravitational influence of gigantic structures like the superclusters and attractors. Here again, their peculiar velocities are several hundred kilometers per second. The expansion of space therefore only acts definitively above local motion starting from a distance of around 100 million light-years, corresponding to a recession speed of 2,000 km/s. One hundred million light-years is the approximate size of our local Supercluster. It is only beyond this limit that the realm of relativistic cosmology really opens up.

Soon even its great age will curl up
until it is smaller by three quarters.

—Alfred Jarry, *Du soleil, solide froid*

33
The Age of the Universe

Under some simplifying assumptions, the value of the expansion rate gives an indication of the age of the Universe. It is easy to understand why a high expansion rate corresponds to a shorter theoretical age of the Universe: a greater speed implies that it began expanding more recently.

The theoretical age of the Universe deduced from the expansion rate, meaning the time taken by the Universe to expand to its present state, assuming the rate of expansion is constant over the course of time, is called the *Hubble time*. For a universe whose matter density is exactly equal to the critical density (the Einstein-de Sitter Euclidean model), the Hubble time is given by the simple formula $2/(3H_0)$. If the expansion rate is as large as 100 km/s per megaparsec, the Hubble time is only 6 billion years. If the expansion rate falls to 50 km/s per megaparsec, the Hubble time reaches 12 billion years.

The Hubble time must now be compared to the experimental age of the Universe, meaning the age of the oldest objects that we know—stars and chemical elements. It is clear that these objects cannot be older than the Universe itself (by definition). This is therefore a crucial test for big-bang models.

To determine the age of a star directly is equivalent to calculating the distance traveled by an automobile starting from the amount of gasoline remaining in the tank: if one knows the capacity of this tank and that it was full to begin with, as well as the gas mileage of the car, it is easy to deduce the total mileage. Applied

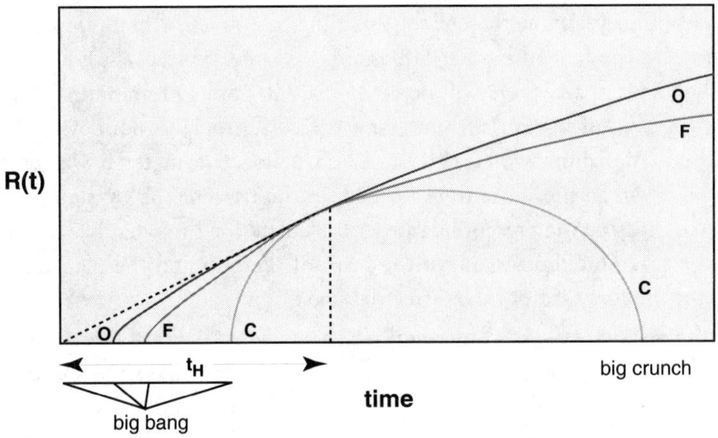

Figure 33.1. The Hubble time. If one knows the present rate of expansion, denoted H, it is not difficult to extrapolate the cosmic past and determine the age of the universe, meaning the time that has passed since the expansion began. This time is necessarily smaller than the Hubble time $t_H = 1/H$, obtained by tracing the tangent to the curve $R(t)$ at the present instant (dotted line). For a flat universe with critical density (F), the age is equal to $\frac{2}{3}t_H$; for a closed universe (C), it is smaller than $\frac{2}{3}t_H$; and for an open universe (O) it is between $\frac{2}{3}t_H$ and t_H.

to a star, the capacity of the tank corresponds to the mass, while the amount of fuel remaining is indicated by the color and luminosity of the star. It therefore remains to determine the efficiency, that is to say the nuclear reactions and the rate at which they liberate energy. This knowledge depends on detailed models of stellar evolution.

In our galaxy, the oldest stars are found in the central clump and in the globular clusters, compact groups of stars containing around a million stars each that gravitate around the center of the Milky Way. To measure the age of the stars starting from the luminosity of globular clusters is a delicate task, since a small error on this luminosity changes the age by more than 20%. The results therefore have some flexibility. The data obtained by the European satellite Hipparcos indicate a range of values between 9 and 14 billion years. One must add about one billion years to obtain the age of the Universe, since this is the approximate amount of time needed for the first stars to be able to form after the big bang.

The Universe also contains radioactive elements, which decay with known lifetimes, since they can be measured in the laboratory. For example, uranium 238 is a radioactive isotope with a half-life of 6.5 billion years that transforms into lead 206. In plain language: let us take one kilogram of uranium 238 and wait 6.5 billion years; at the end of this time, the mass of uranium 238 will be decreased by one half (the other 500 grams will now be lead 206); after another 6.5 billion light years, there will now be only 250 grams of uranium 238, and so on. The dating of terrestrial, lunar, and meteorite rocks is done with this type of method. We think we know how much radioactive material was present at the moment when the rocks were formed, in the nascent solar system, so that it suffices to measure the present quantity to calculate the time that has passed. The results give 4.566 billion years for the birth of the Sun and the planets, and 3.9 billion for the formation of the terrestrial crust.[1]

In the distant stars, the other radioactive elements play the role of long duration chronometers, for example, thorium 232 (observed thanks to spectroscopy), which has a half-life of 14 billion light-years. This method gives a result that is compatible with that of stellar evolution: the age of the Universe is between 10 and 15 billion years.

By comparing with the Hubble time deduced from the simplest big-bang model (Einstein-de Sitter), we see immediately that, if the rate of expansion is too high, it is incompatible with the experimental age. This is the reason why the debate on the value of the expansion rate has been so harsh, in the still quite recent era when two camps were in conflict, one around a high value (100) that was perfectly incompatible with the age of the stars, the other around a low value (50) that was marginally compatible. According to the latest estimates coming from the study of supernovae, the two camps have reconciled around a value very close to 70 km/s per megaparsec, which gives a Hubble time of only 9 billion years.

It is above all the reactionary opponents to the big-bang model, as well as the popular scientific press in quest of scoops, who have exploited this apparent incompatibility between the theoretical and experimental ages. The former had a field day proclaiming the "death of the big bang," taken up immediately by a poorly informed press, who judged it more sellable to make headlines about the death of a popular (though poorly understood) theory than to speak of simplifying hypotheses and experimental uncertainties.

However, in this confrontation, the two other cosmological parameters on which the theoretical age of the Universe depends were purely and simply forgotten. In the first place, the Hubble time depends not only on the rate of expansion, but also on the matter density. If we assume that the Universe is hyperbolic, with a matter density five times smaller than the critical density, the Hubble time is lengthened by 2 billion light-years, and there is no incompatibility with the age of the stars.

[1]Everyone has heard of carbon 14, which allows us to date, within shorter time periods, the remains of organic beings.

But above all, the theoretical age of the Universe, as deduced from general big-bang models (Friedmann-Lemaître solutions), depends on the cosmological constant. In fact, the reasoning of Figure 33.1 is only correct if the expansion is continually slowed down by gravitation, in other words if the Universe is decelerating. Many astronomers have long believed this to be the case, but the situation completely changed over the course of recent years, with new experimental data leaning in favor of an accelerating expansion or, equivalently, of the existence of a 207→
non-zero cosmological constant.

For a given matter density and a given expansion rate, the age of the Universe is always greater with a positive cosmological constant. To tell the truth, the Hubble time is no longer even pertinent for speaking of the age of the Universe. The latter can be as large as one likes if the cosmological constant is judiciously chosen. This therefore easily resolves any age problem. The only thing that the believed incompatibility of theoretical and experimental ages of the Universe signifies is that the ultra-simplified Einstein-de Sitter model, with its critical density and vanishing cosmological constant, must definitively be ruled out. However, general big-bang models do not reduce to this extremely particular case; with a subcritical matter density or a positive cosmological constant they pass the test of age compatibility without trouble.

At any rate, there is nothing new under the sun: such a situation already occurred at the very beginning of relativistic cosmology. The value of the Hubble constant H_0, experimentally determined by Lemaître in 1927 and Hubble in 1929, was, as we have seen, ten times greater than that measured today. The error, due to a bad estimate of the distance of Cepheids, played a retarding role in the 192→
acceptance of the concept of the big bang. Indeed, by adopting this elevated value for the rate of expansion, the age of the Universe, calculated with the assumption of a vanishing cosmological constant, was less than a billion years; at the time, no one knew how to measure the age of the stars, but they were beginning to have an idea of the age of the Earth, thanks to natural radioactivity, which recently had been discovered. This is one of the reasons for which Lemaître, always careful to make theoretical data agree with those of experiment, never wanted to abandon the idea of the cosmological constant, in contrast to Einstein and the great majority of cosmologists.

Measure the sincerity and piety in your heart
and you will know the distances in the heavens.

—Rabbi Akiva

34
Astronomical Distances

To determine the distance to faraway galaxies requires a surveyor's scale that one can unfold step by step, each rung of the scale influencing the higher rungs. The least error in the first induces enormous uncertainties at the top of the scale.

The first rung corresponds to the distances of the nearby stars, which are obtained by a technique of triangulation called *parallax.* Let us hold a finger in front of our eyes and look at it alternately with the left eye and the right eye: it seems to move against the background. The angular distance, or parallax, is a function of the distance between the eyes and the distance to the finger. In the same way, the nearest stars are viewed in different directions when they are observed at six month intervals, on one side and then the other of the terrestrial orbit.[1] The parallax for a star situated one light-year away is the same as that which would be perceived by our eyes in looking alternately at a candle three kilometers away. In this case, it is three arc-seconds.[2]

The first star whose parallax was measured was 61 Cygni. In 1838, Friedrich Bessel found the value of $0.3''$, or a distance of ten light-years.

The method of parallax only works for stars closer than 3,000 light-years; beyond this, the most sophisticated instruments, whether they are on land

[1]In other words, they are observed with "eyes" separated by 300 million kilometers.

[2]The circle is divided into 360 degrees, each degree into 60 minutes and each minute into 60 arc-seconds. An arc-second therefore represents one 1,296,000th of the circumference of a circle.

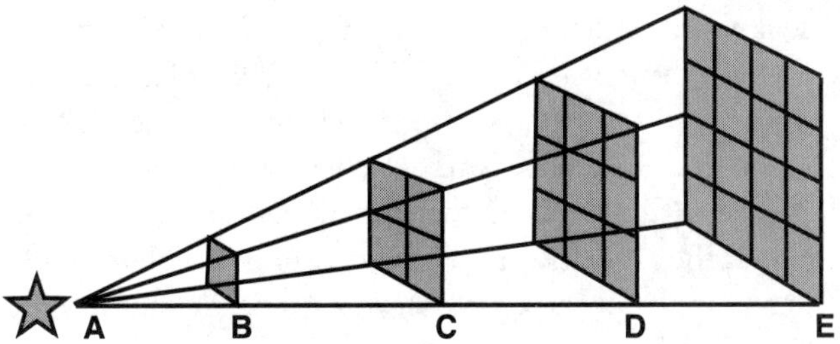

Figure 34.1. Measurement of distance by luminosity. At the distance B from the luminous center A, a certain number of light particles (photons) penetrate the surface B. At twice the distance, namely C, the photons are found to be more spread out, since they penetrate four similar surfaces; at three times the distance, D, they are again spread out to a greater extent since they penetrate nine similar surfaces, and so on.

or traveling by satellite, do not have sufficient resolution to measure angular separations as small as a thousandth of an arc-second.[3] At least, this is true in the optical domain. One hopes that one day, through radio interferometric observations using as a base the Earth and an artificial satellite in orbit around the Sun, we will be able to measure trigonometric parallaxes out to a distance of 300 million light-years. With a first rung extending so far, the cosmological survey will no longer pose problems.

But we are not there yet and, meanwhile, we must use other methods, founded on the brightness of stars. One of the characteristic properties of the propagation of light in space is that the apparent brightness of a light source varies inversely to the square of its distance (Figure 34.1). It follows that the distance of an object can be determined by comparing its apparent luminosity to its intrinsic luminosity.

The apparent luminosity is that which we measure directly. The difficulty therefore lies in estimating the intrinsic luminosity of a light source. How can we determine the proper brightness of an electric bulb, knowing that bulb A, which is four times more powerful than a bulb B, would have exactly the same apparent brightness if it was situated two times farther away? The solution consists in using standard bulbs. If you have bought bulbs at the supermarket, you would certainly know how to tell the difference between a bulb of 100 watts that you observe from far away and that of 25 watts placed two times closer. Galaxies may not be sold at the supermarket, but astronomers have stocked their toolboxes with standard candles, namely Cepheids, type Ia supernovae, and giant galaxies.

[3]The Gaia project of the European Space Agency expects to resolve parallaxes of ten microarcseconds, the equivalent of a hair's breadth viewed from a distance of one thousand kilometers!

A *Cepheid* is a variable star of a particular type (of which the first example was discovered in the Cepheus constellation). The variability of its brightness is due to a regular pulsation that alternately expands and contracts the star. Starting from its period of variation, which is easily measurable, astronomers are able to estimate its proper luminosity. The Cepheids therefore furnish an indicator of distance, on the condition that the period-distance relation can be calibrated with a Cepheid that is sufficiently close for its separation to have been evaluated geometrically. Thanks to this calibration, it is then possible to estimate the distance to Cepheids that are observable in outside galaxies, so distant that the parallax method can not be applied. It is by this technique that Edwin Hubble, in 1925, showed for the first time the extragalactic nature of the Andromeda nebula.[4] Today, thanks to the Hubble Space Telescope, it is possible to see Cepheids in galaxies up to 15 million light-years away.

A *type Ia supernova* (abbreviated SN Ia) is the explosion of a so-called white dwarf star, which occurs when the star goes beyond the critical mass of 1.4 solar masses, by accretion in a double system. Since white dwarfs all have the same mass, their explosions all unleash the same energy and have the same intrinsic luminosity. The measurement of their apparent luminosity therefore gives their distances. Here again, one can calibrate the relation between the distance and apparent luminosity of an SN Ia in a nearby galaxy, by evaluating the distance to this galaxy with the help of the second rung of the scale, namely a Cepheid. Once this calibration is done, the supernovae can be observed at much greater distances than the Cepheids, since their intrinsic luminosity is much larger. Our surveyor's scale therefore allows us to probe the world of galaxies more deeply.

The brightest galaxies of each rich cluster (which are in general elliptic supergiants) are thought to have the same intrinsic luminosity. Of these, 90% have brightnesses that vary by less than a factor of two. In principle, they could also play a role in determining distances, and do so for even more distant regions. But the farther one goes out into space, the more deeply one plunges into the past. Now, each galaxy has a history: in ten billion years, the stellar evolution, as well as possible collisions with other galaxies, have the time to modify its characteristics. The method of giant galaxies is therefore at least as plagued by experimental errors as those of using Cepheids or SN Ias, which shows well the difficulty of directly determining very large distances in astronomy.

---

[4]He did so with an error in calibration that caused him to think that the Andromeda galaxy was two times closer than it really is, an error that carried through to the value for the rate of expansion that he proposed.

And on their aprons, written in abstract script,
formulas are inscribed in advanced math.

Francis Ponge, *Texte sur l'électricité*

35 Cosmic Mathematics

When, in 1927, Albert Einstein became aware of the model of an expanding universe that the young Georges Lemaître had just published, he declared: "Your mathematics is superb, but your physics is abominable." Beyond Einstein's repugnance upon seeing a non-static cosmology—from whence his unjust assessment of Lemaître's physics—this reaction displays the fact that the father of general relativity, one of the greatest physicists of all time, was not a mathematician of the first order: he was more impressed by the mathematical know-how of his colleagues than by their physical reasoning!

In 1912, his school friend Marcel Grossmann, who had become a professor of mathematics at Zurich, explained to him that if he wanted to incorporate gravitation in the special theory of relativity, he had to introduce a new mathematical framework, as it happened, differential geometry in curved space. The tools that allow for the treatment of curved spaces were still of recent invention. Initially constructed by Bernhard Riemann, they had been developed by Elwin Christoffel, Gregorio Ricci, Tullio Levi-Civita, and a few others under the name of *tensor calculus*. Since Einstein had no experience in this area, it was Grossmann who wrote the mathematical part of general relativity.

The theory saw the light of day in 1915. But for many years it was ignored—even rejected with horror—by many physicists because of its unappealing mathematics. Cosmology, a particular domain of application of general relativity, nevertheless allowed a considerable simplification of Einstein's equations and, consequently, of the underlying mathematical formalism. We here examine the nature, or rather the spirit, of cosmic mathematics.

Since works of scientific popularization are in general advised to get rid of every formula, the general public has a tendency to believe that physics, and in particular astrophysics, contents itself with telling stories, conjuring images, and proposing other analogies. This simplification confers upon it a certain similarity ←169 to myths, which in turn lends support to those who deny cosmology the status of a veritable science.

Now, beyond poetic flights about the big bang, the elastic fabric of space-time, and the vacuum energy, the true aim—and the mad ambition—of relativistic cosmology is to account for the structure and density of the Universe in a handful of equations. I think that the majority of readers can sense the beauty of conciseness, like that of the famous $E = mc^2$, but also that of the equations of general relativity. I shall therefore display them here (in simplified form), not so much so that they are understood, but that they are seen. I also have in mind all those high school students who spend their time working on solving equations who often ask themselves what this will be good for later on in life.

In general relativity, the effects of gravitation are not interpreted as a force caused by massive bodies, but are absorbed into the very geometric structure of the Universe. This identification between gravitation and geometry imposes a new mathematical framework: that of a four-dimensional space-time supplied with a non-Euclidean metric.

The metric is a way of measuring the infinitesimal distance between two neighboring points. Step by step, it allows one to evaluate the interval between any two points in space-time, that is to say, between any two events, each one fixed by three position coordinates (x, y, z) and one time coordinate (t). In cosmology, typical events are the emission of a particle of light by a galaxy and its reception on Earth.

The metric generalizes the Pythagorean Theorem, which in planar Euclidean geometry specifies that the square of the distance between two neighboring points is equal to the sum of the squares of the distances projected onto perpendicular coordinate axes: $dl^2 = dx^2 + dy^2$. In a non-Euclidean space-time, the Pythagorean expression is transformed into a more complicated sum, involving all possible products, two by two, of the coordinates, that is to say terms in x^2, y^2, z^2, t^2, xy, xz, xt, yz, yt, zt, with each of these ten products being associated with a coefficient that itself depends on position and time. These coefficients, g_{xx}, g_{yy}, g_{xy}, g_{yz}, and so forth, are called the *metric potentials*, and they became the basic variables of the new theory of gravitation, since they completely define the curvature properties of space-time. One may easily verify that there are ten of them.

$$\begin{pmatrix} g_{xx} & g_{xy} & g_{xz} & g_{xt} \\ g_{yx} & g_{yy} & g_{yz} & g_{yt} \\ g_{zx} & g_{zy} & g_{zz} & g_{zt} \\ g_{tx} & g_{ty} & g_{tz} & g_{tt} \end{pmatrix}$$

Figure 35.1. The metric tensor. In a four-dimensional space-time, the metric tensor possesses sixteen components that can be arranged in the form of a table with four rows and four columns called a matrix. Since this matrix is symmetric (meaning that $g_{xy} = g_{yx}$, etc.), there are in reality only ten independent metric coefficients.

Einstein's equations express the coupling between the geometry of space-time and its material content. They generalize Poisson's equation, which, in classical Newtonian theory, connects the gravitational potential (given by a single, so-called *scalar* quantity) to the density of matter (also scalar). In general relativity, gravitation is described no longer by a single potential but by the ten metric potentials.

Very schematically, Einstein's equations reduce to the equality $\mathbf{G} = \mathbf{T}$, which is to say: geometry equals matter. The symbols $\mathbf{G}$ and $\mathbf{T}$ are written in boldface to signify that these are not simple numbers, but mathematical objects known as tensors, which are tables with sixteen components (matrices with four rows and four columns similar to the squares of a small checkerboard) containing all the information on the geometry and the distribution of matter.

The left side member, $\mathbf{G}$, is constructed from another, so-called Ricci tensor $\mathbf{R}$, which contains all the information about the curvature; its expression was found by Einstein and Grossmann, after many unfruitful attempts, by way of physical reasoning. In the first version of the equations for a gravitational field, given by Einstein in 1915, one had $\mathbf{G} = \mathbf{R} - \frac{1}{2}\mathbf{g}R$, where $\mathbf{g}$ is the metric tensor, combining the metric potentials into a table, and R is the scalar curvature of space-time.

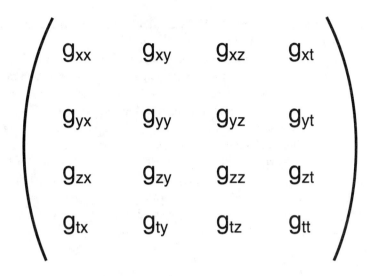

In the second version, given in 1917, Einstein introduced the cosmological constant λ, and consequently found $\mathbf{G} = \mathbf{R} - \frac{1}{2}\mathbf{g}R + \lambda\mathbf{g}$.

The right hand side of the equation, represented by the so-called *energy-momentum* tensor $\mathbf{T}$, contains all the information on the distribution and motion of different forms of energy in the Universe, both matter and radiation. An excellent approximation to the actual state of the Universe is given by a gas with zero pressure whose molecules, without mutual interaction, are the galaxies. To describe the hot primordial Universe simply, one can also add an energy density and a pressure, in the form of radiation. In both cases, the state of matter is that of a perfect fluid described by two numbers that in principle depend on time, the density $\rho(t)$ and the pressure $p(t)$. Note that the "supplementary" term $\lambda\mathbf{g}$ can just as well be considered as being part of the left side term $\mathbf{G}$—in which case it is interpreted as a component of the geometry—as it could be part of the right side term $\mathbf{T}$—in which case it is interpreted as a contribution to the energy (which is ←207 in fact the present preference).

Einstein's equations are impossible to solve in all generality. In order to find exact solutions, one must simplify the left hand side, by assuming particular space-time symmetries, and simplify the right hand side, by reducing the material content of the Universe to its simplest expression: depending on the region of the Universe that one wants to describe, one can assume the vacuum, an isolated mass, a pure electromagnetic field, a pressureless fluid of galaxies, and so on.

Homogeneous Friedmann-Lemaître Models

With this care for symmetry and simplicity, we see the resurgence of a Platonic idea, but here for less noble reasons, since here it is simply done in order to arrive at a solution of the equations! This is in fact precisely what is done to construct a cosmological model that is a solution of general relativity. Fortunately, in the case of cosmology, the simplifications are to a large extent justified.[1] In fact, what do we observe about the real distribution of matter in the Universe on a large scale? First of all, a remarkable symmetry around us, known as *isotropy*: the physical properties of the Universe do not depend on the direction in which we look. A second fundamental symmetry is postulated rather than observed: that of *spatial homogeneity*, according to which the physical properties at a given moment are the same, regardless of which point in space is considered. Taken together, these two postulates—homogeneity and isotropy—form the cosmological principle, which completes the idea of Copernicus, promising an end to anthropocentrism.

[1] This is in contrast to the theory of quantum cosmology, founded on the so-called Wheeler-DeWitt equations, which are just as unsolvable as Einstein's equations. In this case the solutions that have been obtained through severe simplifications, like those that have been proposed by Stephen Hawking or Andrei Linde, are hard to justify.

Mathematically, the cosmological principle translates into the fact that at a given instant, space has the same curvature at every one of its points. The metric then takes a simple form, after bringing out a universal scale factor $a(t)$ that, at each instant, has the same value throughout space, and such that the distance between two arbitrary points varies over the course of time in direct proportion to a. This scale factor is sometimes called the *radius of the universe*, but this term is misleading, since it only takes this meaning in the case of spaces of finite size.

Since the curvature is constant throughout space, there are only three cases to consider, depending on the sign that it takes. Thus, the curvature is given by the expression $k/a(t)^2$, where k can take the values +1, 0, and -1: if the curvature is positive ($k = +1$), the space is spherical and necessarily finite (like the hypersphere); if it is null ($k = 0$), the space is Euclidean; and if it is negative ($k = -1$), the space is hyperbolic. Euclidean and hyperbolic spaces can have either finite or infinite volume; this question concerns topology, not the equations of relativity.

By solving Einstein's equations, after taking into account the preceding simplifications and allowing for a cosmological constant, one obtains a full set of so-called Friedmann-Lemaître cosmological models.[2] A Friedmann-Lemaître model is characterized by its curvature, by its temporal dynamics, and by its topology. Only the first two features are calculable by the equations of relativity. The dynamics is given by the way in which the scale factor $a(t)$ and the density $\rho(t)$ vary with time. When a grows, the Universe is expanding and the distances between faraway galaxies grow with time. In the contrary case, the distances decrease within a contracting Universe. Some models, in fact, exhibit a phase of expansion followed by one of contraction over the course of their evolution. Most Friedmann-Lemaître solutions have a singular beginning, meaning a time t in the past where the scale factor $a(t)$ vanishes; these are the famous big-bang models. The scale factor can also vanish in the future, in which case the Universe ends in a *big crunch*.

Figure 35.2 graphically summarizes all of the possible temporal evolutions of the scale factor $a(t)$, as a function of the two fundamental cosmological parameters, namely, the sign of the curvature and the value of the cosmological constant. Over the course of the eighty-year history of relativistic cosmology, researchers have in turn favored one or the other case in this table. The so-called debate about the big bang that one regularly reads about in the popular press is really only a report on the shifts in consensus that have taken place within the professional community about the accepted values of the cosmological parameters.

[2]These models are also known in English as Friedmann-Robertson-Walker models, adding the names of H. Robertson and A. Walker, since these researchers clarified the mathematical structure of relativistic cosmology in the early 1930s.

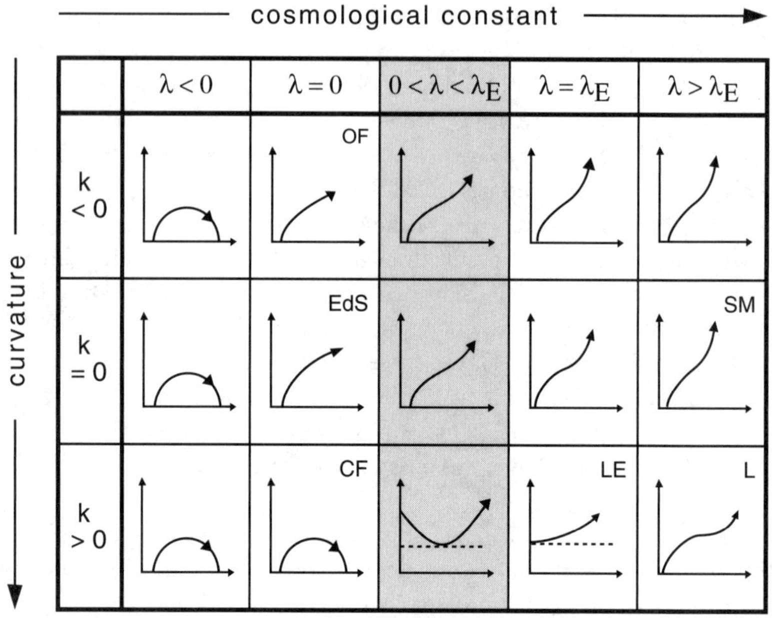

Figure 35.2. Graphical representation of Friedmann-Lemaître universes. The dynamics of Friedmann-Lemaître cosmological models, meaning the variation of the spatial scale factor as a function of cosmic time, is determined by the sign (k) of the curvature of spatial sections and by the value of the cosmological constant λ. Two important particular values of the latter are $\lambda = 0$ and $\lambda = \lambda_E$, where λ_E is the value proposed in 1917 by Einstein in order to assure the static character of his hyperspherical universe. Since a positive cosmological constant is equivalent to a repulsive action at large distances, all of the models with large cosmological constants ($\lambda > \lambda_E$) are *open* in time, meaning that they expand forever, whatever their curvature may be. Inversely, a negative constant $\lambda < 0$ (which is rather unlikely on the physical level) would tend to augment the effective gravity, with the effect that the corresponding universe models would all end up by collapsing in on themselves. In certain cases ($k > 0$ and $0 < \lambda \leq \lambda_E$) the initial singularity might even disappear; therefore not every Friedmann-Lemaître model is a big-bang model.

Only rarely does the discussion move away from the general framework of homogeneous Friedmann-Lemaître solutions, for a simple reason: if one moves away from this framework, it is practically impossible to find exact solutions of Einstein's equations!

Most of the homogeneous relativistic cosmological models proposed since 1927 correspond to one of the cases in this table. Let us survey them in chronological order.

Einstein's static model (1917).

- Curvature: +1
- Matter: $\rho = \text{constant}$, $p = 0$
- Cosmological constant: λ_E
- Dynamics: none

This case is shown as the dotted line in the box labelled LE in Figure 35.2. It is a very particular case of the Friedmann-Lemaître solutions, since the cosmological constant is adjusted very precisely to λ_E in such a way as to impose constancy of the scale factor a and the density of matter ρ over the course of time. The Einstein universe is one of matter without motion.

Closed Friedmann model (1922).

- Curvature: +1
- Matter: $\rho(t)$ variable, $p = 0$
- Cosmological constant: 0
- Dynamics: expansion followed by contraction

This case is shown in the box labelled CF in Figure 35.2. Friedmann also treated solutions with non-zero cosmological constant on the mathematical level, but he remarked that these were superfluous since, without them, one obtained a mathematically coherent model having both a beginning and an end in time.

Open Friedmann model (1924).

- Curvature: -1
- Matter: $\rho(t)$ variable, $p = 0$
- Cosmological constant: 0
- Dynamics: perpetual expansion

This case is shown in the box labelled OF in Figure 35.2. Friedmann also solved the equations with a cosmological constant; his aim was to prove that relativistic cosmology admitted spaces of negative curvature.

Lemaître-Eddington model (1927).

- Curvature: +1
- Matter: $\rho(t), p(t)$ variable
- Cosmological constant: λ_E
- Dynamics: perpetual accelerating expansion

This case is shown in the box labelled LE in Figure 35.2. The cosmological constant is adjusted so that $a(t)$ grows without end, starting from Einstein's static hypersphere at $t = -\infty$. There is therefore neither a big bang nor an age problem for the Universe.

Lemaître hesitating universe model (1931).

- Curvature: +1
- Matter: $\rho(t), p(t)$ variable
- Cosmological constant: $\lambda > \lambda_E$
- Dynamics: perpetual expansion, initially decelerating, then accelerating

This case is shown in the box labelled L in Figure 35.2. Starting from a singular origin called the primeval atom, the Universe passes through a phase of "hesitation," during which it skirts the static Einstein state, before starting to expand again in continual acceleration. This model resolved the problem of the age of the Universe and of the time necessary to form galaxies.

Euclidean Einstein-de Sitter model (1932).

- Curvature: 0
- Matter: $\rho(t)$ variable, $p = 0$
- Cosmological constant: 0
- Dynamics: perpetual decelerating expansion

This case is shown in the box labelled EdS in Figure 35.2. This ultra-simplified solution served as a standard model for 60 years, to the point of blocking out research on other models. Reinforced in the 1980s by the inflationary model, it has today been abandoned, primarily because it gives an age for the Universe that is too short.

Model with sub-critical density (circa 1990).

- Curvature: -1
- Matter: $\rho(t), p(t)$ variable
- Cosmological constant: 0
- Dynamics: perpetual decelerating expansion

This case is also as shown in the box labelled OF in Figure 35.2. Measurements of the density of dark matter using the dynamics of galaxy clusters suggested a density that is much smaller than the critical density, thus favoring a hyperbolic model.

The provisionally standard model (2000).

- Curvature: close to 0
- Matter: $\rho(t), p(t)$ variable
- Cosmological constant: yes ("dark energy")
- Dynamics: perpetual accelerating expansion

This case is shown in the box labelled SM in Figure 35.2. This is a return to a model close to that of Lemaître imposed by the latest experimental data, but with dynamics dominated by the vacuum energy, the modern interpretation of the cosmological constant. The range of uncertainty covers the three possible cases of curvature, -1, 0, and +1, but many theorists, remaining attached to the inflationary model, adjust the cosmological constant in such a way as to obtain an exactly Euclidean universe.

Variants of the Homogeneous Friedmann-Lemaître Models

While continuing to assume the homogeneity and isotropy of the Universe, it is permissible to make exotic hypotheses on the nature and distribution of matter and energy that are no longer those of a perfect fluid. The three solutions below have drawn much attention in their times, but have been (provisionally?) abandoned.

Static de Sitter model (1917).

- Curvature: +1
- Matter: $\rho = 0, p = 0$
- Cosmological constant: yes
- Dynamics: none

Since the density of matter is assumed to be zero, the de Sitter model cannot represent the complete evolution of the real Universe. However, it has the advantage of showing the effect the cosmological constant has on the expansion of space. The de Sitter universe is one of motion without matter. It has been taken up again as support for the steady state theory, and more recently to describe the phase of inflation experiencing very rapid expansion, which is believed by many researchers to have taken place during the very first instants of the Universe.

Oscillatory Einstein-Tolman universe (1931).

- Curvature: +1
- Matter: $\rho(t)$ variable
- Cosmological constant: yes
- Dynamics: endless cycle of expansions and contractions

Normally, the closed Friedmann model begins in a big bang and finishes in a big crunch, but Friedmann suggested the possibility of a cyclical solution where the closed Universe successively expands and contracts a great number of times. If one accepts certain exotic equations of state for the matter, one in fact obtains a solution where the radius of space oscillates between a minimum value (the state of maximal compression) and a maximum value (state of minimal compression). According to Richard Tolman, the singularities must be replaced by very small and very dense universes, similar to the primitive atom. Eddington responded that he found it rather stupid to repeat the same thing endlessly. Lemaître baptized this the *phoenix universe* model and showed that the singularities were inevitable, unless one introduced into general relativity quantum corrections whose effect was to "dull the point." The phoenix universe still found advocates as prestigious as George Gamow and John Wheeler. The philosophically motivated forms of the phoenix universe go back very far. In Brahman cosmology, the *kalpa* (a day of Brahma) was a cosmic period equivalent to around 4 billion years, ending with total destruction and a period of non-creation (a night of Brahma), before being

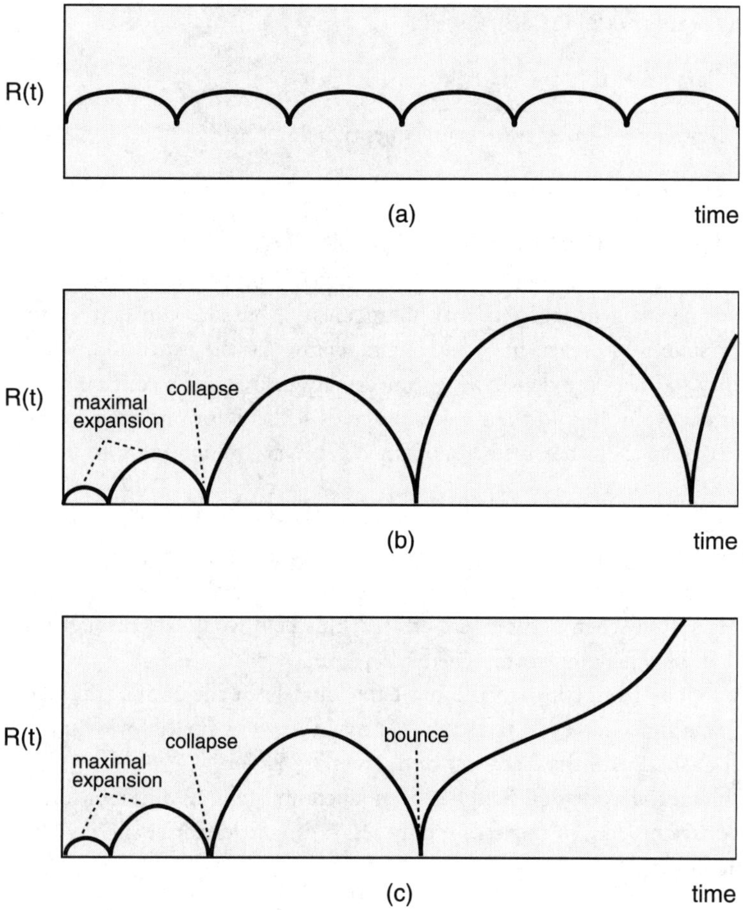

Figure 35.3. Three phoenix universe models. (a) In this model, the periodic universe imagined by Friedmann endlessly repeats the same cycle of expansion and contraction. (b) In the model studied by Lemaître, the maximal duration and radius of each cycle grows over the course of time. (c) In this model, the creation of entropy from one cycle to another is such that at the end of a small number of cycles of expansion and contraction, the Universe experiences a final bounce (our big bang) and enters into perpetual accelerating expansion, in which we presently find ourselves.

reborn in an infinite succession of cycles. Analogous speculations were common in antiquity, the "great year" symbolizing the vast period after which the history of the world must repeat itself.

Steady state model (1948).

- Curvature: 0
- Matter: $\rho(t)$ constant + exotic energy field
- Dynamics: expansion

In 1948, Thomas Gold and Hermann Bondi wished to apply the perfect cosmological principle, according to which the physical properties of the Universe are not only constant in space (spatial homogeneity), but also constant in time (stationary state). However, in order to compensate for the expansion, which has a tendency to lower the density, it was necessary to imagine a continuous creation of matter. According to their calculations, it sufficed for one atom of hydrogen per liter of space to be created spontaneously every billion years to preserve the steady state.

Fred Hoyle showed that this could be realized within the framework of a de Sitter model, with the condition of introducing a *creation field* into the energy-momentum tensor. Conceived as a reservoir of negative potential energy that manifests itself when particles are created, this field would act throughout a perpetual and unchanging cosmic history.

A serious competitor to the big-bang models in the 1950s (the name "big bang" was given derisively to Lemaître's rival theory by Fred Hoyle himself), this model clashed with the discovery of the fossil radiation.

The idea of a creation field has been taken up again, in different form, by the inflationary models; in this case it only acts during a very brief and early epoch of cosmic history.

Nothing can be created from nothing.

—Lucretius

36 Cosmic Repulsion

Since 1927, theory and observations have shown that the Universe is progressively expanding. Most astronomers believed that the expansion of the Universe, governed by the attractive force of gravity, must evolve by slowing down, meaning 184→ that the rate of expansion should decrease over the course of time.

Some recent research has contradicted this simplifying thesis, which at any rate only corresponded with very particular cases of the cosmological solutions of general relativity.

The first chapter of this history begins with Albert Einstein and a "blunder" (as Einstein put it) he committed in 1917. Einstein quite naturally wished to use his brand new theory of general relativity to describe the structure of the Universe as a whole. The cosmological conception which then prevailed was that the Universe must be static, that is, invariable in time. Einstein expected that general relativity would support this opinion. However, to his great surprise, this was not the case. The model universe that he developed, filled uniformly with matter and having the geometry of a hypersphere, did not have a constant radius of curvature: the inexorable force of gravity, acting on each celestial body, had a tendency to destabilize it.

For Einstein, the only remedy for this dilemma was to add an ad hoc but mathematically coherent term to his original equations. This addition corresponds to some sort of antigravity, which acts like a repulsive force that only makes itself felt at the cosmic scale. Thanks to this mathematical trick, Einstein's model remained as permanent and invariable as the apparent Universe. Einstein represented this term by the Greek letter λ and named it the cosmological constant. It was indeed a constant, since it had to keep exactly the same value in space and time. Formally, it could take any value whatsoever a priori, but the value selected by Einstein led to a static Universe.

In the same year, 1917, Willem de Sitter built another model for a static relativistic universe, which was very different from Einstein's. It assumed, in effect, that the density of matter remained vanishing over the course of time; in compensation, he had to introduce a positive cosmological constant: in the absence of matter, and therefore of gravity, only a cosmological constant can determine the curvature of space.[1] The possible role of the cosmological constant had expanded: in de Sitter's model, the trajectories followed by "galaxies" (of zero mass, since matter is absent by assumption) rapidly separate from each other over the course of time, as if there were a general expansion of space. This meant that in the absence of matter, the cosmological constant has a particular influence on the structure of space. In Einstein's eyes, de Sitter's solution nevertheless reduced to a simple mathematical curiosity, since the real Universe indeed has a mass.

A first sudden turn of events took place in 1922, when the Russian mathematician Alexander Friedmann studied cosmological models without making the assumption of a static Universe. He then discovered some solutions of general relativity where the radius of curvature of space varies with time, whether or not there is a cosmological constant. The static solutions previously found by Einstein and de Sitter now appeared to be no more than very particular cases of more general solutions, which were dynamic.

A few years later, Edwin Hubble's observations indicated that galaxies do in fact have the tendency to separate irresistibly from each other over the course of time. At the same time, the British theorist Arthur Eddington reexamined Einstein's model and discovered that, like a pen balanced on its point, it is unstable: with the least perturbation, it begins either expanding or contracting.

All of these new facts came together to prove that Einstein's model is not a good model for the Universe, and sparked a new debate on the cosmological constant: not only did it fail to perform its function, which was to stabilize the radius of the Universe, but it seemed that, in any case, this radius varies over time. In these conditions, what good did it do to keep this superfluous term? Einstein was the first to criticize himself; after some hesitation, he ended up admitting the physical relevance of Friedmann's solutions and declared that, by

[1]If the cosmological constant and the density of matter are both vanishing, the Universe reduces to the static and flat Minkowski space-time, the special case used in special relativity.

introducing the cosmological constant, he had committed the greatest blunder of his life. In the new relativistic model that he proposed in 1932 with his colleague de Sitter—a Euclidean model with uniform density that expanded eternally—the term disappeared!

Naively, the debate on the cosmological constant seemed closed. However, one man did not agree, namely, Georges Lemaître. He began his cosmological research in 1925, and had understood the dynamical role played by the cosmological constant in de Sitter's model. In 1927, he rediscovered the dynamic solutions discovered by Friedmann, and for the first time made the connection between the 143→ expansion of space and the apparent recession of galaxies.

For Lemaître, the cosmological constant remained an essential ingredient of relativistic cosmology. A first point that he used to support his argument comes from cosmic mathematics: in 1922, Élie Cartan demonstrated that the most general form of the equations of relativity must necessarily include a cosmological 195→ constant. This did not, however, suffice to convince Einstein. "Einstein's prestigious authority naturally found more than one echo. But to suppress the cosmological term from the equations, is this really a solution to the difficulty? Is it not rather the politics of the ostrich?" wrote Lemaître, sorry for the attitude of the father of relativity, who had renounced one of his greatest contributions to science.

Further, Lemaître insisted: "I am led to come around to a solution of the equation by Friedmann where the radius of space starts from zero with an infinite speed, slows and passes by the unstable equilibrium ... before expanding once again at accelerated speed. It is this period of slowing which seems to me to have played one of the most important roles in the formation of the galaxies and stars. It is obviously essentially connected to the cosmological constant" [Lemaître 60].

The verbal and written exchanges between Einstein and Lemaître concerning the cosmological constant were rather colorful. During one of their encounters in Pasadena, California, the journalists who reported their conversation spoke of a "little lamb," which followed the two physicists everywhere they went.[2] More seriously, Lemaître considered the cosmological constant to be both a logical and observational necessity. He developed a premonitory argument, in three points:

- The cosmological constant is necessary to obtain a scale for the lifetime of the Universe that definitively moves away from the limit imposed by the duration of geological ages. Big-bang models without a cosmological constant in fact predict that the theoretical age of the Universe, calculated

[2]The pun on *lambda* makes reference to a nursery rhyme well-known in American pre-schools, but completely unknown to Lemaître, who accordingly did not understand the joke!

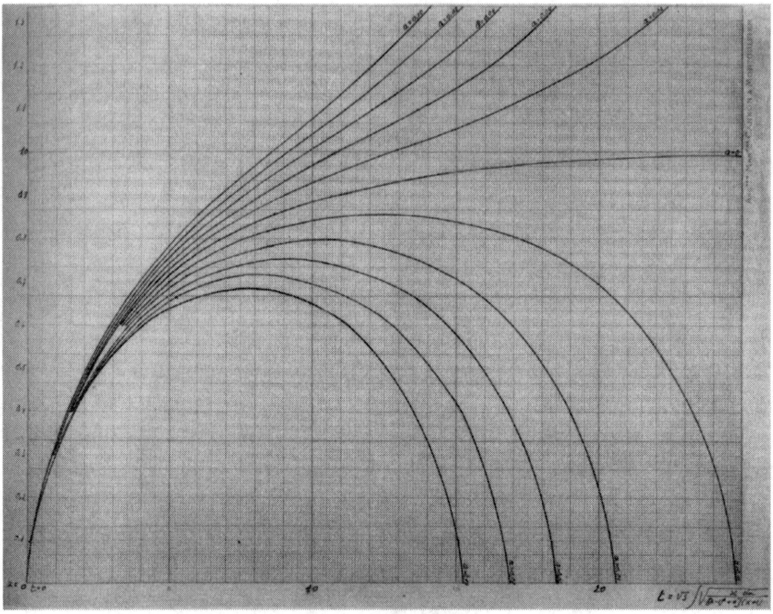

Figure 36.1. Lemaître's graphs. This extraordinary diagram, drawn by Lemaître in 1927 but never published, follows different possible temporal evolutions of the radius of the universe for different cosmological constants, in a space of positive curvature. All of the models begin with a singularity at ($x = 0$, $t = 0$). For a sufficiently large cosmological constant, the universe becomes open. The most recent data are compatible with a Lemaître solution with accelerating expansion (uppermost curves). (Image courtesy of the Lemaître archives, University of Louvain.)

starting from its rate of expansion, is shorter than that of the stars, which is impossible. The term λ helps to suppress this contradiction: a Universe with a cosmological constant is older than a Universe without a cosmological constant; it starts off more slowly, and it takes more time to reach the
←188
rate of expansion measured today.

- The instability of the equilibrium between gravitational attraction and cosmic repulsion is the only way in which to understand evolution on the scale of stars and galaxies during a lapse of time corresponding to about a billion years. This argument poses the basic problem, still not resolved in 2007, of the time scale necessary for the formation of galaxies.[3]

[3]This is all the more so since the observations from WMAP in 2003 suggest an epoch of galaxy formation even more precocious, on the order of 200 million years!

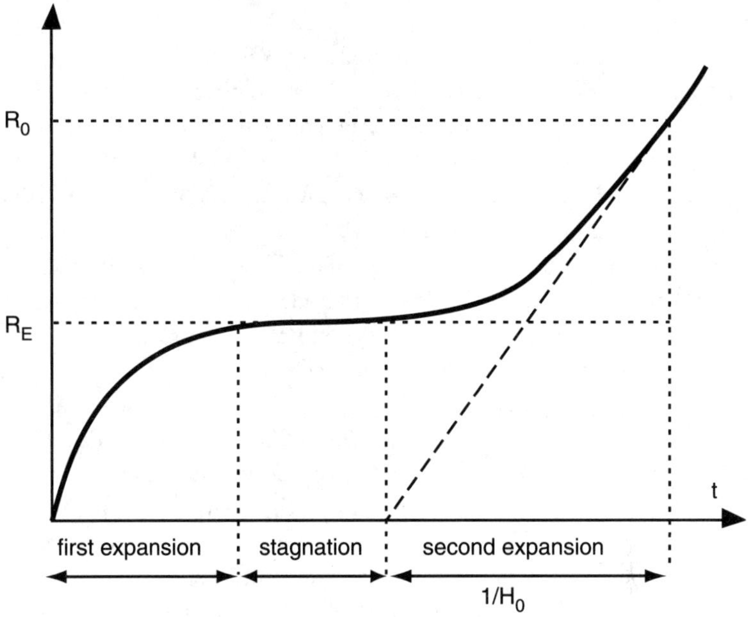

Figure 36.2. Lemaître's hesitant universe. A model universe with accelerated expansion, such as that calculated by Lemaître in 1931, allows for a simple resolution of questions about the age of the universe and the time taken to form galaxies.

- The energy is only defined up to an additive constant, since general relativity does not provide a method of adjustment to compensate for an arbitrary change in the zero level of the energy. According to Lemaître, the cosmological constant allows one to make this adjustment. He thus anticipated the link between the cosmological constant and quantum mechanics: "The cosmological constant is certainly superfluous for explaining gravitation, but gravitation is not all of physics. From a purely geometric point of view, the radius of the Riemann space was also an unnecessary constant; it was an indication that geometry must not be sufficient in itself, but must be founded on a vaster synthesis, more precisely the geometric theory of gravitation. Is not the cosmological constant which appears in the latter theory an indication of a later enlargement of the theory, to a theory which would succeed in uniting and synthesizing the points of view of the theory of relativity and that of quantum mechanics where a characteristic constant also figures: Planck's constant, which is perhaps not without relation to the cosmological constant."

In Lemaître's eyes, the cosmological constant was therefore equipped with all the virtues. All the same, he never succeeded in convincing Einstein, or the rest of the community of cosmologists. It is true that no direct astronomical observation came to establish this mysterious term, and that the disavowal of the father of relativity came down with all its weight. The cosmological constant was almost completely abandoned by astronomers for a half century, between 1940 and 1990.

In the interval, theoretical physicists explored the idea advanced by Lemaître in his third argument: the link with quantum mechanics. With the development of the theory of quantum fields, they began to realize that the notion of empty space presents greater subtleties than they had thought. Space is not an empty receptacle filled with matter and radiation; it is a physical and dynamical entity that has "flesh." This flesh is the vacuum energy. Physicists such as Paul Dirac and Richard Feynman made the hypothesis that what we call empty space is in fact filled with *virtual particles*, which only manifest themselves as a material reality in an ephemeral fashion. Take a cubic meter of space, and take out all matter and all radiation. What would remain? Most of us would respond: nothing. On the contrary, these physicists said, there is still energy there, and it confers upon the vacuum a sort of latent pressure. In other terms, the vacuum is not nothingness; it is potentially crammed with energy. This vacuum energy is not only implied by quantum mechanics, but has been experimentally demonstrated through the Casimir effect.[4]

Applying this concept to cosmology, the Russian Yakov Zel'dovich showed in 1962 that the term λ was equivalent to a latent energy field in empty space. The cosmological constant is in a sense the dynamite of the vacuum!

Astronomers continued to ignore these theoretical developments until the mid-1990s. However, in a series of developments, the cosmological constant suddenly made a strong comeback. Astronomers all of a sudden rediscovered its virtues, which had previously been advocated by Lemaître. It is not that they had finally read the prophetic articles of the Belgian physicist. More prosaically, new observational data indicated an acceleration of the cosmological expansion. By measuring, with unequaled precision, the remains of fourteen supernovae situated at distances varying between seven and ten billion light-years, astronomers discovered that these moribund stars are ten to fifteen percent farther away than they should be. The calculations, performed with data collected by the Hubble

[4]In 1948, Hendrik Casimir had predicted that two parallel and closely separated conducting plates should be attracted to each other. It took until 1958 for the first experimental verification to be seen, and until 1997 for a rigorous verification. The general explanation that is advanced is that the mere presence of the plates extracts energy from the quantum vacuum, in the form of electrons that induce an attractive electric field.

space telescope and then by telescopes placed in Hawaii, Australia, and Chile, were published in 1998. They shocked more than one scientist, and went on to be debated in numerous seminars.[5]

This discovery gave rise to the usual types of announcement in the popular press: "Our model of the Universe put into question," "All of the foundations established by cosmology could be overturned," could be read in the newspapers. In January of 2000, even the Davos "World Economic Forum," which gathers together the political and economic leaders of the planet every year, organized a session dedicated to the subject! Invited to the debate, I tried to temper the media frenzy a little bit, by explaining that the acceleration of the cosmic expansion did not necessitate putting the big bang into doubt, and was only a surprise for those who do not know the subtleties. It was not that simple, in front of an audience that was three-quarters Anglo-American, for whom the history of modern cosmology reduces to the single name of Hubble, to explain that Lemaître, in French and 40 years ahead of time, had already given all the ins and outs of the cosmological constant!

The experimental indications of a non-zero cosmological constant come not only from type Ia supernovae, but also from independent measurements of the fluctuations in the cosmic microwave background radiation. What is the value of this constant?

The cosmological constant plays the role of an energy density. Its value can therefore be expressed in the same units as the matter density, for example, in grams per cubic centimeter, or, even better, as a factor Ω_λ, which is highly convenient for appraising at a glance the contribution of the cosmological constant to the total density of the Universe, Ω_{tot}. The two methods (supernovae and background radiation) suggest that Ω_m (m for matter) is close to 0.3, whereas the Universe is Euclidean. This latter case corresponds to $\Omega_{\text{tot}} = 1$. To arrive at a flat Universe, one must therefore add a contribution of $\Omega_\lambda = 0.7$. We thus see that the measurement of λ remains indirect. There are, at any rate, other possible explanations than a cosmological constant that can account for this dark energy.

In the present situation, it would seem that everyone could agree on the existence of a cosmological constant, the fruit of a logical imperative as well as an observational necessity. But is the value deduced from astronomy compatible with that calculated by theoretical physicists? This is where things become terribly complicated, and are the subject of one of the most important challenges of present-day cosmology. Let us see why.

[5] Since 2001, the observations of type Ia supernovae have been pursued and completed on statistically significant samples. The acceleration of the cosmic expansion has been confirmed, and one can even date the epoch when the rate of expansion ceased to slow down in order to begin growing under the influence of the dark energy: this *inflection point* was produced about 6 billion years ago.

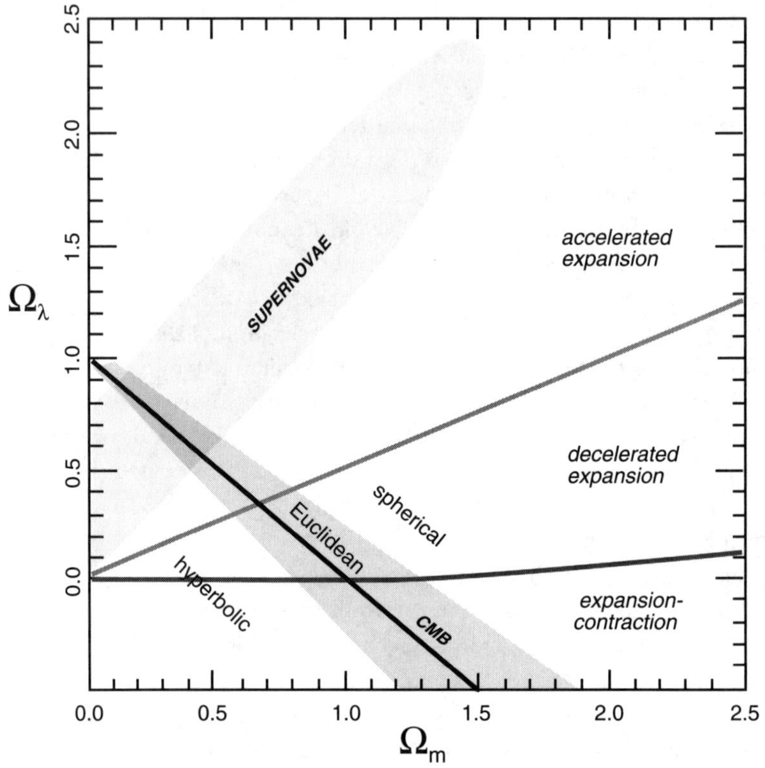

Figure 36.3. The cosmic square. The cosmic square presents the two cosmological parameters that determine the curvature and dynamics of the universe: Ω_m is the contribution of matter to the average density of the Universe and Ω_λ is the contribution of the cosmological constant. The parameter Ω_m is necessarily positive, but Ω_λ can be negative. The black line (marked "Euclidean") corresponds to a space of zero curvature, and separates the spherical spaces (positive curvature) from the hyperbolic spaces (negative curvature). The dark gray line separates open universes (in perpetual expansion) and closed universes (in expansion followed by contraction). The light gray line separates universes whose rates of expansion accelerate and those whose expansion rates slow down. The shaded zones delimit the experimental constraints presently imposed by the observations of supernovae (in light gray) and of the cosmic microwave background (in medium gray). The two zones intersect in one region (in dark gray), clearly corresponding to a universe in perpetual accelerated expansion, with $0.1 < \Omega_m < 0.4$ and $0.4 < \Omega_\lambda < 0.9$, without however determining the value of the curvature.

Physicists who reflect on the concept of the vacuum energy have a more grandiose project in mind: the unification of the fundamental interactions. For that purpose, they are obliged to add terms to their equations that represent entirely new natural fields. The field concept was invented in the nineteenth century by mathematicians in order to express how a given quantity can vary from one point to another in space. Physicists immediately adopted this idea in order to describe quantitatively how forces, such as gravity and electromagnetism, change as a function of the distance with respect to a source.

The fields put into play in the theories of unification are very different from those that we already know about. For example, the Higgs field, which is so actively sought, was introduced by Sheldon Glashow, Abdus Salam, and Steven Weinberg in their electroweak theory, which unifies the electromagnetic and weak nuclear interactions. Taking into account the experimental success with which this theory was met (the mediating particles of the electroweak interaction, called intermediate bosons, were actually detected at CERN), we expect that the Higgs field is not a simple theorist's whim, but shall very soon be put into evidence thanks to physical experiments at very high energy.[6]

Before this electroweak unification, the weak interaction, responsible for the radioactive decay of certain particles, as well as the electromagnetic interaction, responsible for the mutual influences between charged particles and for the movement of magnetized needles, were considered as forces of completely distinct natures. By combining their mathematical descriptions in a common language, Weinberg, Salam, and Glashow showed that this distinction was not fundamental. A new natural field, the Higgs field, causes these two interactions to act differently at low temperature, meaning in the present day Universe. But at temperatures higher than a billion billion degrees (10^{18}), the weak and electromagnetic interactions become virtually identical in the way in which they affect matter.

There is a price to pay for the introduction of new quantum fields: they confer upon the vacuum state an enormous latent energy, which behaves exactly like the λ of cosmological models. Here the cosmological constant is brilliantly resuscitated! Nothing remains but to calculate the value of λ in the framework of these unified theories. There's the rub. There are several possible models, but the value

[6]The CERN electron-positron collider (LEP) was recently able to push its experiments to energies higher than expected, and in autumn 2000 the effects due to the Higgs field would have been detected for a mass of 114 GeV, equivalent to a temperature of 1.32×10^{15} K. The future Large Hadron Collider (LHC) at CERN, which will start functioning around 2008, will have the necessary resources to confirm these indications.

in most cases is 10^{122} times greater than the limits prescribed by astronomical observations![7]

This difference is so immense that our minds have difficulty grasping it. However, we can easily see that it is unpleasant, in the eyes of theoretical physicists, that their best unified models, which are supposed to make the correct predictions in the domain of elementary particles, lead to such aberrant cosmological consequences.

The astronomers are probably right, and it is up to the physicists to better understand the unified theories, as well as the true nature of the vacuum energy. In so-called quintessence theories, the cosmological constant is replaced by a field that varies over the course of time, being very strong in the primordial phases of the Universe, in agreement with the calculations of physicists, but falling to very low values in the later stages of cosmic evolution, conforming to the value measured today by the astronomers.[8]

Must one really go that far? The theories predicting a high value of the cosmological constant are purely speculative. None of them have been the object of the least experimental verification. Their approach is perhaps entirely mistaken. Even worse, new programs that have as their ultimate goal the unification of gravity and quantum mechanics, such as string theory, have as a secondary objective the prediction of a strictly zero cosmological constant. If the astronomical measurements are confirmed, the mystery of the cosmological constant will therefore continue to deepen, since no theory at present sets forth an explanation as to why its value is neither zero, nor very large. Like a phantom, the cosmological constant signals us that something essential is still missing in our understanding of the physical Universe, and may serve as a guide to how we can put the finishing touches on a good theory of unification.

It is a singular homage to the genius of Einstein and Lemaître to see how the "little lamb" that they politely debated in Pasadena, more than 50 years ago, has followed a tortuous path whose end is not yet in sight.

---

[7]The cosmological constant is comparable to the inverse of the square of a length. For the physicists of the infinitesimally small, this length is interpreted as the distance scale at which the gravitational effects due to the vacuum energy become manifest on the geometry of space-time. They estimate that this scale is the Planck length, or 10^{-33} centimeters. For astronomers, the cosmological constant is a cosmic repulsive force that affects the rate of expansion on the scale of the radius of the observable universe, or 10^{28} centimeters. The ratio between these two lengths is 10^{61}, which is in fact the square root of 10^{122}.

[8]The quintessence field would naturally evolve toward an attractor giving it a low value, whatever its original value may be. Physicists thus estimate that a great number of different initial conditions would lead to a similar universe—precisely the one that is observed.

[In the universe] there are stars that are solitary and others accompanied by satellites, bodies of light and masses of shadows.

—Georges Buffon, *Première vue de la Nature*

37 Dark Matter

Space is populated by mysterious bodies. Some form of dark matter, so-called because it is inaccessible to telescope observation, dominates it. Its nature is enigmatic. Does it consist of bodies made of ordinary matter that nevertheless remain invisible—small stars, gaseous clouds, or black holes—as astronomers propose? Or, does it rather consist of exotic particles created during the big bang, or even still stranger energy fields, as some theoretical physicists suggest?

Beyond its nature, the quantity of dark matter is crucial: it is this upon which the curvature of space and the fate of the Universe depend.

Measurement of the Invisible

From the solar system to the observable universe as a whole, astronomical structures are dynamically governed by gravity. This puts bodies into motion by imparting accelerations to them, and therefore velocities. To measure the speeds of visible celestial bodies amounts to estimating the mass of the other bodies, either luminous or not, around which they gravitate.

The procedure is well-known. Kepler's laws, reformulated by Newton according to his theory of universal attraction, allow one to calculate the mass of the planets starting from their orbital velocities. It was by measuring the observed anomalies in the movement of distant planets that Joseph Le Verrier and John

Adams, in 1845, predicted the existence, position, and mass of Neptune. It was by analyzing the irregular movements of the stars Sirius and Procyon that Friedrich Bessel, in 1884, predicted that they must have invisible companions of comparable mass; the white dwarfs Sirius B and Procyon B were in fact discovered, in 1862 and 1896, respectively. The existence of black holes—invisible bodies par excellence!—was also inferred from the motion they impart to surrounding matter, either gas or stars. Finally, the *exoplanets* that are spoken of so often today remain invisible most of the time, but are indirectly discovered thanks to the faint dance that they impart upon the star around which they gravitate.

These three celestial objects are in fact plausible candidates for constituting a fraction of the dark matter: giant gaseous planets, white or brown dwarf stars, and small or large black holes.

The discovery of invisible matter on the scale of galaxy clusters is owed to the astrophysicist Fritz Zwicky. In 1933, he measured the average speeds of galaxies in the Coma Berenices cluster, situated one hundred million light-years away. He was perplexed because the galaxies seemed to move too rapidly within the cluster: the visible mass of Coma, estimated from its luminous components, could not by itself explain such high local speeds. Zwicky concluded from this that a missing mass must be added, which could explain Coma's dynamics.[1]

A similar phenomenon was observed on the scale of individual galaxies. In spiral galaxies, the rotational speeds of stars around the galactic center seem too high to be explained solely by its luminous mass. One can therefore infer that these galaxies are surrounded by a spherical halo of dark matter, around ten times more massive than the stars and shining gas in the galactic disk.

Out of this, the hunt for invisible matter was launched. The term *missing mass* fell into disuse since, in truth, nothing is missing: the dark matter is really there, and it governs the overall organization of the cosmos!

The Cosmological Challenge

Although galaxies have a tendency to be dispersed through the action of cosmic expansion, the contrary effect of gravity is to reassemble them. In other words, the gravitational attraction that the combined masses of the Universe exert upon each other tends to slow down the expansion. The competition has begun. Is there a sufficient amount of matter for the expansion one day to cease?

In general relativity, energy in all its forms dictates the curvature of space. If the average density of energy passes a certain critical threshold, space has positive

[1]This is so if one assumes the hypothesis that Newtonian dynamics is satisfied at the scale of the cluster. Certain alternative theories of gravity modify the Newtonian law in such a way that it is no longer necessary to invoke a missing mass.

curvature. In the contrary case, space has negative curvature or, precisely at the critical limit, it has an average zero curvature.

Astronomers denote the value of the real density divided by the critical density by Ω, so that a discussion on the curvature of space reduces to how Ω compares to 1. If Ω is greater than 1, space is spherical and finite. If it is smaller than or equal to 1, space is hyperbolic or Euclidean (whether it is finite or infinite depends on the topology: this is the subject of this book!).

Moreover, if one neglects the particular forms of energy that work against gravity, such as the cosmological constant, the curvature of space also dictates the long-term evolution of the Universe. If Ω is greater than 1, the expansion that the Universe is presently experiencing will be interrupted in the future to make way for an overall contraction. In the contrary case, the expansion will continue eternally, slowing down all the while.

The cosmological challenge raised by dark matter is therefore considerable. Its total quantity dictates the curvature of space, and its distribution between gravitating matter and antigravitating energy governs the dynamics of the Universe.

Amount of Luminous Matter

It is not very difficult to estimate the number of luminous stars in a galaxy, and thus their total mass, by measuring the amount of light that the galaxy emits. To calculate the mass of interstellar and intergalactic gas is also not an insurmountable task. The cold gas emits radio waves, which can be collected on Earth by radio telescopes, while the hot gas emits x-rays. The atmosphere acts as a screen for the latter, but they can be detected by instruments on board rockets or satellites.

By adding up the masses of all the luminous bodies, one obtains a value Ω_v (v for "visible") no greater than 0.005—very much smaller than the amount needed to close the Universe. Dark matter will necessarily raise this figure, but by how much?

Amount of Dark Matter in Galaxies and Galaxy Clusters

In a spiral galaxy, we see stars, gas, and dust assembled into a thin disk, in rotation around an axis passing through the center. Most of the visible mass is concentrated in a large central bulge, the galactic bulb; on the level of celestial mechanics, the galaxy should behave like a Keplerian system, meaning a massive central body surrounded by bodies of smaller mass whose rotation speeds decrease rapidly with their distance from the center. One can measure the speed of rotation of the stars and the gas (more precisely, its component projected in the direction of the observer) thanks to the Doppler effect, which shifts the spectral lines toward

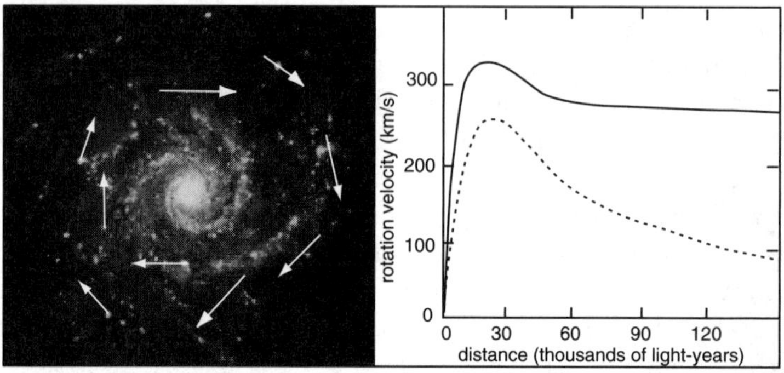

Figure 37.1. Rotation curve for a spiral galaxy. At left, a view of a spiral galaxy. The arrows represent the speed of rotation of the stars around the center. At right, the solid line is the rotation speed as a function of the distance from the center. The dotted curve corresponds to the predicted behavior if the galaxy only contained the visible matter.

the red (respectively blue) proportionally to the velocity of recession (respectively of approach) of the source. The rotation curve is the graph of this speed as a function of the distance from the center (see Figure 37.1). We see that the observed velocities are abnormally high far away from the center, contrary to the Keplerian case where these velocities would only depend on the visible mass. To explain the difference between the two curves, astronomers calculate the mass necessary to confer such motion to the stars and gas. By also calculating the theoretical rotation curve, obtained in the case that the galaxy only contained visible matter, one sees that one needs at least ten times more dark matter than luminous matter.

Moreover, many astronomers believe that the dark matter cannot be distributed in the disk, but must be in an extended spherical halo encompassing the visible galaxy.

For elliptical galaxies, which do not have a disk, such measurements are impossible, but one can evaluate the local stellar speeds as well as the temperature of the galaxy's internal gas. Here again, through the help of some models, one concludes that elliptic galaxies must be at least ten times more massive than their visible part.

Galaxy clusters are made up of hundreds of galaxies grouped together by gravity. Three methods are available in astronomy to evaluate their dynamical mass.

Galaxies move within the cluster under the influence of the general gravitational field. By modeling this, one can calculate the total mass of the cluster as a function of the local speeds of the galaxies. The latter can be supplied through spectroscopic observation.

A second method is based on analysis of the distribution and temperature of the extremely hot intergalactic gas that bathes these clusters. It is observed in the x-ray domain thanks to astronomical satellites like Rosat or Chandra.

The third method is based on gravitational mirages. Light rays are deflected 25 → by the masses of celestial bodies. A gravitational lensing effect is obtained when a galaxy cluster—the lens—is situated in the foreground of a very distant light source—a galaxy or a quasar. Several dozen of these mirages have been studied carefully. When they are suitably configured, their analysis allows one to reconstitute the total mass of the lensing cluster.

All three methods converge toward the same result: galaxy clusters contain between ten and one hundred times as much dark matter as luminous matter! Their contribution to the density parameter is presently estimated at $0.05 \leq \Omega_d \leq 0.3$ (d for "dynamic").

Dark Matter and the Primordial Universe

One of the successes of big-bang models is the explanation of the observed abundances of the light chemical elements (deuterium, helium, and lithium) by a brief phase of primordial *nucleosynthesis*, dating from the first three minutes of the Universe.

For example, deuterium is an isotope of hydrogen that takes part in hydrogen fusion. However, as the most fragile of all nuclei, it is destroyed by nuclear reactions. The more protons and neutrons there are in a given volume of space, the more reactions there are that destroy deuterium, and the less the deuterium can form in abundance. Since there has been no new creation of deuterium since primordial nucleosynthesis,[2] its present abundance reflects its past abundance. We understand how the quantity of deuterium, evaluated today with the help of spectroscopy, is connected to the density of protons and neutrons at the epoch when they were formed.

It is the same for the other elements formed in the primordial Universe: helium 3 (two protons and one neutron), helium 4 (two protons and two neutrons), and lithium 7 (three protons and four neutrons). Their presently measured proportions give the value of the density of protons and neutrons in the primordial Universe. By the law of expansion,[3] this density gives the present density of protons and neutrons, which contributes to the parameter Ω. The theoretical calculation of abundances, carried out within the framework of big-bang models, is in agreement with observations if the present matter density is around 0.05.

[2]The new deuterium formed in stars by nuclear reactions is destroyed right away by other reactions.

[3]To be more precise, we assume that the density remains constant per unit of comoving volume. This volume grows over the course of cosmic history like the cube of the scale factor $a(t)$. The density therefore dilutes very quickly over the course of time.

Can we conclude from this that we live in a Universe of largely sub-critical density, that is to say, of negative spatial curvature and in perpetual expansion? No, since the above reasoning only applies to so-called baryonic matter. The *baryons* are the class of particles made up of quarks. Protons and neutrons are the most stable of these and, because of this, they are the constituents of atomic nuclei: the nucleons. Baryonic matter can therefore simply be called ordinary matter;[4] it is, in fact, this which makes up the stars, the planets, pebbles, living beings, automobiles, etc.

Exotic Dark Matter

Let us review the situation:

- The estimate from visible matter (galaxies, gas) gives the value $\Omega_v \approx 0.005$.
- The dynamical analysis of galaxies and galaxy clusters gives $0.05 \leq \Omega_d \leq 0.3$.
- The calculation of primordial nucleosynthesis imposes $\Omega_b \approx 0.05$ (b for "baryonic").

The conclusion that one could draw from these comparisons is that, not only does the Universe contain at least 90% dark matter (since $\Omega_d \geq 10\Omega_v$), but this dark matter is not composed uniquely of ordinary matter (since $\Omega_d > \Omega_b$).

This last point suggests that there must be "extraordinary" matter. Theoretical physicists prefer to call it exotic matter. From where could this come? To justify its existence, we must open the Pandora's box of particle physics.

For certain cosmologists, the credibility of exotic matter is reinforced because they believe that a Universe having exactly the critical density $\Omega = 1$ would be more satisfying from an aesthetic point of view. Their arguments are highly debatable, because on the one hand they are based on the idea that the corresponding cosmological equations are simpler, and on the other hand they support themselves with theories of high energy physics applied to the description of the primordial Universe. According to these cosmologists, the Universe could have ←245 experienced an extremely short phase of so-called inflation, during which its rate of expansion would have grown to a remarkable extent.

One of the consequences of inflation would be to irresistibly flatten the Universe (a little bit like the surface of an immensely inflated balloon would tend to

[4]Ordinary matter (e.g., atoms, molecules) is made up not only of protons and neutrons (baryons), but also of electrons (leptons), in numbers nearly equal to those of the protons (if not, the Universe would not be electrically neutral). But the electrons are extremely light—1,836 times lighter than the protons—so much so that their contribution to the mass of the Universe is completely negligible.

become flat), in other words, to lead the average density of the observable universe to the critical value $\Omega = 1$.

If such were the case, the totality of cosmic energy would be distributed as 0.5% luminous matter and 99.5% dark matter and, within the latter, 5% ordinary dark matter versus 95% exotic energy! Let us examine in more detail what types of constituents could contribute to one or the other of these forms of dark matter-energy.

The Nature of Dark Matter

Let us review the possible forms of matter:

- Visible matter, meaning ordinary matter composed principally of protons and neutrons, is found in the form of stars, dust, and interstellar gas.
- Baryonic dark matter, meaning ordinary matter that is too dim to be observed, could be composed of invisible bodies collectively called *machos* (e.g., black holes, giant planets, or brown dwarfs), or of non-emitting cold gas.
- Nonbaryonic dark matter is made up of exotic particles collectively called *wimps*; it is subdivided into two types: cold dark matter and warm dark matter.
- Nonparticulate dark energy, if it exists, is necessarily comprised of cosmological nature, and probably comes from the vacuum, from topological defects, or from quantum fields.

By all logic, baryonic dark matter is at least in part made up of rather small celestial objects (a sufficiently large star is luminous) that populate galaxy halos. They are therefore called machos, an acronym for massive astrophysical compact halo objects. Nonbaryonic dark matter could be made up of massive elementary particles, produced at very high energy in the first instants of the big bang, which interact more or less weakly with known matter (from which arises the difficulty in detecting them). These are called weakly interacting massive particles (wimps). The astrophysicists have summed up the problem of dark matter with a dubious joke: is the Universe dominated by machos or by wimps? More seriously, let us examine the experiments dedicated to the detection of machos and wimps.

Macho Detection

Visible baryonic matter, which can be directly detected by astronomers, is that which emits substantial amounts of radiation (like stars and gas clouds) or that absorbs radiation (like interstellar dust and cold clouds). By all evidence, many of the bodies in the Universe do not belong in these categories.

Compact, massive, and invisible bodies are either black holes, paragons of dark matter, or stars too small to sustain nuclear reactions and shine by themselves, meaning either dwarf stars or giant planets.

Although the nature and existence of black holes raises fascinating questions, it is not likely that they can bring a significant contribution to the hidden mass. In fact, we know that black holes of stellar mass can only be created by more massive stars, which are very rare. Moreover, giant black holes, which seem to be present at the center of certain galaxies, cannot exist in great number in the halo of spiral galaxies, since they would then impart to the stars in the disk local speeds that are too high with respect to what is observed. Finally, the small black holes formed in the primordial Universe remain highly hypothetical. In total, the mass density contained in black holes of any sort is certainly negligible.

Hydrogen and helium are the two most abundant chemical elements in the Universe, making up 90% of stars. Nevertheless, hydrogen and helium stars whose mass is smaller than 8% of the solar mass will never become hot enough to begin nuclear reactions, and will never shine. These failed stars are called brown dwarfs. We know that they exist, and we know of at least one that is found in our own solar system: Jupiter. One can, in fact, consider this body as a brown dwarf rather than as a planet since, although its mass is a thousand times smaller than that of the Sun, Jupiter radiates in the infrared: at high pressure but low temperature, the hydrogen gas of which it is composed crystallizes at the center of the planet and liberates energy.

How do we detect such failed stars? Above, I evoked the phenomenon of gravitational lensing on the scale of galaxy clusters. The same phenomenon can occur on a smaller scale, when the light source is a star in a nearby galaxy, for example, the Large Magellanic Cloud or the Andromeda galaxy, and the lens a brown dwarf situated in the halo or the bulge of our own galaxy. One then finds oneself in the presence of a *microlens*. As remarked for the first time in 1986 by the Polish astrophysicist Bohdan Paczynski, over the course of its galactic motion a brown dwarf has a non-zero probability of momentarily passing through the line of sight of a star in the Large Magellanic Cloud, situated in the background. Calculations show that at a given moment about one star of the Large Magellanic Cloud out of two million should experience an amplification of its luminosity greater than 30%.

Things become complicated when one takes notice of the fact that, out of the millions of stars in the Large Magellanic Cloud, several dozens of thousands are intrinsically variable. How do we make a distinction between an intrinsic variability and a chance variability due to a microlensing effect? The latter case in fact exhibits three distinctive characteristics that facilitate its identification. In the first place, the event, because it is very rare, must be nonrepetitive for a given star; also the light curve should be achromatic, meaning that it is independent of the observed wavelength; finally, the form of the light peak should be symmetric in time and of a duration of approximately 30 days (this last figure is obtained by considering a brown dwarf of one-tenth of a solar mass situated 30,000 light-years away and having a transverse velocity of 160 km/s).

With the observational program thus posed, the practical realization of it remains, which implies billions of measurements of stellar luminosities over several years. Indeed, if the galaxy's halo is populated by ten thousand billion brown dwarfs (one hundred times more than the luminous stars) that are able to play the role of microlenses, only the light from a handful of stars among the ten million contained in the Large Magellanic Cloud can vary in a recognizable way.

In 1990, three programs, baptized EROS[5] (French), MACHO (American and Australian), and OGLE[6] (American and Polish), launched ambitious experiments surveying millions of stars in the directions of the Large and Small Magellanic Clouds (for EROS and MACHO) and of the Andromeda galaxy (for OGLE), in order to track possible microlenses situated in the halo or bulge of our galaxy.

The first events were detected in September 1993. Since then they have been reported almost daily, tracked 24 hours a day by sites spread out in Chile, South Africa, or Australia, regions of the globe from which one can see both the galactic center and the Magellanic Clouds. Large field photographic plates, taken at the southern European observatory in Chile for EROS and in Australia for MACHO, give instantaneous images of the Magellanic Clouds as a whole. A regular surveillance, 40 shots per night for several years, allows the gradual reconstitution of the light curves of each star in the Magellanic Clouds. A similar procedure is used for the OGLE program and the stars of the Andromeda galaxy.

It is always interesting when experiments deliver results different from those for which they were designed. This is the case for the microlensing observation programs. In 1998, the combined analysis of the results allowed one in the first place to exclude the possibility that dwarfs of mass smaller than one-thousandth of a solar mass contribute more than ten percent of the mass of the galaxy. Secondly, the events detected in the direction of the Large Magellanic Cloud are compatible with a small population of galactic machos of half a solar mass; these are therefore not brown dwarfs, but normal dwarf stars. One must doubtless look elsewhere for an explanation of the dark matter.

---

[5]EROS is the acronym for *expérience de recherche d'objets sombres*, whose English translation is "experiment for the search for dark objects."

[6]OGLE is the acronym for optical gravitational lensing experiment.

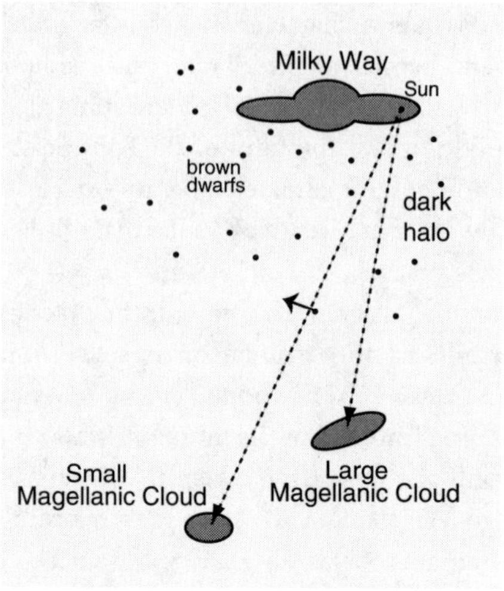

Figure 37.2. The search for gravitational microlenses. The Large and Small Magellanic Clouds are two satellite galaxies of our own Milky Way; their stars serve as background for detecting, through the microlens effect, the brown dwarfs in the halo.

The Detection of Cold Hydrogen

We have spoken of failed stars. Could one imagine failed galaxies? Yes, in the form of immense clouds of cold hydrogen molecules[7] that were never able to condense into stars. These are serious candidates for the baryonic dark matter. Nevertheless, recent (2000) data coming from the far ultraviolet spectrographic explorer (FUSE) satellite lead one to think, rather, that essentially all of the cosmic hydrogen is at high temperature, nearly a million degrees.

The spectra supplied by FUSE have in fact detected a particular ion in their hydrogen clouds, oxygen VI. This strongly ionized atom has the particularity of being formed at a minimum temperature of a few hundred thousand degrees. This means that the vast clouds of gaseous hydrogen that shelter it are much hotter than predicted. Before this, some scientists had believed that cold molecular hydrogen might explain baryonic dark matter. The FUSE data refute this hypothesis.

As for warm hydrogen, since it can be detected in the ultraviolet, it enters the category of visible matter!

[7] As soon as atomic hydrogen is heated, for example by stellar radiation, it loses its electron and is ionized. When it is sufficiently cold, on the contrary, it combines into H_2 molecules, made up of two hydrogen atoms that share their electrons.

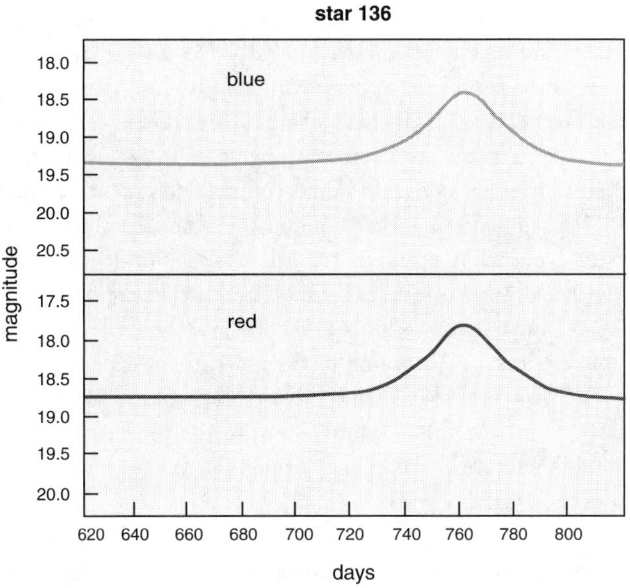

Figure 37.3. Light curve of a gravitational micromirage. This diagram represents the light curve of star 136 of the LMC. The x-axis is the observation time measured in days and the y-axis is the apparent magnitude of the star. Above, the observation is made in the color blue, below, in the color red. There is no apparent difference. The luminosity peak is indeed symmetric, and the duration of the event is around 30 days, in conformity with expectations for the microlensing effect.

The Detection of Wimps

Since the machos were no help, let us now interest ourselves in the wimps. The supposed existence of these exotic particles results from various unified theories of the fundamental interactions (electromagnetism, strong and weak nuclear interactions, and gravitation). Since unification can only be realized at very high energy, these hypothetical particles could only have formed during the big bang. It could be that billions of them cross through our bodies every second, but since they are wimps, they do not interact with the baryonic matter of which we are composed, and we cannot sense their existence.

The candidate particles for forming nonbaryonic dark matter could have been baptized by Tristan Tzara and the surrealists: axion, majoron, higgsino, photino, sneutrino, paraphoton, gravitino, axino, vorton, neutralino, cosmion, flatino, magnino, gluino, wino, bino, dino, preon, pyrgon, maximon, and so on. We are only missing the casino!

These particles are classed into two categories: hot and cold. The term *hot* does not really refer to a temperature, but rather to the fact that these particles would have emerged from the big bang with speeds close to that of light. There exists at least one particle of this type: the neutrino. One even knows that there is on average 400 of them per cubic centimeter of space (as many as there are photons, the particles of light). It would suffice if the neutrinos had a mass as small as 10^{-32} grams (one hundred million times smaller than the mass of the proton) for the Universe to attain its critical density. But does the neutrino have a mass, or on the contrary is it, like the photon, of strictly zero mass? The subject is under debate, since theory admits both possibilities. The last word belongs, as always, to experiment. According to the most recent experiments, completed in 1998 at Superkamiokande (Japan), neutrinos should have a mass, but of no more than 10^{-35} grams. Their contribution to the total energy density of the Universe should therefore be completely negligible, representing at most only one thousandth of the critical density.

There could, nevertheless, be other hot, massive wimps that are not neutrinos. One constraint on their number comes from the problem of galaxy formation. High speed particles would slow down the process of galaxy formation, by having the tendency to fragment the vast agglomerates of matter. Calculations indicate that, if hot wimps dominate the dark matter, objects the size of galaxy clusters would be the first to form in the history of the Universe and would then be fragmented into smaller masses. However, this is in conflict with observations, in particular those of the WMAP telescope, which show instead that the galaxies must have formed relatively early, less than one billion years after the big bang.

Cold wimps would be those that would have emerged from the big bang at speeds much smaller than that of light. These particles would have aggregated into galactic masses more rapidly than hot matter, with the effect that the galaxies would have formed before the clusters. Concerning possible scenarios of galaxy formation, the hypothesis of cold dark matter has therefore for a long time seemed more plausible. However, although numerous unified theories set forth what they could be, no particle of this type has been discovered. Moreover, some of the most precise data on the distribution of large structures in the Universe go against a dark matter explained exclusively with the help of cold wimps. A mix of various forms of dark matter seems to work better.

To prove the existence of wimps, the ideal would be to construct particle accelerators powerful enough to create these very high energy particles. Nevertheless, the race toward gigantic accelerators is presently approaching its limits, and one cannot conceive of using this method in the near future. We are left with the Universe, which *has* been, in its distant past, a prodigious natural particle accelerator, capable of producing every possible species of particle, as massive as they may be.

The wimp hunters have therefore turned toward cosmological observations in the hope of pinpointing their quarry, in the form of fossil particles from the first instants of the Universe. The difficulty lies in constructing detectors capable of collecting cosmological wimps. A number of experimental projects are in the works. They are based on massive target composite bolometers coupled to a temperature sensor. In plain language, a bolometer is an electrical resistance thermometer that is able to measure small flows of heat. When a wimp meets a nucleus in the target, its energy is changed into heat. The detector sensitive to the heating of the target converts this heat into an electric signal. To be efficient, the ensemble must function at extremely low temperatures, smaller than 0.1 Kelvin (or $-273°$ C). The first detection campaigns have just started up; thus, detectors have been installed in the subterranean laboratory of Modane, France, in order to be isolated from the background noise produced by cosmic radiation. Whatever the future result—either positive or negative—of these experiments may be, the innovative detectors designed for the occasion will help advance research in other disciplines.

The Detection of Dark Energy

And what if the majority of the dark matter was not matter? What if it was found in the form of a pure energy field, not made of material particles? A cosmological constant, and more generally a vacuum energy, ideally simulates the presence of dark energy. But this dark energy has a nature that is distinct from that of matter. In relation to the curvature of space, the effects of dark energy would be identical to those of an equivalent amount of matter by virtue of the relation $E = mc^2$. For example, if the Universe had a total matter density (including visible, baryonic, and nonbaryonic forms) of $\Omega_m = 0.3$ and a corresponding cosmological constant with an energy density $\Omega_\lambda = 0.7$, it would have zero curvature, and thus be Euclidean. On the other hand, for cosmological dynamics, the effects of dark energy would be completely different: acting in a repulsive fashion, like a sort of antigravity, the cosmological constant would provoke an acceleration of the cosmic expansion. The cosmological constant in fact confers an energy density to space while exerting, not a compressive pressure, but an expansive tension that is equivalent to a negative pressure.

In the same way as for other forms of dark matter, the dark energy could just as well play a specific role in the process of formation of large-scale structures. In 1989, astronomers discovered an enormous grouping of galaxies called the Great Wall, measuring about 500 million light-years long, 200 million light-years in width, and 15 million light-years in thickness (Figure 9.1). Gravity alone does not seem to be sufficient to form a concentration of density this enormous over

a duration as brief as that granted by the Hubble time. As Lemaître had fore-
←188 seen, the cosmological constant helps to resolve this problem by favoring, over the course of certain past phases, the coalescence of large aggregates of matter.

We see how, without direct observations on the nature of dark matter, the constraints imposed by the formation of galaxies can guide cosmologists' choices. The fabrication of a galaxy resembles a delicate cooking recipe. For it to succeed, one must know how to select the ingredients, how to incorporate them in the right proportions and in the right order, and how to cook them for an adequate amount of time. For the moment, it is still short-order cooking: a large dose (70%) of vacuum energy, a good part (25%) of non-baryonic dark matter, a dash (5%) of ordinary dark matter and a smidgen ($<$ 1%) of visible matter. One once thought that nature abhorred a vacuum. Today, if one believes this recipe, we must admit that the physical world is filled to the brim: the vacuum pasta would totally dominate the universal energy, topped here and there by a foam wherein all the material reality of the world would take refuge.

We are perhaps missing a cosmic Bocuse, capable of concocting a galactic soufflé for us that would rate several stars.

Further Reading

Iain Nicolson. *Dark Side of the Universe: Dark Matter, Dark Energy, and the Fate of the Cosmos.* Baltimore: Johns Hopkins University Press, 2007.

Ken Freeman and Geoff McNamara. *In Search of Dark Matter.* New York: Springer-Praxis, 2006.

Heat! One might say that you don't have to do anything but bend down to pick it up, in space. Yes, heat at 270 degrees below zero.

—Louis Auguste Blanqui, *L'Éternité par les astres*

38
The Cosmic Microwave Background

The Discovery of the Cosmological Blackbody

If you enter into a room whose fireplace has recently welcomed a fire, you can learn much about it simply by observing the coals: if they are bright red, the flames have only recently gone out; if they are orange, the fire finished burning a half-hour ago; if they are gray or if they no longer give off any visible light, but if you can still feel their heat on your hands, then the fire is even older, but not completely extinguished. In this progression, the coals emit a radiation going from visible light to the infrared, meaning radiation whose wavelength becomes larger and larger, or the energy more and more weak, as time goes by.

The first stages of the big bang can be thought of as a great fire, of which the Universe itself would be the coals. The more that it expanded, the more the Universe cooled, and it produced a radiation whose wavelength became more and more stretched toward the red. Today, 14 billion years after the cosmic fire, its radiation is found in the microwave range, similar to the radiation used in microwave ovens. There is a basic difference between the big bang and the chimney fire: on the one hand, we do not contemplate the cosmic coals from the sidelines since we are part of the cooling blaze, and on the other hand, there is no chimney!

This type of reasoning was first presented by Georges Lemaître in 1931. As a point of departure for the cosmic expansion, he imagined the *primeval atom*, a giant ultra-dense nucleus, comparable to ordinary atomic nuclei but much larger, whose fundamental instability would have caused the expansion: "The atom-world was broken into fragments, each fragment into still smaller pieces.... The evolution of the world can be compared to a display of fireworks that has just ended: a few red wisps, ashes, and smoke. Standing on a cooled cinder, we see the slow fading of the suns, and we try to recall the vanishing brilliance of the origin of the worlds."[1] In cosmic rays, those massive particles that penetrate Earth's atmosphere at very high speed, Lemaître thought he had found vestiges of the primitive Universe; however, it was necessary to look for thermal radiation instead.[2]

It was the task of George Gamow, who had learned relativity in Saint Petersburg under the direction of Friedmann, but then emigrated to the United States to flee the communist regime, to enrich Lemaître's hypothesis considerably by adding the notion of temperature. Lemaître had imagined that the Universe at the beginning should have been very dense. In 1948, Gamow and his collaborators, Ralph Alpher and Robert Hermann, specified that it should equally well have been very hot. Assuming a primitive mixture of nuclear particles called *ylem*, a Hebrew term referring to a primitive substance from which the elements are supposed to have been formed, they were able to explain the genesis of the lightest nuclei (deuterium, helium, and lithium) during the first three minutes of the ←188 Universe, at an epoch when the cosmic temperature reached 10 billion degrees.

They then calculated that, at a slightly later epoch, when the Universe had cooled to a few thousand degrees, it suddenly became transparent, and allowed light to escape for the first time. Alpher and Hermann calculated that one should today receive an echo of the big bang in the form of blackbody radiation.

Blackbody radiation is a universal form of light whose spectrum (the distribution of energy as a function of the wavelength) is uniquely determined by a temperature. The black body represents an idealized physical situation in which the radiation is imprisoned within an isothermal wall. This would be the case of a perfectly hermetic oven (which therefore serves as the blackbody) that you make red hot, heating it to a temperature of, say, 3,000 degrees. If you make a small hole in the surface of the oven, then radiation of the given color (red, in our case) comes out, independently of what type of objects you might have put inside—a key, a multicolored ceramic pot, or black and white pebbles. This means that the radiation emitted by the blackbody does not depend on its constituents, since they have attained perfect thermodynamic equilibrium. The surface of a star like

[1] *Revue des questions scientifiques*, November 1931.

[2] Nevertheless, the recently discovered ultra-high-energy cosmic rays might indeed have been created in the big bang.

the Sun emits radiation close to that of a blackbody; in practice sunlight is yellow in color, since it corresponds to a temperature of nearly 6,000 degrees—the temperature found on the surface of the Sun.

For the cosmological blackbody, Gamow and his collaborators calculated a residual temperature of 5 degrees above absolute zero (5 Kelvins or 5 K), or −268° C. Their prediction did not elicit excitement. They refined their calculations several times until 1956, without arousing any more interest; no attempt at detection was made.

In the middle of the 1960s, at Princeton University, the theorists Robert Dicke and James Peebles studied oscillatory universe models in which a space in expansion followed by contraction, instead of being infinitely crushed in a big crunch, passes through a minimum radius before bouncing into a new cycle. They calculated that such a hot bounce would cause blackbody radiation detectable today at a temperature of 10 K. It was then that they learned that radiation of this type had just been detected, at the Bell Company laboratories in New Jersey. There, the engineers Arno Penzias and Robert Wilson had been putting the finishing touches on a radiometer dedicated to astronomy, and they had found a background noise that was higher than expected. To put the finishing touches on a radiometer means to eliminate background noise from the antenna, or at least filter it out. The brouhaha of conversations in a large ballroom is an acoustic background noise, which is measured in decibels. In radiometry, the background noise is due to the agitation of the electrons in the antenna and to external interference, and is expressed by a temperature. When Penzias and Wilson pointed their antenna toward the zenith, they found a background noise temperature of 6.7 K; after subtracting the antenna noise and absorption by the atmosphere, there remained an excess of 3.5 K. The engineers of the Bell Company spent an entire year looking for possible causes, including the pigeon droppings that dirtied the horn of the radio telescope. The temperature excess was still there. Not coming from the equipment, not depending on the seasons, captured in every direction of the sky and having the characteristics of blackbody radiation, this background noise had to be of cosmic origin. It was the fossil radiation!

The teams of the Bell Company and Princeton University published their articles separately in the same July 1965 issue of the *Astrophysical Journal.* Penzias and Wilson only gave the results of their measurements, while Dicke, Peebles, Peter Roll, and David Wilkinson gave their cosmological interpretation. None of them mentioned the predictions of Alpher and Hermann, much less those of Lemaître. Penzias and Wilson obtained the Nobel prize in physics in 1978. However, at the moment of their discovery, they believed instead in the steady- 195→
state theory, rival to that of the big bang. In fact, the detection of the fossil radiation practically signed the death sentence of the steady-state model!

Recombination

What physical mechanism is at the origin of the fossil radiation? The early Universe did not contain any stars or any galaxies, just a mixture of charged particles and radiation, made up of the nuclei of light elements, electrons, neutrinos, and photons, all in thermodynamic equilibrium. In other terms, the entire Universe was filled with matter similar to the outer layers of our modern-day Sun. Like solar matter, the primordial plasma was too hot to welcome stable atoms: as soon as a positive nucleus and a negative electron grazed each other, their speed from thermal agitation was so large that they could not associate into a neutral atom. In consequence, photons, the particles of light, had a very hard time propagating; they were inevitably scattered by free electrons. The early Universe was opaque. In everyday life, things like metal, wood, and stone are opaque precisely because they are filled with free electrons that intercept light, while transparent materials, like glass and water, possess very few free electrons.

As the expansion continued, the primordial plasma cooled. After about 300,000 years, its temperature fell to 3,000 K, a sufficiently low level for the electrons to lose their liberty: they combined with atomic nuclei in a process called *recombination*. The plasma has now condensed into a neutral gas of hydrogen and helium. This condensation is analogous to the evaporation of mist in Earth's atmosphere. The mist is made of minuscule droplets of suspended water that diffuse the photons of light; it is therefore opaque. But when the atmosphere is reheated by the Sun, the drops evaporate, transforming into water vapor, and the air rapidly becomes transparent. An analogous phenomenon took place in the early Universe, when the plasma condensed into gas (even though it was cooling). Having lost its free electrons, the Universe became transparent. The photons, which had been scattered ceaselessly by the charged particles of the plasma (from whence the name of *surface of last scattering* given to their region of departure), suddenly[3] became free to voyage without constraint through space, which they have been doing ever since.

The temperature of 3000 degrees is the one found today on the surface of stars like red giants. At recombination, the primitive plasma was therefore a gigantic luminous object occupying all of space, shining as brightly as a star. If we do not see it at night (much less during the day!) in visible light, it is because the energy of this radiation has diminished over the course of time. In fact, in 14 ←156 billion years, the temperature has fallen by a factor of 1100, since the radius of curvature of the Universe has in the same time grown by the same factor. This

[3] The Universe did not become transparent instantaneously. The surface of last scattering is, more properly speaking, a spherical layer of non-zero thickness; however, this thickness is less than 1% of its radius. This is the reason why, to simplify while not introducing too much error, cosmologists often treat the layer of last scattering like an infinitely thin spherical shell.

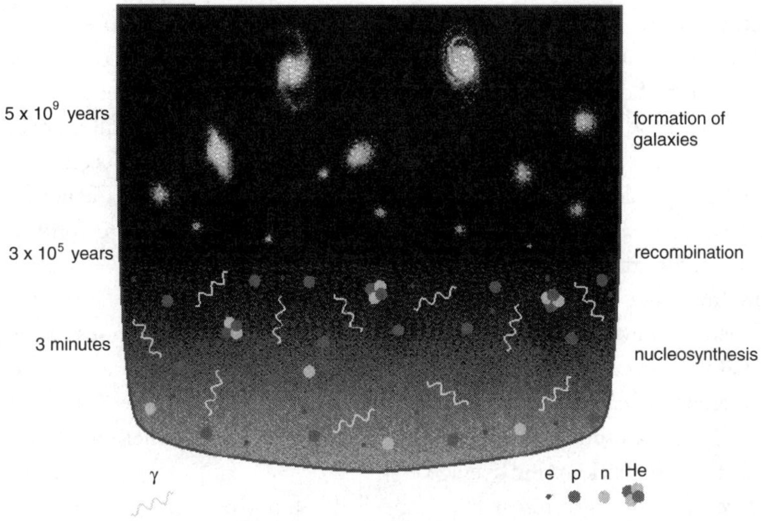

Figure 38.1. Recombination and transparency. Starting from protons (p) and neutrons (n), the first helium nuclei (He) were formed after three minutes, during nucleosynthesis. It was only after 300,000 years of expansion that the negatively charged electrons (e) combined with the nuclei to form electrically neutral atoms. The Universe then became transparent, and the cosmological radiation (γ) began to propagate.

is required by cosmic thermodynamics. The fossil radiation captured today has a blackbody temperature of only 2.7 K and is only "seen" with wavelengths between a few centimeters and a few millimeters—in the radio and microwave range. In spite of this, the photons of the fossil radiation are quite numerous: there are on average 400 per cubic centimeter of space. When your television is not tuned to a regular channel, you see the "snow" of the screen; remember to tell yourself that a few percent of the snow comes from the photons of the fossil radiation!

Spots in the Fossil Radiation

It was the condensed gaseous masses coming out of recombination, made up of three quarters hydrogen and one quarter helium, which necessarily formed the first galaxies and the first stars. But to do this, two things were needed: lumps, meaning excesses of density within the primitive soup, and a sufficiently long time for the lumps to grow under the action of gravity. Without these lumps, the universe would have remained in the form of a diffuse and perfectly homogeneous gas forever. It would contain neither stars, nor galaxies, nor planets—nor, a fortiori, living beings, astronomers, or readers.

It is clear that the observed galaxies and galaxy clusters could only have been created if the primitive plasma was denser in certain regions. The zones presenting an excess of density with respect to the average exerted a greater gravitational attraction on the surrounding matter, attracting it and becoming still more dense, reinforcing their attractive power, and so on. Through a series of complex processes, gravity caused the seeds of the celestial bodies to sprout. It is not known if it was the stars, galaxies, or galaxy clusters that were the first to form, but it is estimated that a minimum of a billion years had to pass after recombination before these first bodies appeared.

Scientists have tried to find the traces of these density fluctuations in the primordial Universe. They should translate into small temperature variations in the fossil radiation, called *anisotropies* since they depend on the direction of observation. These anisotropies look like spots, some slightly warmer, others slightly cooler, than the background average.

Twenty-six years of research were needed before they were detected, in 1991, by the differential microwave radiometer (DMR) launched on board the cosmic background explorer (COBE) satellite by NASA.

The detection of the anisotropies was arduous in more than one way. First of all, the microwaves are a range of the electromagnetic spectrum that are difficult to observe. Only the part with the lowest energy reaches the ground, which in fact caused Penzias and Wilson to detect them in the radio range. But the part at higher energy, which contains more information on the warm and cool spots of the fossil radiation, is absorbed by the atmosphere. One must therefore launch microwave telescopes either on board satellites, or on stratospheric balloons that fly sufficiently high.

In addition, the traditional technology of radio antennas becomes badly adapted and is not sensitive enough as soon as one goes up in energy, from large waves to microwaves, and as soon as one wants to measure the background cosmic noise with great precision. Rather than detecting the waves that make the antennae vibrate, one measures the heat that they transport. The way to detect this ←217 minute heat consists in using thermometers sensitive to low temperatures, called *bolometers*. These are cooled to the neighborhood of absolute zero. For example, the COBE detectors were at 2 K, which in fact allowed for the capture of the signal at 3 K as well as its small variations.

Moreover, contrary to initial expectations, the temperature differences from one point to another of the sky are smaller than one ten-thousandth of a degree, a level of fluctuation comparable to that which would be represented on the Earth's surface by a mountain or valley no greater than 100 meters in height.

Finally, the data are drowned under the antenna noise and under interfering perturbations coming not from the fossil radiation, but from other astronomical

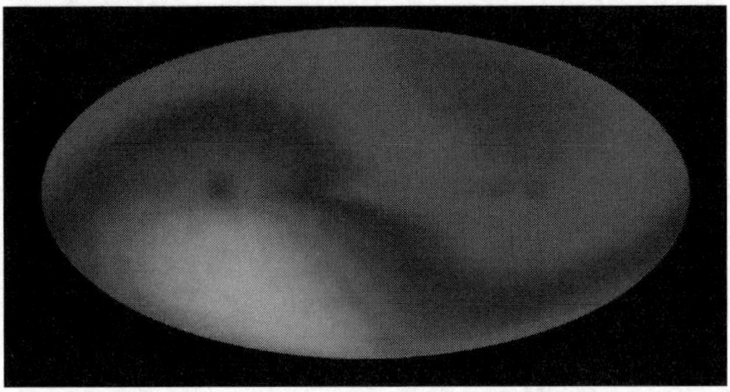

Figure 38.2. Dipole anisotropy. This false-color map of the fossil radiation shows the dipolar temperature variation due to the Doppler shift. The red hemisphere presents a maximal excess of temperature of 3 millikelvin; the warmest point gives the direction and speed of motion of the Earth with respect to the fossil radiation. (See Plate XII. Image from NASA COBE.)

sources. To access the true noise of the cosmic background, one must subtract all these interfering contributions.

First of all, it was noticed that the temperature of the fossil radiation, measured in a precise direction pointing toward the Aquarius constellation, is 0.2% higher than in the opposite direction. This dissymmetry, called *dipolar anisotropy*, translates a local effect, the movement of Earth with respect to the distant space-time reference frame represented by the fossil radiation. The relative motion of the Earth with respect to the source of radiation translates, through the Doppler effect, into a slight augmentation of the energy received in the direction of motion, and to a lessening in the opposite direction. The dipolar anisotropy measured by COBE shows that the Earth is moving at a speed of 370 km/s toward Aquarius. This speed is the sum of all the particular speeds of the Earth: that of our planet around the Sun, that of the Sun around the center of the galaxy, that of the galaxy with respect to the center of gravity of our local Cluster, that of the Cluster with respect to the local Virgo Supercluster and, finally, that of our Supercluster with respect to a Great Attractor.

One must then remove the contribution of our galaxy, which emits primarily microwaves in the galactic plane coinciding with the Milky Way.

When all these effects are subtracted, the residual temperature of the sky's background shows itself to be equal everywhere to 2.728 K (−270.43°C), with fluctuations no greater than one hundred-thousandth of a degree. This extraordinary uniformity attests to the fact that the Universe was well and truly homoge-

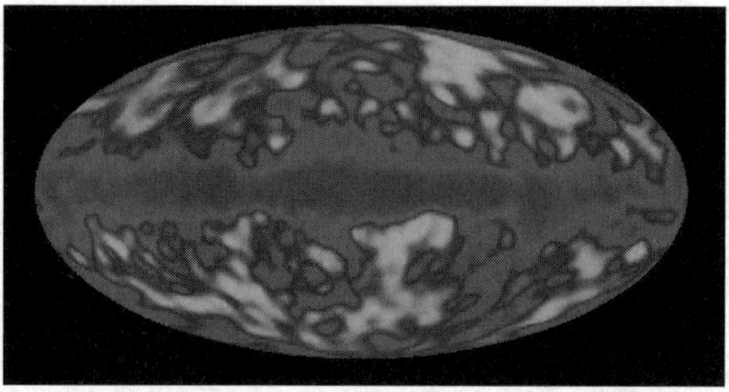

Figure 38.3. Microwave emission from the Milky Way. When the dipole anisotropy is subtracted, the residual fluctuations of the fossil radiation are jumbled up with the microwave signal emitted by the disk of our galaxy (in red). (See Plate XIII. Image from NASA COBE.)

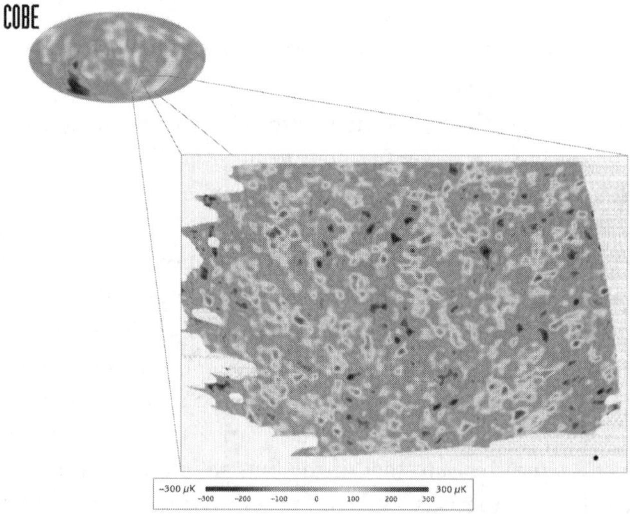

Figure 38.4. Fluctuation map measured by Boomerang. Compared to the map of fluctuations measured by the COBE satellite in 1992 (Figure 21.2), the one obtained by the Boomerang experiment in 1999 had a resolution 70 times better but coverage of the sky that was 25 times narrower. The hot and cold spots observed in the fossil radiation are encoded by colors. These are the seeds of the large-scale structures that we see today in the universe: galaxies, galaxy clusters, and superclusters, corresponding to primordial anisotropies that were later amplified by gravity. (See Plate XIV. Image © Boomerang Collaboration.)

neous at the epoch of recombination, as the big-bang models foresaw. The temperature map established by COBE is a projection of the surface of last scattering onto a plane, analogous to a map of the world, where the temperature variations are encoded by colors (Figure 21.2).

The density variations of the primitive plasma at the surface of last scattering are not entirely responsible for the temperature variations. The speed of the electron and photon plasma and the variations of the gravitational potential of the Universe at the last scattering surface, as well as the variations of the gravitational potential along the photon trajectories, also induce temperature anisotropies. We see how a very detailed study of the fluctuations can provide a veritable mine of information on the conditions that held in the early Universe, particularly on cosmological parameters like the curvature, the density of dark matter, the cosmological constant, and so on. But how do we separate all these effects?

Acoustics of the Big Bang

Cosmologists have found it convenient to represent the spots by an angular spectrum, which plots the amplitude of temperature fluctuations as a function of a wave number l. This wave number is nothing more than $180°/\theta$, where θ (theta) is the angle at which the fluctuation is seen. The temperature fluctuations, for a mathematician, are expressed by a real valued function on the sphere. As such, they can be decomposed into an infinite series of spherical harmonics, just like a real-valued function on a circle can be decomposed into an infinite series of sines and cosines, called a Fourier series decomposition. And in the same way that the Fourier coefficients of a sound wave furnish all the useful information about a sound, allowing one to recognize, for example, if it is an E flat played by an oboe, the Fourier coefficients of the fossil radiation reflect the values of the rate of expansion, the matter density, the cosmological constant, and other parameters of the Universe.

If you sprinkle sand on the surface of a drum and then gently vibrate the drum skin, the grains of sand will assemble into characteristic patterns; the detailed study of these patterns contains information on the size and shape of the drum, on its elasticity, the physical nature of the drum skin, and so on. The analysis of the spots in the fossil radiation resembles the analysis one could perform on a vibrating spherical drum. The primordial Universe was traversed by veritable acoustic waves.

An acoustic wave is the propagation of a vibration within a medium. The speed of propagation depends on the characteristics (temperature, density, etc.) of the medium crossed. For example, sound propagates at 331 m/s in air at 0°C, at 1,435 m/s in water at 8°C, and at 5 km/s in steel.

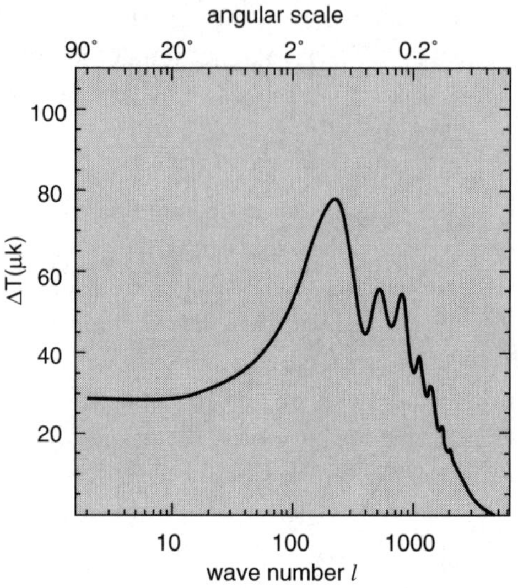

Figure 38.5. Acoustic peaks. This sketch shows the prediction of the model for a Euclidean universe filled with cold ordinary matter. The amplitude of temperature fluctuations (in millionths of kelvin) is given as a function of the angular scale at which one looks (in degrees) or as a function of the wave number l, which amounts to the same thing. The first peak corresponds to the sonic horizon. The secondary oscillations depend on parameters like the spatial curvature, the cosmological constant, the rate of expansion, the ratio of photons to electrons, and so on.

The primordial plasma behaved like air. Gravitational vibrations propagated in the form of acoustic waves at a fairly large speed, although smaller than that of light. After 300,000 years, the acoustic waves had transported the vibrations over a distance called the *sonic horizon*. We do not hear these primordial acoustic waves, but since the photons recorded all of the modifications of the plasma, we see them, in precisely the form of the warm and cool spots that cover the last scattering surface.

We expect that these fluctuations would be maximal at an angle of about one degree. This scale corresponds to the angle subtended by the sonic horizon, meaning the distance covered by an acoustic wave between the big bang and recombination. Since fluctuations have maximal amplitude at the sonic horizon, a principal *acoustic peak* appears in the angular spectrum of fluctuations, corresponding to a fundamental vibrational frequency, followed by secondary peaks similar to musical harmonics, meaning they vibrate at whole multiples of the fundamental frequency.

The position of the first peak depends on the size of the sonic horizon at the moment of recombination. This size serves as a standard ruler for measuring the geometry of the Universe. In fact, in a space of zero curvature, the angle subtended by the sonic horizon at recombination is around one degree; in the fluctuation spectrum, this angle corresponds to a wave number $l = 180$. If space is hyperbolic, that is if it has negative curvature, the photons move more rapidly along diverging trajectories, so much so that the dominant scale of the angular fluctuations becomes smaller and corresponds to a larger wave number. For a positively curved spherical space, it is the inverse.

Most of the detailed cosmological information is encoded in the first few peaks of the angular spectrum. Unfortunately, they were inaccessible to the DMR 245→ aboard COBE. The reason for this is that the DMR was only capable of resolving structures on angular scales of seven degrees, which corresponds to an apparent diameter of fifteen full moons laid end to end. If COBE had looked toward the Earth and not toward the sky, it could only have distinguished spots that were greater than 1,000 km in diameter: it would have seen in a rather blurry manner the edges of continents, but would have made no distinction, for example, between France and England. We can here plainly see the difference between the sensitivity and resolution of a detector. The sensitivity allows us to measure a change in temperature of one hundred-thousandth of a degree, the resolution allows us to say that this change is localized in a more or less crude pixel. Thus, with only the data from COBE, one cannot be certain of having correctly recognized a particular spot on the map of the fossil radiation. All that one can say is that, statistically, some of the spots that one sees as warm on the map are in fact warm, and some that one sees as cold are in fact cold.

In terms of the wave number, COBE could not collect any information beyond $l = 25$, which is therefore very much below the first acoustic peak. This is the reason for which, after COBE, scientists have sought to improve the resolution and to reduce the noise of their instruments. New low-temperature technology allowed for such progress. In fact, to measure the temperature anisotropies with a precision of one millionth of a degree, one must be able to cool the detector to only a fraction of a degree above absolute zero.

Some of these instruments are already in operation. Boomerang-98 is an experiment launched aboard a stratospheric balloon to measure the anisotropies of the cosmological blackbody, by bolometers cooled to 0.3 K. This Italian and American experiment performed its first long duration flight (more than ten days around Antarctica) between December 1998 and January 1999. The data collected have been exceptional: on the one hand, the bolometers' sensitivity has allowed one to reach a millionth of a degree; on the other hand, their angular resolution (10 arc-minutes, 70 times better than COBE) allowed one to reach wave

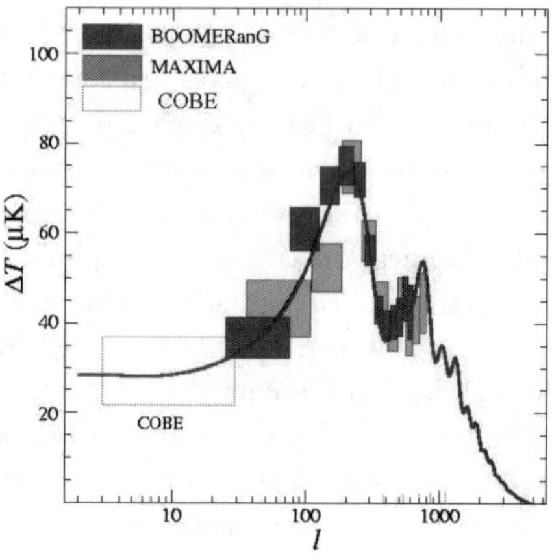

Figure 38.6. The measurement of acoustic peaks. The data on fluctuations of the fossil radiation, obtained by the COBE, Maxima, and Boomerang experiments, are included within error boxes. Cosmologists attempt a best fit adjustment of the corresponding spectrum of angular fluctuations (black curve).

numbers of $l = 800$. In 190 hours of observation, four percent of the sky was covered (see Figure 38.4).

Maxima-1, another balloon experiment, took flight in August of 1998 and flew ten hours at an altitude of 39 kilometers above Texas. It was followed by Maxima-2 in June of 1999, with comparable performance.

The first acoustic peak is clearly found in the data, there where it was expected, at a wave number of around 200. The deciphering and interpretation of the other peaks are less obvious. The results were announced in April 2000 during a spectacular press conference, after which it was a bit prematurely trumpeted that the Universe was flat. Constraints on the cosmological parameters are determined by making adjustments for a best fit between the observed angular spectrum and the predicted spectrum, averaging out the variations of a good dozen parameters, all while remaining in a very particular class of models that from the very beginning favor the theory of inflation, to the detriment of other possible models. It is therefore no surprise to find the conclusion from the hypotheses put forth: ←245 inflation in fact implies a space that is very close to being Euclidean!

Despite everything, Boomerang and Maxima have entered us into an era of high precision observational cosmology. What about the *topological parameters*, which here interest us first and foremost? We must confess that, whatever precision may be brought by balloon flights that finely measure the curvature, they prove to be ineffective for testing the topology of the Universe. They only cover a few squared degrees of the sky. However, the topology is a property of the Universe on a very large scale. To detect or eliminate it in a convincing manner, we need very high resolution maps of the entire sky. As in cosmic crystallography, the topological mirage created by the ultimate wrapped shape of space is global, 116→ and not localized. The method of matched circles predicts temperature correlations in very different directions. Instead of sending an armada of balloons, with each one carrying a bolometer destined to calibrate a very small section of the sky, and then reassembling the data into an immense patchwork to establish a complete map, the best method is to place bolometers on a satellite; there, the time of observation and the orientability are sufficient to cover the entirety of the sky. Only satellite experiments can achieve a complete covering of the sky, with the exception of the zone occupied by the Milky Way.

The scientific community expects much from two satellites designed to this effect. The first, the microwave anisotropy probe (MAP), has been constructed by a team of scientists from several American institutions, under the aegis of Princeton University and NASA. Launched in 2001,[4] MAP will be placed 1.5 million kilometers from the Earth, in a precise orbital position directly opposite the Sun, called the Lagrange point. It is an ideal placement, for the three brightest sources in the sky (which therefore cause the most interference in the microwave range), namely, the Sun, the Earth, and the Moon, will all three be grouped together at the back of the instrument.

MAP will carry an improved version of the DMR used on COBE, consisting of two microwave telescopes whose signals will be electronically subtracted to give the temperature differences between two directions of the sky. The differences will be measured for wavelengths between three and fourteen millimeters. Since MAP will rotate around itself, the directions in which it will point its telescopes will change, because of which over the 24 months expected for the mission, the temperature differences between several million pairs of directions will be measured, and the data will be periodically transmitted to the Earth. The scientists of the MAP program can then reconstruct a high resolution map of the temperature fluctuations of the background of the sky, with a sensitivity of 0.00002 kelvin per pixel of 0.3 squared degrees.

PLANCK is a mission of the European Space Agency, planned for 2007. Its objective is to measure the anisotropies of the fossil radiation on all angular scales greater than ten arc-minutes. To take up the geographic analogy again, this amounts to measuring details of about fifteen kilometers on the surface of the Earth. The project in its present state calls for a 1.5 meter radio telescope

[4]Note for the present edition: MAP, renamed WMAP in honor of David Wilkinson, one of the initiators of the project, who passed away in September 2002, was in fact launched in June 2001. It has produced extraordinary results, discussed in the afterword.

launched on a space vessel that will also be placed at the Lagrange point. However, in contrast to MAP, which will only measure temperature differences, PLANCK will measure absolute temperatures. It will thus directly furnish the map of the background sky.

PLANCK will have access to a larger range of wavelengths at higher energy than MAP, between 0.3 and 10 millimeters. Its angular resolution, its coverage in frequencies, and the sensitivity of the maps obtained will be better than that of MAP, which is the least that one can expect from a mission planned six years later. As a bonus, PLANCK will be the first to finely measure the polarization of the fossil radiation and to give information on the gravitational waves emitted by the primordial Universe.

Further Reading

Marc Lachièze-Rey and Edgard Gunzig. *The Cosmological Background Radiation.* New York: Cambridge University Press, 1999.

George Smoot and Keay Davidson. *Wrinkles in Time.* New York: William Morrow, 1993.

The ignorant believes it flat, I deepen its roundness.

—Jean de La Fontaine, *Un animal dans la lune*

39 Is the Universe Flat?

"The Universe is flat." This announcement made the front page of the newspapers in the spring of 2000. It earned a nearly unanimous ovation from the scientific community for the researchers at the origin of this experimental result, deduced from new data on the spots in the cosmic microwave background published in the journal *Nature*. However, this announcement was incorrect in more than one way.

To begin with, it was false in its formulation: the Universe cannot be flat. In fact, all modern discourse on the Universe comes from relativistic cosmology. Now, in relativity, the Universe does not reduce to just the cosmic space that contains the stars, the galaxies, the vacuum energy, and so on: the Universe is a space-time. In this regard, the major result of relativistic cosmology is that space-time possesses a dynamics (including expansion, either decelerating or accelerating; contraction; or oscillations), which translates automatically into a space-time curvature. The only relativistic model of the Universe that is really flat is Minkowski space-time; but since this is empty of matter and gravity, it cannot in any case 195→ represent the observed Universe. The static models of Einstein and de Sitter are *temporally flat*, in the sense that their radius of curvature remains constant, but their spatial curvature is positive. These models are, moreover, invalidated by the observation of cosmological redshifts. The good relativistic solutions are the Friedmann-Lemaître models. They all have a space-time curvature, and their purely spatial part is itself curved, save for the extremely particular case of the Euclidean model, of zero curvature.

It is precisely this latter case that was promoted by the announcement: space (and not the Universe) would be flat. But this corrected formulation is hardly more satisfying. The adjective *flat* is only applicable to surfaces, and not to volumes. It would therefore have been more correct to say "space has zero curvature" or even "space is Euclidean."

This distinction between *Universe* and *space*, and *flat* and *Euclidean*, may seem like an affectation of rigorous science, completely worthless in the field of popularization. It is nothing of the kind! If one wants the public to understand what science returns, one must be very attentive to the choice of terms. The words *Universe* and *flat* have precise meanings in the dictionary, which do not correspond to those implicitly given to them in the statement "the Universe is flat." As a consequence, instead of informing the reader, this announcement misinforms him. After the diffusion of this scoop in newspapers and on television, a good number of people had an entirely false image of the shape of the Universe, as I was able to witness many times. Some believed, for example, that all the galaxies were laid out in an infinite plane. Others thought that Einstein and his theory of general relativity, according to which the Universe is curved by matter, were contradicted by new astronomical observations. Their reasoning was the following: matter curves space; more matter implies more curvature (a black hole in fact curves its surrounding space more strongly than the Sun); if space is flat, that means it has no curvature, meaning little or no matter; the Universe would therefore be too empty to be curved on the scale of billions of light-years.

This reasoning is false. If cosmic space has an average zero curvature, it is not because it is too empty, but on the contrary because it is on average "full." More precisely, it is because its density parameter Ω, defined as the ratio between the real density (all forms of matter and energy together) and the critical density, is equal to 1. A truly "empty" space, let's say with Ω much smaller than 1, is *strongly* curved; this curvature is negative, of course,[1] but it is still a curvature!

To conclude this first point, it is the responsibility of the popularizer and of the scientific journalist to be particularly vigilant on the choice of words—even if it means correcting the sometimes poorly rigorous declarations of the scientific professionals themselves.

As a second point, to believe that space is flat is logically faulty. Strictly speaking, this is equivalent to saying that the total density of matter Ω is equal to a theoretically well-defined value, with infinite precision. Now, it is obvious that no physical experiment could ever obtain a result measured with infinite precision. Every measurement is associated with an error bar. It is unfortunate that the scientists at the origin of the announcement declared that they were able "to

[1] In this case, space has a hyperbolic geometry.

say with great certainty that the Universe is flat." They have sacrificed rigor for the scoop. It would be forever impossible to prove (in the experimental sense of the term) that space is flat.

In fact, this is the only case that could never be proved! It would proceed differently with a spherical space (of positive curvature) or a hyperbolic space (of negative curvature), for these solutions correspond to intervals of possible values of the density parameter ($1 < \Omega < \infty$ in the spherical case, $0 < \Omega < 1$ in the hyperbolic case). Let us imagine that, in the future, the precision of experimental data gives an error bar entirely contained within one of these two intervals (for example, $1.05 < \Omega < 1.23$): one would in fact have experimental proof that space is curved (positively, in my example).

The real experimental situation is the following.[2] The data on the spectrum of the fossil radiation, obtained by the balloon experiments Boomerang-98 and Maxima-1, and interpreted according to certain theoretical models, give an error bar of $0.95 < \Omega < 1.25$ with a confidence level of 95%. Starting from there, one cannot conclude anything much about the curvature of space, only that the Euclidean case $\Omega = 1$ is found within the error bar, without being particularly in the middle. This simply implies that observable space has a radius of curvature greater than that of the cosmological horizon, and that it seems to us nearly "flat," just like a balloon that is several meters in diameter would seem flat to a bacterium whose horizon goes no farther than a few microns. Would we say because of this that the balloon is flat and that its volume is infinite? In believing that the Universe is strictly flat, the public has a tendency also to deduce from this that space is infinite (generally ignoring the possibility of a wraparound Universe). However, it would suffice to have $\Omega = 1.00001$ for space to be spherical, and thus necessarily of finite volume. The distinction between finite and infinite is nevertheless huge!

231→

In reality, since the Boomerang and Maxima experiments only map out a small part of the background sky, they say nothing about the global geometric properties of space. One might be tempted to rectify the announcement by saying that "space is weakly curved," since the density parameter Ω is close to 1. This would once again be a crude error! Weakly curved with respect to what reference scale? In fact, cosmologists normalize the curvature of space to three whole numbers, $k = -1$, $k = 0$, or $k = 1$, according to whether the space is hyperbolic, Euclidean, or spherical. If $\Omega = 1.00001$, then $k = +1$, and one can certainly not say that +1 is slightly larger than 0. The ratio between these two numbers is in fact infinite!

It would be another error to believe that it is because space has zero curvature that the apocalyptic end of the cosmos, announced by the partisans of the big

[2] See the afterword for readjusted data in light of the observations of the WMAP telescope.

crunch hypothesis, would not take place. To avoid a big crunch only requires that the contribution of the cosmological constant would be sufficiently large to accelerate the expansion. Even if the total amount of energy contained in the Universe was larger than the critical density, and in consequence space was closed into the shape of a hypersphere or any of its multiply-connected variants, it would all the same be in perpetual expansion because of the cosmological constant, and not because of its curvature.

To conclude this second point, a correct formulation of the famous announcement would be: "The spatial part of the observable universe is approximately Euclidean." I recognize that this is not an announcement that would make headlines.

Let us push the analysis further. Why did most scientists themselves, who are normally rigorous, give themselves over to the game of this abusive simplification? A deeper examination of the affair opens interesting ideas on the sociology of science, unknown not only by the public, but usually also by the mediators of these ideas, the scientific journalists. Science, and in particular cosmology, is not sheltered from a true curse: groupthink. Science functions by provisional consensus, meaning opinions shared by the majority. Of course, contrary to other domains like philosophy or politics, these opinions do not arise arbitrarily; they result from a satisfying agreement between theory and experiment. However, a consensus that survives long enough has the tendency to be transformed into orthodox thought. It then begins to exercise a veritable campaign of intellectual intimidation against all contrary thought. In the history of cosmology, Aristotelianism is the best example of a vision that was initially original and innovative, but that, subsequently, ossified to the point of halting the development of the discipline for nearly two millennia.

On a different scale, the model of inflation has for 20 years played the role of an original cosmological concept that has transformed into a tyrannical thought, which one however disguises under the nobler term of *paradigm*. Inflation is a model coming from high energy physics, according to which the Universe would have experienced, in the first fractions of a second of its evolution, a frenetically elevated rate of expansion, to the point that space would have become flat. At first sight, therefore, the model of inflation seems to be favored by the latest observations. The shouts of the press attest to it: "victory of inflation" and "inflation confirmed" could be read in a number of newspapers. This is obviously what the inflation lobby would like to have people believe. Is it really the case? To answer this, one must enter into the meat of the subject and closely examine the way in which the conclusion $\Omega = 1$ was obtained from the raw data of the Boomerang and Maxima experiments.

In exchange for introducing a certain number of hypotheses, some of them plausible, a whole range of angular spectra of the fluctuations of the cosmic microwave background are theoretically calculated by varying a dozen cosmological parameters. 231→ They are then compared to the observed spectrum, and a best fit adjustment is made to deduce the various parameters. However, as we enter into an era of high precision experimental cosmology, it becomes more and more difficult to understand how the cosmological parameters are extracted from the observational data. The temperature anisotropies are influenced by a great many factors: e.g., the rate of expansion, the curvature of space, the energy density of the vacuum, the baryonic density, the number of different species of neutrinos, the amplitude and spectrum of the perturbations in the primordial density, and so on. Many different causes produce comparable effects. The ranges of constraints given with great fanfare should therefore be taken with a lot of precautions.

The well-informed reader will have every reason to be perplexed if he delves into a detailed reading of scientific articles addressing the question. A close analysis of the angular spectrum of fluctuations suggests in fact that the most likely value for Ω is 1.2, and not 1. It would be logical to conclude from this that the balance of experiment leans in favor of a spherical space, of finite volume, although with a very large radius of curvature and in perpetual expansion thanks to the cosmological constant. This would be a singular return to the models proposed be- 195→ tween 1927 and 1931 by Lemaître (see Figure 36.1), which would however render the announcement of a cosmological revolution a little bit overrated. Nevertheless, it is exactly this that the inflation lobby has a hard time admitting, especially since not one simple inflationary model predicts $\Omega > 1$. One can see it in the extraordinary way that the three teams of scientists responsible for the COBE, Boomerang, and Maxima experiments came together in July 2000 to write a consensus article on the latest values of the cosmological parameters (as one speaks of the latest Parisian fashions). Their way of proceeding was somewhat surprising: their aim is no longer to find the best adjustment of the various parameters (which would in particular give $\Omega = 1.2$ for the total density), but to calculate the values of the other cosmological parameters while assuming at the outset that $\Omega = 1$. In other terms, the scenario of a fluctuation spectrum caused by inflation is *already* found in the hypotheses. It is not astonishing that it is found again in conclusion! This type of biased analysis is rendered possible by the fact that $\Omega = 1$ is indeed compatible with the data. However, if the data on the cosmic microwave background obtained by the MAP or PLANCK space missions confirm the spherical model, the champions of inflation will have to seriously revise their copy.

It is not a question of minimizing the importance of the data collected by the Boomerang and Maxima experiments. Never has observational cosmology attained such precision. But the error bars and the pile of implicit hypotheses leave the door open to original interpretations quite different from that of a senselessly flat Universe.

The sense of the symmetrical is an instinct which may be depended on with an almost blindfold reliance. It is the poetical essence of the Universe—of the Universe which, in the supremeness of its symmetry, is but the most sublime of poems.

—Edgar Allan Poe, *Eureka*

40 Symmetry

Symmetry is present everywhere in nature, from our faces to the wings of a bird, from the leaves of a tree to atomic crystals. It is seen also in human achievements. The art and architecture of all civilizations show us masterpieces of simplicity and equilibrium thanks to the precepts of symmetry used in their development.

In geometry, an object possesses a certain symmetry if different viewing angles produce indiscernible images of the object. A symmetry is therefore a mathematical operation that leaves a shape globally invariant.

Let us draw a circle on a page, then trace a circle of the same radius on a transparent page and superpose the two figures. There are an infinite number of ways to place the tracing paper so that the two circles remain identical: the circle is symmetric with respect to all permutations on both sides of all its diameters. In geometry, this property is formalized by saying that the circle is invariant under all angular rotations whatsoever around its center.

We intuitively feel that the circle possesses a sort of perfect symmetry. It is clear that other geometric figures possess regularities, while being less symmetric than the circle. Let us, for example, trace two equilateral triangles on our two pages; there are only six different ways of moving the tracing paper so that the two figures coincide: three rotations of 120 degrees each, or three flips of the page with respect to the bisectors (Figure 40.2).

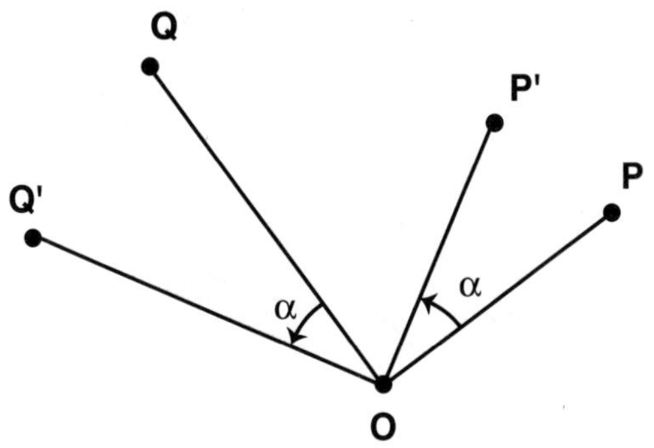

Figure 40.1. Rotation. A rotation by an angle α in the plane around a point O transports a point P into a point P' such that $OP = OP'$, and the angle formed by OP and OP' is equal to α.

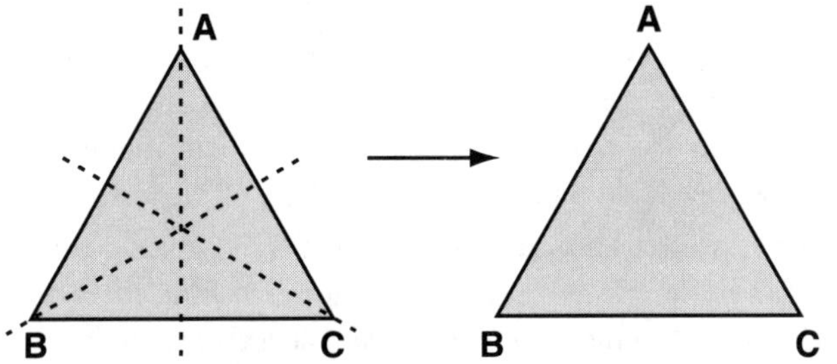

Figure 40.2. Symmetries of an equilateral triangle. Six elementary operations leave an equilateral triangle invariant: three rotations of $120°$ around the center, and three reflections with respect to the bisecting lines (dotted lines).

In the same way, a square is only invariant under rotations by a quarter turn, the pentagon under rotations by a fifth of a turn, and so forth, as well as by certain flipping operations called *reflections*. A reflection is the geometric transformation that, at a point, makes a correspondence to another point situated exactly on the other side of a straight line called the axis of symmetry, at the same distance and in a perpendicular direction. The concrete experiments that we have described with mirrors of course come from this operation of reflection, with the mirror playing 10 →
the role of the axis of symmetry.

The shape of a planar figure or of an object of finite extent, for example the leaf of a tree, can be completely explored in its symmetrical elements solely with

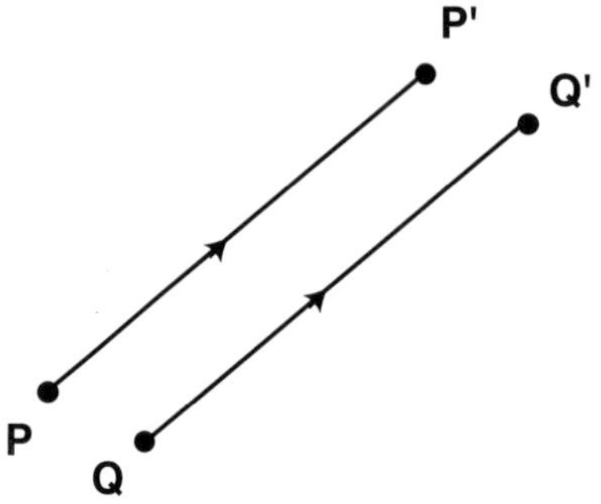

Figure 40.3. Translation. A translation by a length L in the plane transports the points P and Q into P' and Q', such that $PP' = QQ' = L$.

the help of reflections and rotations. On the other hand, the plane in its entirety presents other types of symmetry, which manifest themselves as soon as one tries to cut it into patterns repeated out to infinity. These pavings, or tessellations, are found everywhere in decorative art: e.g., tapestries, floors, mosaics, and so on. To analyze them, we need additional symmetries. One of these is *translation.*

A translation is a rectilinear displacement taken in a given direction for a given distance. It is a very simple transformation, which has little effect on objects: if all the points in a figure are displaced by the same length and in the same direction, the figure remains identical to itself; it has simply glided a bit further on, as if on rails.

Combinations of symmetry operations are also interesting. To begin with, it is easy to compose symmetries of the same nature. Two combined rotations give a new rotation, whose angle is equal to the sum of the angles of the two original rotations. Two successive translations are equivalent to a translation whose length is the sum of the lengths of the two translations. In a kaleidoscope formed from mirrors with multiple orientations, the images result from a combination of reflections associated to the various mirrors.

This property, according to which the combination of two symmetries gives a symmetry of the same nature, characterizes a fundamental mathematical structure, the *group* structure. The theory of groups is nothing other than the mathematical formalization that allows one to classify symmetries in spaces of an arbitrary number of dimensions. In this regard, as Henri Poincaré stated, “the theory of groups is, so to speak, the entirety of mathematics stripped of its matter and reduced to a pure form.”

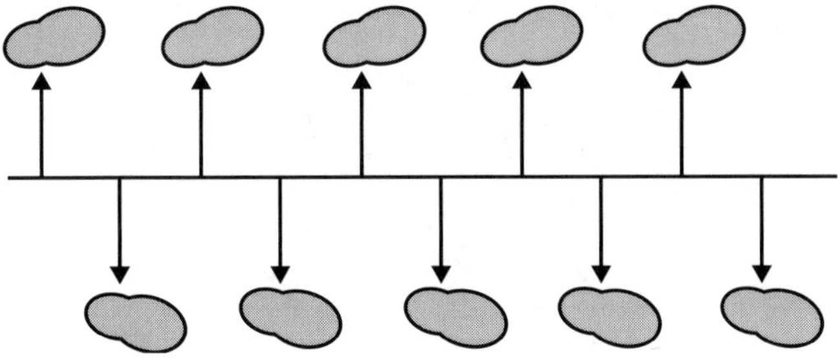

Figure 40.4. Glide reflection. A glide reflection is the combination of a translation and a reflection.

It is just as interesting to combine symmetries of different natures. For example, a *glide reflection* is the result of a reflection with respect to a given straight line, followed by a translation along the same line. Footprints in the snow, if they are truly regular, present this type of symmetry around the central path (Figure 40.4).

These four symmetry operations in the plane—rotations, reflections, translations, and glide reflections—possess an important common property: they preserve distances. In other words, any given line segment, if it is transformed by means of any one of these operations or any of their combinations, keeps the same length. Because of this, the term *isometry* (from the Greek *isos*, meaning equal, and *metron*, meaning measure) is given to these transformations. Since the procedure that allows one to measure the distance between any two points, called the metric, is one of the essential characteristics of geometry, the isometries play a privileged role in the study of the global forms of spaces and of their topologies.[1]

The classification of isometry groups allows in particular for a listing of all regular tilings of the plane starting from a fundamental cell, on which one performs various translations, rotations, reflections, glide reflections, and their possible combinations. For example, by transposing a parallelogram in two different directions by way of two independent translations, we cover the plane with an infinite network, with each parallelogram capable of being considered as the fundamental cell. However, the choice of the cell is not unique. An arbitrary parallelogram can be chosen, and even shapes that are no longer parallelograms. One could pave a surface just as well with squares as with hexagonal tiles.

[1]An example of a planar symmetry that is not an isometry is homothety: a point is transformed into another by seeing its distance to a certain fixed point multiplied by a constant number. A square transformed by a homothety of factor two gives a square with edges two times as long, and thus a surface four times larger.

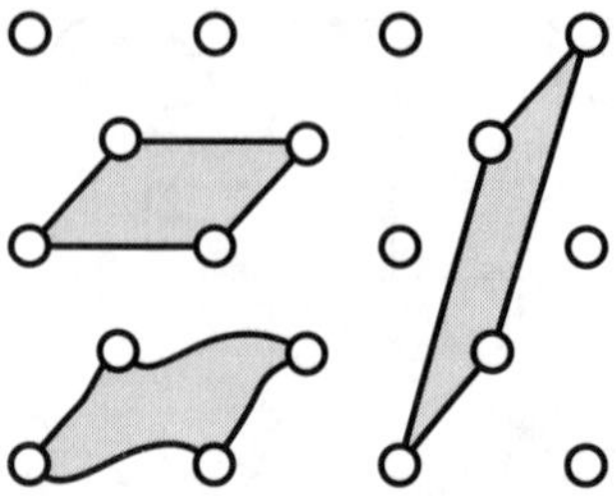

Figure 40.5. Different cells for a single network.

In 1891, the Russian mineralogist Yevgraf Stepanovich Fedorov demonstrated that the number of symmetry groups allowing one to regularly tile the plane is equal to 17. In 1922, the archeologist Andreas Speiser remarked that these 17 groups were discovered empirically 4000 years earlier in decorative art. While studying Greek weavings, the pavings of Egyptian temples, and the mosaics of the Alhambra in Granada, Spain, he noticed that they were composed of identical patterns, combined by simple or composite symmetries, with all possible operations reducing to the 17 groups identified by Fedorov. The innumerable variety of planar decorations can therefore be reduced to an exhaustive mathematical description.

The freedom of choice in selecting the fundamental cell allows one to tile the entire plane with patterns of various shapes. This precept was applied by the Dutch engraver Maurits Cornelis Escher. In 1936, the young artist traveled to the Alhambra of Granada, where he was fascinated by the Moorish tilings. Soon after this visit, he read a popularizing article that the Hungarian mathematician George Pólya had published in 1924 on the symmetry groups of the plane. Without understanding the abstract aspect, Escher was able to extract the 17 symmetry groups that were described there. Between 1936 and 1941, he applied his new knowledge in an impressive series of engravings presenting all possible periodic tilings. Taking the opposite approach from Islamic art, which had to confine itself to purely geometric patterns, Escher used animal or human forms: butterflies, birds, fish, lizards, and imps. He entered into contact with renowned mathematicians like Donald Coxeter and Roger Penrose, and worked in collaboration with them. By introducing colored patterns into his engravings—a supplementary dimension that was not taken into account in Fedorov's classification—Escher opened up a new field of geometry, the theory of polychromatic symmetry groups, subsequently studied by Coxeter.[2]

[2]See, for example, [Ernst 76, Escher and Locher 71].

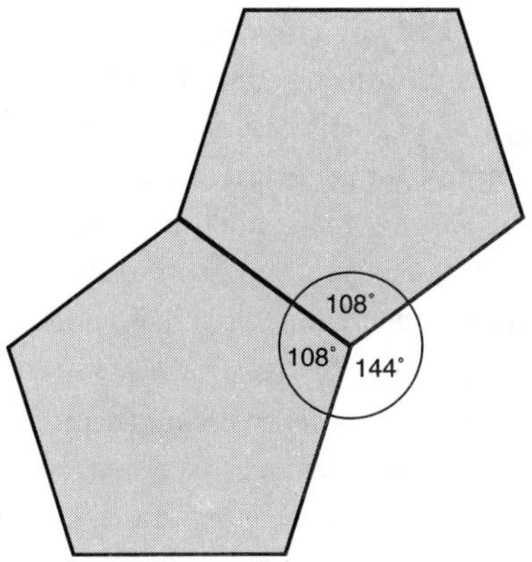

Figure 40.6. The impossibility of a pentagonal tiling.

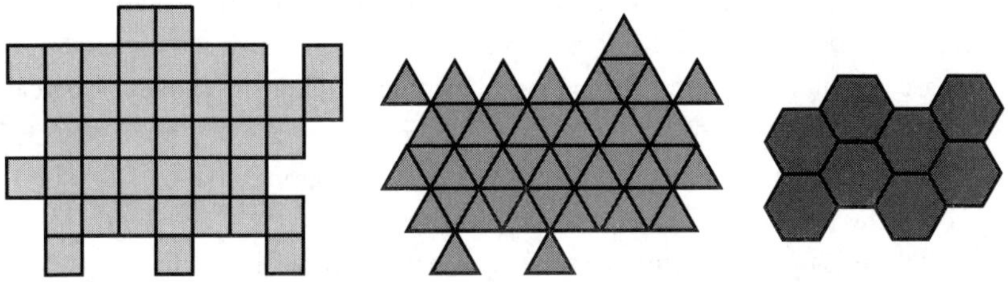

Figure 40.7. The three regular polygonal tilings of the plane.

One cannot produce a tiling with pentagons. However, semi-regular tilings, made with the help of different patterns, are also a fascinating domain of study, which was formalized by the mathematical works of Roger Penrose. Here again, semi-regular tilings were first explored empirically by artists. Origami, the traditional art of paper folding so valued in Japan, is very precisely tied to semi-regular tessellations of the plane; as anyone can verify by making a paper hen, the tesserae 264→
are polygons such as triangles, rectangles, and so forth, that are not all identical.

Still working in two dimensions, but now considering non-Euclidean geometries, the classification of isometry groups and the list of regular tilings precisely determine all possible surface topologies.

What about the symmetries of three-dimensional space? The rotational symmetry is manifest: the motion of the potter's wheel or that of our planet around itself are particular examples defining our ordinary Euclidean world.

The sphere is invariant under all spatial rotations around its center, but a cube must be turned by 90 degrees (or an integer multiple of 90 degrees) for its exterior appearance to not change. The cube is one of the regular polyhedra, figures that identically reproduce themselves under given rotations whose angles can only take a finite number of values. Among the isometries of three-dimensional Euclidean space, in addition to rotations, we also find translations, reflections, and new operations that do not exist in the plane, like screw symmetry.

The symmetries that allow one to tile space in a regular fashion are made manifest in crystals, atomic arrangements reproducing themselves with a certain periodicity in space. We have seen that there are 17 symmetry groups in the plane; in three-dimensional Euclidean space, one counts 230. This implies that the thousands of different crystals that are present in nature or that can be synthesized in the laboratory have a structure that reduces to one or the other of these 230 groups, for this reason known as the crystallographic groups.

One never finds pentagonal shapes in crystals.[3] There nevertheless exist alloys of aluminum and manganese that possess pentagonal symmetries, without being able to tile space in a regular fashion. These *quasicrystals*, which have special properties, are the subject of intensive research, as much theoretical as technological.

The first crystallographer in history was Johannes Kepler. In a marvelous little treatise from 1610, *La Strena*, he remarked on the hexagonal shape of snowflakes and made the connection with the shape of certain crystals [Kepler 66]. He accounted for this by premonitory geometric reasoning, by asking himself, for example, how to pack regular solids in the most compact possible way. Nature produces these types of packings: pomegranate seeds, of rhombohedric shape, occupy the least possible space in the fruit, and bee honeycombs have a hexagonal shape that allows them to contain the greatest amount of honey. But Kepler above all caught sight of the underlying symmetry principles that preside in the ordering of the world, on all distance scales, from crystals to planetary orbits and the cosmos as a whole. Of a mystical temperament, Kepler believed that geometric symmetry was the natural language with which God expressed himself in Creation.

Symmetry is so present around us that one cannot stop oneself from thinking, following Kepler, that it occupies an important place in the explanation of the world. In antiquity, Plato was the first to want to apply general principles of symmetry to describe the general organization of the cosmos, its shape, and its content. He elected the sphere, the symmetric figure par excellence, to symbolize the cosmic architecture, supposed to reflect divine perfection and immutability. In order to complete the geometrization of the world, Plato wanted to represent the

---

[3]This is true even though pentagonal symmetry is common in the living world, e.g., in primrose or starfish.

elements (earth, water, air, and fire) by symmetric figures. He chose the regular polyhedra, which are less "perfect" than the sphere, since the elements are exposed 259→ to corruption, alteration, and change.

Fundamentally, nothing has changed in contemporary physics. The physicist geometers of today make an inventory of the symmetries of nature in order to harmonize with the structural properties of matter (the verb *harmonize* also evokes musical harmony). They still feel that the way in which diverse physical phenomena and the properties of matter express themselves can be put in correspondence with geometric symmetries. The difference is that the symmetries sought today are much subtler than invariance under translation or rotation, the only ones known to the ancients.

Symmetry's role in physics was better understood thanks to the work of Évariste Galois, in 1832, and of Emmy Noether in 1916. The latter showed how the conservation laws of certain physical quantities are connected to geometric symmetries. For example, the well-known principle of conservation of energy is a consequence of the fact that, when one examines fragments of matter at different times, they conserve exactly the same energy, in the same way that a cube is conserved when one subjects it to a rotation by an angle of 90 degrees. The difference is that the symmetry that comes into play in the conservation of energy is a temporal translation instead of a spatial rotation. The invariance of a material system under spatial rotation leads to the conservation of another physical property, called the *angular momentum*. It is this that explains why an ice skater who spins around slows down her speed by extending her arms (as the mass of the arms moves away from the axis of rotation, the speed must decrease for the amount of angular momentum to remain constant) and turns more quickly by bringing her arms close to her body. It is the same for pulsars, stars that have considerably increased their initial rotation speed by contracting in on themselves.

The concept of symmetry and of a group of transformations that leave an object invariant, whether applied to a geometric figure, a material body, or a conceptual object, has extended itself to the very structure of space-time. The principle of relativity stipulates that the laws of nature keep the same form in all inertial reference frames. In Galileo's mechanics and Einstein's special relativity, the inertial reference frames are in uniform translation with respect to each other; in general relativity, the inertial reference frames are those that fall freely in a gravitational field. The geometric operations that allow one to calculate the positions, times, speeds, and so forth, in passing from one inertial reference frame to another, are equivalent to symmetries. These form a group: the group of relativity. Symmetry is therefore omnipresent both in classical mechanics, with the Galilean group in which, for example, speeds add together, and in relativistic mechanics, with the Poincaré group.

After relativity, symmetry was extended to quantum mechanics and particle physics. Physicists use it to classify the elementary particles and understand the nature of the fundamental interactions, through so-called gauge symmetries. These gauge theories have been crowned with success, since the particles predicted with the help of such concepts have in fact been detected in the laboratory.

Today, symmetry covers the entire field of physics, to the point of becoming its foundational pillar. For 30 years, researchers have tried to unify the forces and particles that make up our material universe, in other words to find a common mathematical description. Such a *theory of everything* would take into account not only all known and unknown forms of matter, but also the four fundamental interactions, namely, gravitation, electromagnetism, and the strong and weak nuclear interactions. These unified theories are still varied: superstrings, M-theory, loop quantum gravity, and so forth, but their common basic hypothesis is that nature operates according to a collection of mathematical rules that boil down to symmetries.

Although the symmetries of nature are presently hidden in our low energy Universe, they would reveal themselves at very high temperature and can be studied in particle accelerators.

The real aim of unified theories is, in fact, twofold: it is not only the search to discover the underlying symmetries of the early Universe (at very high temperature), but also to find physical mechanisms capable of breaking these symmetries when, over the course of its expansion, the Universe descended to low energy. After all, we live in a Universe that has become complex, filled with particles and interactions so diverse that they do not lend themselves well to an overly symmetric description. The complexity of the world can therefore be expressed through the departure from perfect symmetry. Physics studies precisely these *symmetry breakings* and shows that they play a role in nature at least as fundamental as the symmetries themselves.

It is remarkable that this approach is mirrored in art and aesthetics. Here symmetry is omnipresent, but the (subjective) notion of beauty is moreover connected to a slight break from symmetry, rather than perfect symmetry. The most beautiful faces are not exactly symmetric, and the most successful architecture mixes symmetry and surprise.

Further Reading

Hermann Weyl. *Symmetry*. Princeton, NJ: Princeton University Press, 1983.

Mark Armstrong. *Groups and Symmetry*, 2nd edition. Berlin: Springer-Verlag, 1997.

At twenty years of age, upon learning that there were only five perfect regular polyhedra that could be inscribed in a sphere, five and not one more, I told myself that the universe could only be limited.

—Salvador Dali

41 Polyhedra

A regular polyhedron is a solid figure in space whose faces are regular and identical polygons, and which is constructed in such a way that two faces and two faces alone meet at every edge. Thus, an ant that would like to pass from one face to another only needs to cross one single edge; any shortcut passing through the interior of the polyhedron is forbidden.

All in all, there are only five regular polyhedra. These are:

- the tetrahedron, more commonly called the pyramid, whose four faces are equilateral triangles;
- the hexahedron, more commonly called the cube, whose six faces are squares;
- the octahedron, with eight equilateral triangle faces;
- the icosahedron, with twenty equilateral triangle faces; and
- the dodecahedron, with twelve pentagonal faces.

The five regular polyhedra have been known since antiquity. After having been studied by Theaetetus and by Plato, they formed the crowning achievement of Euclid's *Elements*. The scholium of Proposition 18 of the thirteenth and final book shows why there can be no more than five regular polyhedra.

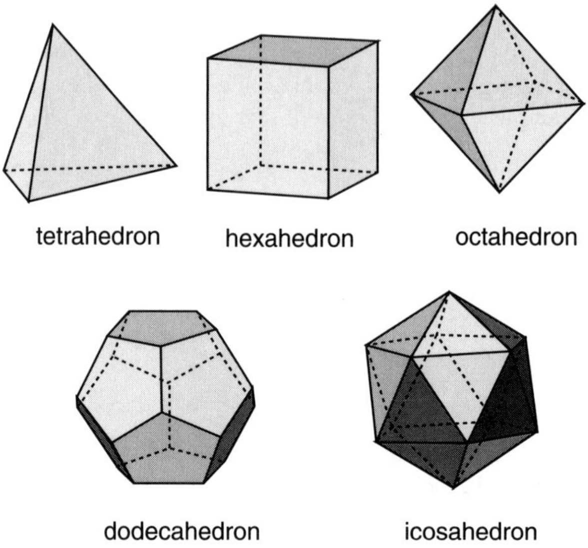

Figure 41.1. The five regular polyhedra. The regular polyhedra, invoked long ago by Plato to geometrize the "elements," serve today to represent certain multiply connected spaces, with the added condition that one considers the faces to be identified in pairs by specific geometric transformations.

Plato, who had engraved over the door of his philosophical school, the Academy, the inscription "Let no one enter who is not a geometer," used them to
←143 describe the elements (earth, water, air, and fire) in terms of pure geometry—
←250 from whence the names "Platonic bodies" and "Platonic solids" often given to the regular polyhedra.

Earth is associated to the cube, water to the icosahedron, air to the octahedron, and fire to the tetrahedron. Plato's argumentation is rather flavorful. For example, a stone in cubic shape is more difficult to put in motion than a stone having any other polyhedral shape. As a consequence, it is the heaviest and most inert figure, and in this regard, it should be associated to the terrestrial element. The icosahedron should be associated to water because, in the series of polyhedra, it is the one that possesses the greatest number of faces: five triangles meet at every point, which makes its structure seem relatively round and fluid. The rest were determined similarly.

As for the fifth regular polyhedron, the dodecahedron, Plato and Aristotle used it to complete the geometric representation of the world by putting it in correspondence with a fifth element: the quintessence.[1] With its 12 pentagonal faces, the dodecahedron is in fact the figure that is closest to the sphere, whose perfect roundness symbolizes celestial perfection.

According to Iamblichus, it was the pythagorean Hypasus of Metapontum who for the first time constructed the figure of the dodecahedron. In his *Life of Pythagoras*, he related the fact that divulging in writing how to construct a quasi-perfect sphere starting from 12 pentagons was such an impious act that Hypasus perished in the sea. Iamblichus added that Hypasus had boasted of a discovery whose merit went back to the supreme master himself, Pythagoras! The latter would have been intrigued by the shape of the pyrite crystals that were to be found in Sicily, where he lived. Pyrite presents 12 slightly irregular pentagonal faces, and this shape would have led Pythagoras to interest himself in the golden ratio, which can be deduced from the construction of regular polygons. Since then, the dodecahedron has been charged with a heavy symbolism. The number five, associated to the number of sides of its polygonal faces, plays a particular role in occultism: the pentacle or five-pointed star, inside of which are inscribed letters, words, and signs, is supposed to translate a universal structure. The number 12, which is the number of faces, put it naturally in correspondence with the 12 signs of the zodiac, the 12 months of the year, the 12 apostles, and so forth. Bronze artifacts of gallo-roman origin have been found in dodecahedral form.

Numerous artists have drawn polyhedra, including Piero della Francesca, Leonardo da Vinci, Dürer, and Escher (whose brother was a professor of crystallography at Leiden University). Salvador Dali, in particular, painted the impressive *Sacrament of the Last Supper*, which takes place in a room of dodecahedral shape, surrounded with large pentagonal bay windows. But it is above all in cosmology, in crystallography, and in topology that the polyhedra have revealed all of their explanatory potential.

In 1596, in his *Mysterium Cosmographicum* (*The Secret of the Universe*), Kepler interested himself in the regular polyhedra in order to resolve, in quite an original way, the problem of the relations between planetary orbits. According to him, the five Platonic solids should correspond exactly to the five intervals between the six known planets: Mercury, Venus, Earth, Mars, Jupiter, and Saturn. He showed mathematically that the Platonic solids fit within each other in a unique way, into an architecture reproducing that of the solar system.

When, following the new astronomical data of Tycho Brahe, Kepler had to abandon this model in favor of ellipses in order to describe the shape of planetary orbits, he lost none of his fascination for these nearly perfect figures. He closely

[1]In *Epinomis*, a later text than *Timaeus*, Plato calls it *ether*, whose etymology signifies "always moving," like the heavens.

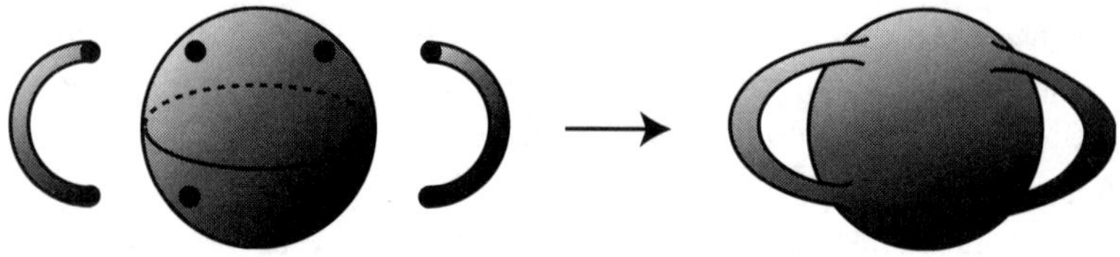

Figure 41.2. The multiply handled sphere. The Euler-Lhuilier formula applies to all closed surfaces of genus H, equivalent to a sphere equipped with H handles.

studied the semi-regular convex polyhedra, known as Archimedean polyhedra since they were studied by the geometer from Syracuse (rhomboids, prisms, antiprisms, and so on). Then, by abandoning the hypothesis of convexity, Kepler constructed the first *stellated* polyhedra, made of assemblages of multi-branched ←250 stars. He used them to model the structure of snow crystals and lay down the basics of crystallography.[2]

The polyhedra, regular or not, finally played an essential role in the development of topology. In 1750, in a letter addressed to his colleague Christian Goldbach, Leonhard Euler cited a certain number of properties of general polyhedra. Among them was a formula of extremely simple appearance, which however ended up revealing itself to have considerable depth, $F - E + V = 2$, where F is the number of faces, E the number of edges, and V the number of vertices of the polyhedron.

For example, a cube has six faces, twelve edges, and eight vertices. One can verify without too much effort that $6 - 12 + 8 = 2$. It is the same with the other regular polyhedra: the tetrahedron (4, 6, 4), the octahedron (8, 12, 6), the dodecahedron (12, 30, 20), and the icosahedron (20, 30, 12). However, the formula is true for any type of polyhedron whatever its shape or size may be! After resolving the problem of the Seven Bridges of Königsberg, Euler thus reached a new stage in the progressive liberation from the notion of distance in geometry.

What happens if one digs one or more holes in a polyhedron? Simon Lhuilier reflected on this strange question and demonstrated, in 1813, that Euler's formula could be generalized to an arbitrary polyhedral figure, either pierced or not: $F - E + V = 2 - 2H$, where H is the number of holes. This number as well appears as a topological invariant, independent of the size of the polyhedron. It can also be defined for any type of closed surface, and it is then called the *genus*. The genus of the torus is 1, that of a sphere is zero, that of a sphere equipped with H handles is H.

[2]In 1809, the mathematician Louis Ponsot, unaware of Kepler's results, rediscovered the regular stellated polyhedra and made a complete classification of them.

Another topological invariant is the dimensionality of space. It is only in three dimensions that the number of regular polyhedra is equal to five. In the two-dimensional plane, one can obviously construct as many regular "polyhedra" (meaning regular polygons) as one wants; it suffices to increase the number of sides.

Mathematicians, always curious, worked hard to see what became of the regular polyhedra in spaces of higher dimension. They found, for example, that in a space of dimension four, one could construct six regular polyhedra; one of them, called the *hypercube*, is a solid figure bounded by eight cubes, in the same way that a cube in ordinary space is bounded by six square faces. In dimension five, one can imagine the ultracube, bounded this time by a hypercube on each of its ten faces! But the most talented geometer living in a five-dimensional universe could never assemble more than three regular polyhedra: the icosahedron and the dodecahedron have no equivalent in five-dimensional space.

The number of regular polyhedra therefore appears as a characteristic of the dimensionality of the space within which these polyhedra are embedded. Some even see this as a rigorous demonstration of the three-dimensional nature of our space, because in real space one can in fact construct the five regular polyhedra in stone, wood, or paper!

Speak to what stays silent
Under the surface

—Michel Cassé, *Varech primordial*

42
The Classification of Surfaces

I shall here use the word *surface* to mean a space of two dimensions that has no border. A disk or a square, which have borders, are therefore not surfaces. In general, a surface possesses a curvature that varies from one point to the other (Figure 42.1). When a surface has no cusps, meaning points where the curvature becomes infinite, it has the property of being homogenizable. This means that a certain change of coordinates will transform its metric of variable curvature to a metric of constant curvature.

The complete classification of homogenizable surfaces therefore reduces to the classification of surfaces of constant curvature. The three cases to envisage are Euclidean surfaces (zero curvature), spherical surfaces (positive curvature), and hyperbolic surfaces (negative curvature).

Euclidean Surfaces

Among the Euclidean surfaces, only one is simply connected: the usual plane E^2. The others, which are multiply connected, are characterized by their fundamental polygon and their holonomy group. The fundamental polygon is therefore either a *digon* (a polygon of two sides that only meet at infinity, meaning a band

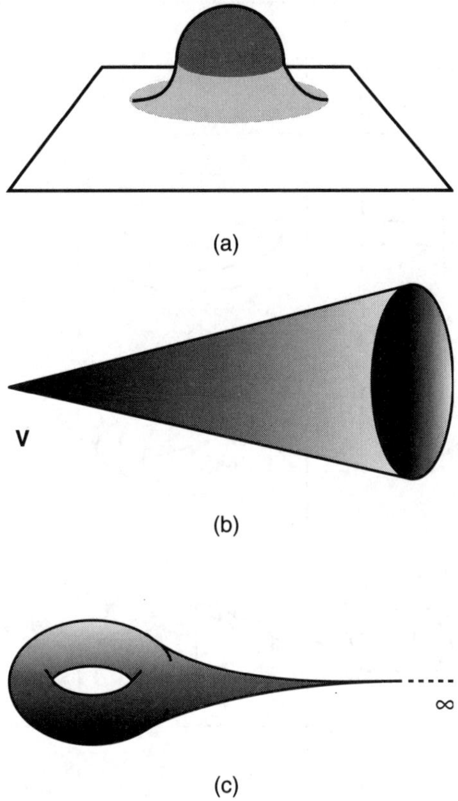

Figure 42.1. Three surfaces of variable curvature. (a) Homogenizable surface of varying curvature. A plane supplemented with a bump is split into three regions, one of zero curvature (in white), one of negative curvature (in light gray), and another of positive curvature (in dark gray). It is possible to stretch this surface continuously and flatten it completely, that is to say, to give it a metric with zero curvature everywhere. In a certain sense, in a lumpy plain, the regions of positive curvature compensate the regions of negative curvature. (b) Conic surface. This surface has a curvature that is zero everywhere except at one point (the vertex V of the cone), where the curvature is infinite. It is not homogenizable. (c) Surface with a cusp. This nearly toric surface, although of finite area, possesses a cusp extending out to infinity. It is not homogenizable.

with parallel edges of infinite length), a quadrilateral, or a hexagon. As for the holonomies (possible point identifications), these are either translations or glide reflections. The number of different possible combinations turns out to be very few: there are all in all four multiply connected Euclidean surfaces (Figure 42.2).

name	fundamental polygon and identifications	shape	closed	orientable
cylinder		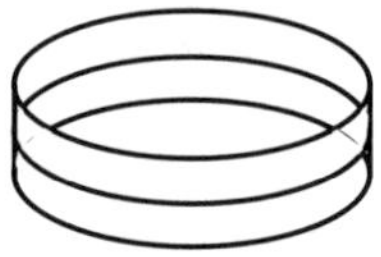	no	yes
Möbius band		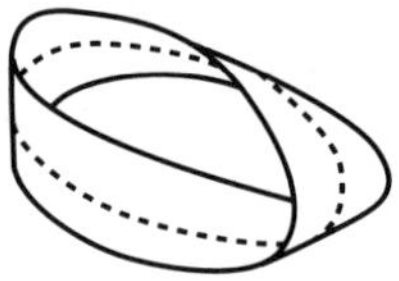	no	no
torus	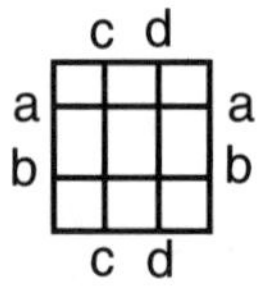	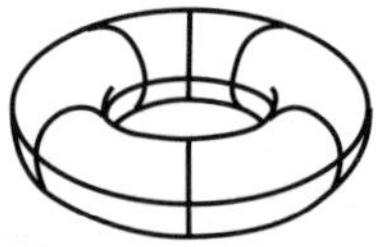	yes	yes
Klein bottle	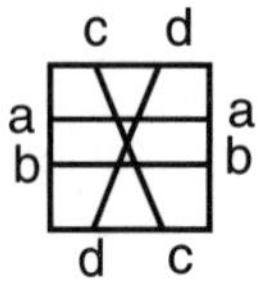	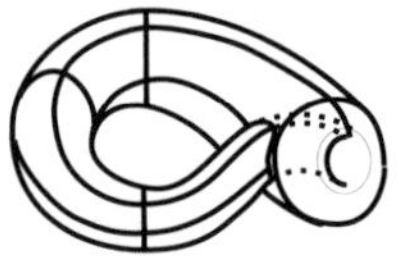	yes	no

Figure 42.2. The four multiply connected Euclidean surfaces. The second column shows the shape of the fundamental polygon and possible point identifications, the fourth column indicates the closed (finite area) or open (infinite area) character of the surface, the last column whether it is orientable or not. When the fundamental polygon is a digon (as for the cylinder and the Möbius band), the surface is open; if not, it is closed.

It is easy to obtain finite sections of the cylinder or Möbius band (Figure 42.3) starting from a piece of paper. On the other hand, it is impossible, even with ←62 elastic material, to construct a Euclidean (or *flat*) torus T^2 or a Klein bottle K^2 (Figure 42.4).

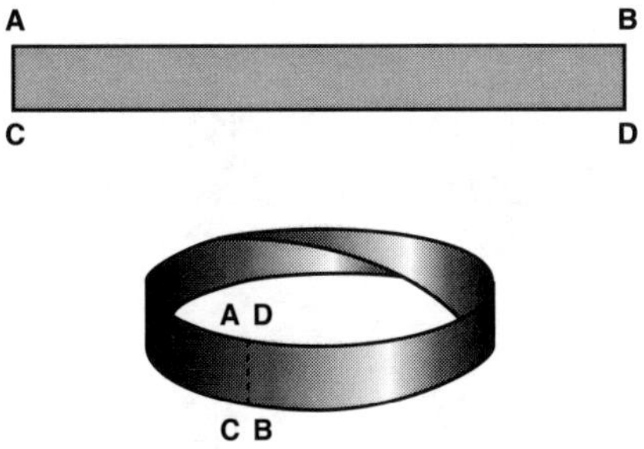

Figure 42.3. Construction of a Möbius strip. The Möbius band is of infinite size, but one can remove a section of finite size called the "Möbius strip." Discovered in 1858 by Johann Benedict Listing and, independently, by August Ferdinand Möbius in 1865, the Möbius strip can be easily constructed in three-dimensional space: it suffices to cut a long strip of paper and to glue one of the ends, after twisting it, to the other. The Möbius band is not an orientable surface, in other words a surface that can be painted a different color on each side: this is because it is a unilateral surface, meaning one that only has a single side.

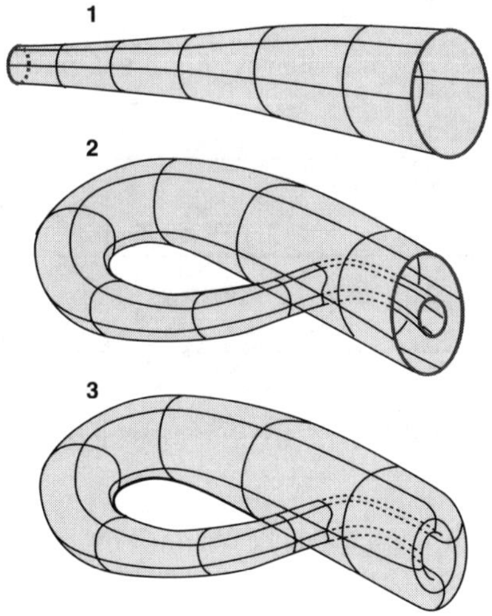

Figure 42.4. Idealized construction of a Klein bottle. The Klein bottle (discovered in 1882) can be obtained from an ordinary bottle by lengthening the neck, curving it around, making it pierce the surface (it is this crossing that makes the construction physically unrealizable), and reconnecting it inside with the bottom. Like the Möbius band, this surface is not orientable.

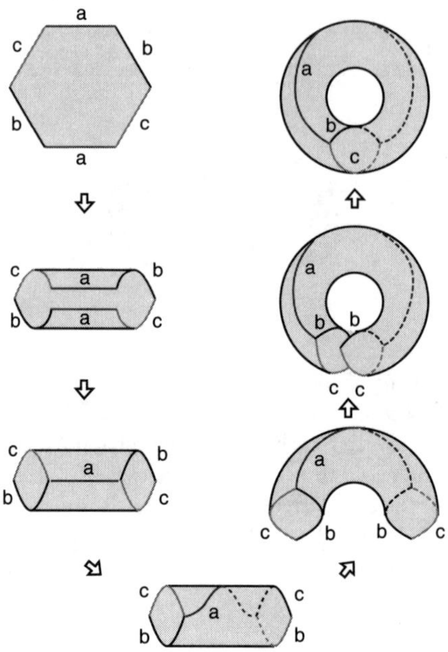

Figure 42.5. Construction of a flat torus from a hexagon.

sphere $\mathbf{S}^2$		simply connected	closed	orientable
projective plane $\mathbf{P}^2$	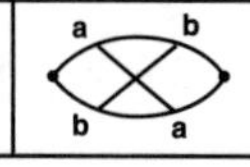	multiply connected	closed	non-orientable

Figure 42.6. The two surfaces of constant positive curvature.

The flat torus T^2 can alternatively be obtained by identifying the parallel sides of either a square or a regular hexagon (Figure 42.5). These two equivalent de-
←250 scriptions are intimately connected to the two possibilities for tiling the Euclidean plane by squares or by hexagons.

Spherical Surfaces

Among the two-dimensional spaces of positive curvature, only the sphere S^2 is simply connected. From the topological point of view, it is representable by a closed digon (Figure 42.6).

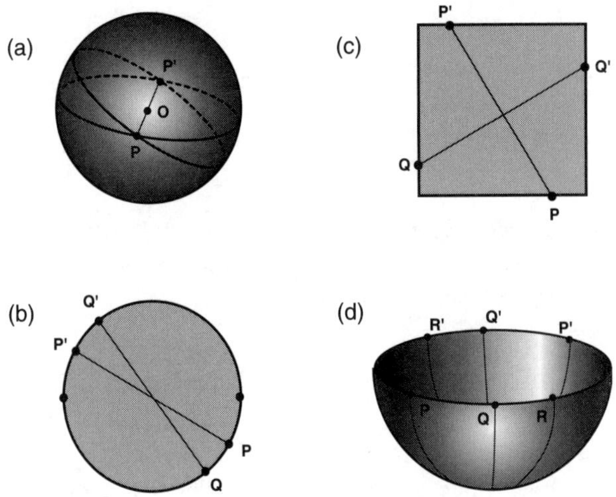

Figure 42.7. Different constructions of the projective plane. (a) When the antipodes P and P' (diametrically opposite points on the sphere) are considered as one and the same point, one obtains the projective plane, equipped with an elliptic geometry. (b) The fundamental polygon of the projective plane is a closed digon—for example, two semicircles placed side by side—whose diametrically opposite points are identified. (c) An equivalent representation is that of a quadrilateral whose opposite sides are identified with the help of two glide reflections (translation followed by a rotation of 180 degrees). (d) The projective plane can also be represented by a hemisphere. The points P, Q, and R are identified with the points P', Q', and R'.

This map of spherical space is familiar to us through world maps. When, on a map of the world, we follow a circumnavigation, we move along toward the west, for example, until we reach the left edge of the map, nevertheless we know very well that we have not reached the end of the Earth, but a point that is represented two times on the two opposite borders of the map. The digon therefore does not really have a boundary; the two points represented on its border are completely indistinguishable from the other points.

There is only one multiply connected variant of spherical space: the projective plane (also called the elliptic plane) P^2. It is obtained by identifying, two by two, diametrically opposite points of the sphere. Like the sphere, the projective plane is finite and has no boundary; on the other hand, it is not orientable. P^2 can be constructed in multiple ways[1] (Figure 42.7).

[1] Among these representations, *Boy's surface* allows one to visualize the projective plane P^2 embedded in the Euclidean space E^2; like the Klein bottle, this representation necessarily leads to self-intersections.

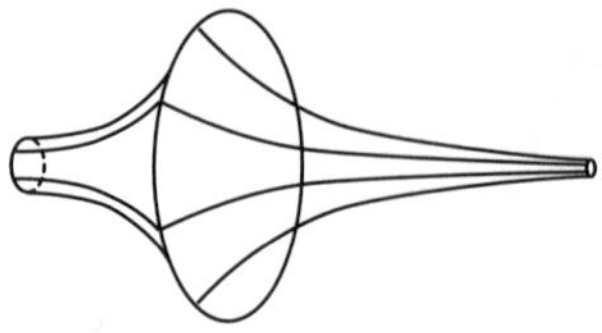

Figure 42.8. The pseudo-sphere. This surface, also called a tractroid, is a model of Lobachevsky's hyperbolic plane: it has a constant negative curvature, and there are an infinite number of parallels to a given "straight line."

In projective space, a straight line is closed, but two straight lines only intersect at one point (while on the sphere they intersect at two points). By tracing along one of the lines, one passes from one side to the other of the other line without meeting it. The straight line is therefore a closed line that does not, however, cut the projective plane into two pieces.

Hyperbolic Surfaces

The surfaces of negative curvature admit as a simply connected prototype the hyperbolic plane (or Lobachevsky plane) H^2. Like the Euclidean plane, the Lobachevsky plane is an open surface: if one moves in a straight line while following the contours of the surface, one never returns to the starting point, but moves out to infinity.

As soon as one adds a handle to the Lobachevsky plane, one obtains a new open hyperbolic surface, which is now multiply connected. One can equally well construct multiply connected closed hyperbolic surfaces, by starting with a fundamental polygon of least eight sides, and by identifying the sides in pairs, with the aid of holonomies forming a subset of the transformations that leave the hyperbolic plane invariant. Since a regular polygon can have an arbitrarily large number of sides, one immediately deduces that there are an infinite number of closed hyperbolic surfaces: the pretzels. The simplest of these closed hyperbolic surfaces is the double torus; it has two holes, and its topology is represented by an octagon, ←62 or by the connected sum of two tori $T^2 \# T^2$ (where the symbol # designates the connected sum). More generally, the n-torus—the connected sum of n tori—is a hyperbolic pretzel with n holes, representable by a polygon with $4n$ sides.

Now we have classified all homogeneous surfaces. Five are Euclidean, two are spherical, and all the others, which are infinite in number, are hyperbolic.

If one is interested only in closed surfaces, the classification is even simpler—it reduces to playing Legos with three basic elements: the sphere S^2, the torus T^2, and the projective plane P^2. To make a connected sum of two or more of these

	0	1	2	3	...
0	S^2	P^2	$P^2 \# P^2$	$P^2 \# P^2 \# P^2$	...
1	T^2	$T^2 \# P^2$	$T^2 \# P^2 \# P^2$	$T^2 \# P^2 \# P^2 \# P^2$	...
2	$T^2 \# T^2$	$T^2 \# T^2 \# P^2$	$T^2 \# T^2 \# P^2 \# P^2$	...	...
3	$T^2 \# T^2 \# T^2$	$T^2 \# T^2 \# T^2 \# P^2$	...	...	...
...	...	...	...	...	...

Figure 42.9. Construction of closed surfaces. The first row counts the number of projective planes and the first column the number of tori that participate in the construction of the surface.

positive curvature	S^2	P^2	
zero curvature		$P^2 \# P^2$	T^2
negative curvature		$P^2 \# P^2 \# P^2$ $P^2 \# P^2 \# P^2 \# P^2$ ...	$T^2 \# T^2$ $T^2 \# T^2 \# T^2$ ...

Figure 42.10. Complete classification of closed surfaces.

elements amounts to stacking Lego pieces. For example, the Klein bottle reduces to the connected sum of two projective planes ($K^2 = P^2 \# P^2$). The sphere plays a particularly important role, that of the neutral element. In fact, the connected sum of two spheres remains a sphere ($S^2 \# S^2 = S^2$), the connected sum of a sphere and a torus remains a torus ($S^2 \# T^2 = T^2$), and so forth. Figure 42.9 gives the list of all possible closed surfaces, obtained by combining the particular number of projective planes and tori that participate in their construction.

However, this list is highly redundant since, with the exception of the elements in the first column and the first row, it contains duplicated elements; for example, $T^2 \# P^2 = P^2 \# P^2 \# P^2$.

A complete list, without duplications, of all closed surfaces finally reduces to:

- the sphere;
- the projective plane;
- the torus;
- all connected sums of projective planes;
- and all connected sums of tori.

Further Reading

Jeffrey Weeks. *The Shape of Space*, 2nd edition. New York: Marcel Dekker, 2001.

And drink, like a pure and divine liquor
The clear fire which fills the limpid spaces

—Charles Baudelaire, *Elévation*

43 The Classification of Three-Dimensional Spaces

The passage from two dimensions (the classification of surfaces) to three dimensions (the classification of spaces that are likely to be used as geometric models for physical space) in no way reduces to a simple generalization, but leads to the appearance of radically new properties. ←264 We have seen that every regular surface can be homogenized so as to be described by a metric of constant curvature. This means that there are only three prototypical simply connected surfaces (which serve as universal coverings), to which all other surfaces are necessarily related. Things are not the same for three-dimensional spaces: there are eight possible universal covering spaces. Only three of these are homogeneous and isotropic; ←38 the remaining five are homogeneous but not isotropic, meaning that at a given point the measurement of the curvature depends on direction.

Three-dimensional cylinders are some relatively simple examples of these. In the same way that the usual cylinder can be considered as the product $S^1 \times E$ of a circle S^1 and a straight line E (in the sense that if one slides a circle along a straight line perpendicular to its center one creates a cylinder), the three-dimensional spherical cylinder can be pictured as the product $S^2 \times E$ of a sphere S^2 and a straight line E. However, while the cylindrical surface could be described with the metric of the Euclidean plane E, the cylindrical-spherical space is fundamentally distinct from the Euclidean space E^3. The curvatures measured are different depending on the orientations of the referential planes used to cut it. Similarly, the cylindrical-hyperbolic space $H^2 \times E$, obtained by stacking Lobachevsky planes, is fundamentally distinct from E^3.

The aim of this book being to apply the mathematical theory of spaces to the description of the physical universe, and the latter seeming remarkably isotropic, I shall here restrict myself (and in a very succinct manner, given the complexity of the subject) to the classification of homogeneous and isotropic spaces. These alone are capable of describing the spatial geometry of big-bang models. They are all related to the three universal covering spaces of constant curvature: S^3 (positive curvature), E^3 (zero curvature), and H^3 (negative curvature).

Euclidean Spaces (Zero Curvature)

Simply-connected Euclidean space, E^3, with uniformly zero curvature, is infinite in every direction. Its multiply connected variants are characterized by their fundamental polyhedra and their holonomy groups. The fundamental polyhedron is either a finite or infinite parallelepiped, or a prism with a hexagonal base, corresponding to the two ways of tiling Euclidean space (see Figure 14.3). The holonomies (possible point identifications) are either translations, glide reflections, or screwings (combinations of a rotation and a translation parallel to the axis of rotation). The various different combinations reduce to 17 distinct multiply connected Euclidean spaces.

Seven of these spaces are open (of infinite volume); see Figure 43.1. Two of these, called *slab spaces*, are infinite in two directions and "compactified" in the third. The five others, called *chimney spaces*, are infinite in a single direction and compactified in the two others.

Ten other spaces are closed (of finite volume); see Figure 43.2. Among these, six are orientable. It is these six spaces that present a particular interest for cos- 195→
mology, since they can perfectly model the spatial part ofthe Einstein-de Sitter universes. One can in particular distinguish the hypertorus T^3, constructed by 73→
identifying the opposite faces of a parallelepiped by translations.

Spherical Spaces (Positive Curvature)

The simply connected spherical space S^3, with positive curvature, is the hypersphere. Einstein attempted to give an intuitive image of such a finite yet limitless

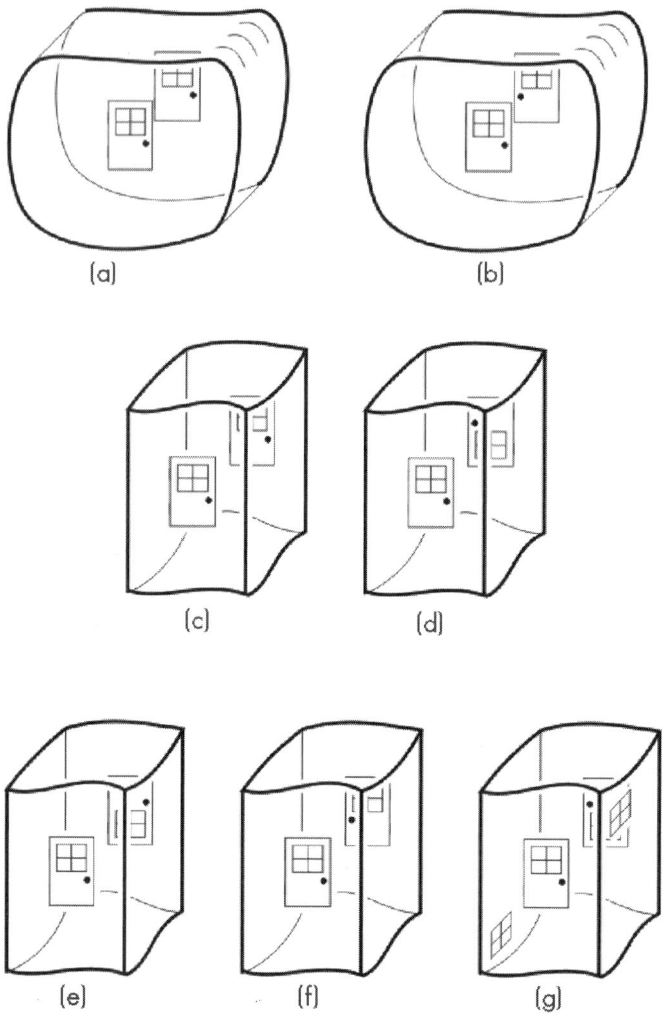

Figure 43.1. The seven multiply-connected infinite Euclidean spaces. The slab space (a) is made of a slab that extends infinitely in two directions, but has finite thickness. The two ends are identified by a translation. Its variant, (b), identifies the walls after a rotation of 180°; shown by the orientation of the doors. The chimney space (c) is made of a rectangular chimney of infinite height, whose front and back (and left and right) surfaces are identified by a translation. The four variants (d)–(g) of the chimney space identify the front and back faces in the way indicated by the orientation of the doors. The left and right surfaces are directly connected, except in case (g), where they are glued after a rotation. (Image courtesy of Adam Weeks Marano.)

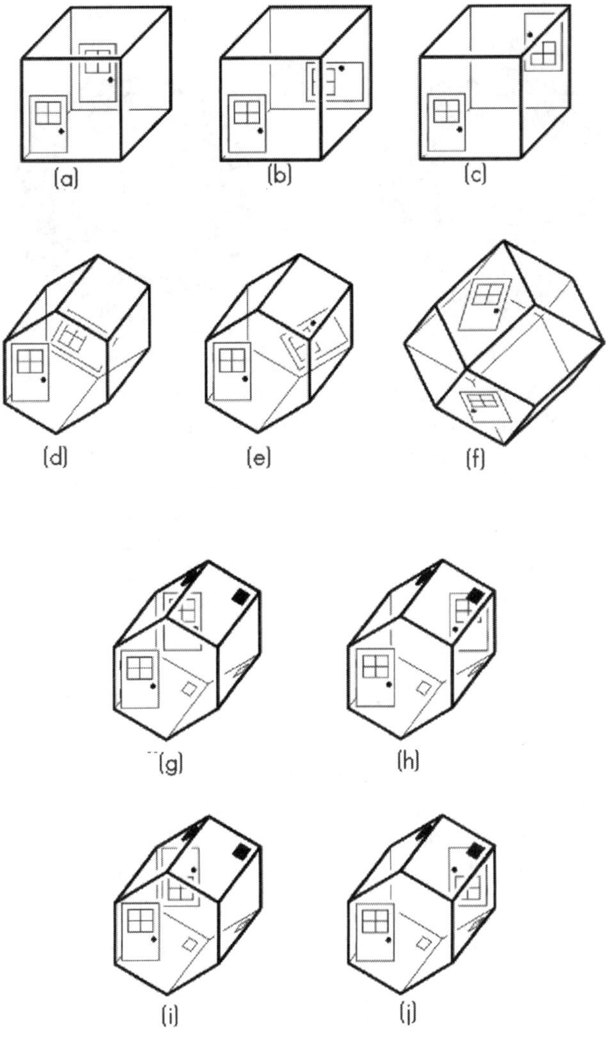

Figure 43.2. The ten multiply-connected finite Euclidean spaces. The unmarked pairs of walls are glued by simple translations. The others are glued with the orientations shown by the doors. The first six spaces are orientable hypertori. The other four are three-dimensional, non-orientable generalizations of the Klein bottle. The representations are as follows: (a) hypertorus, (b) hypertorus with a quarter turn, (c) hypertorus with a half-turn, (d) hypertorus with a one-sixth turn, (e) hypertorus with a one-third turn, (f) Hantzsche-Wendt space, (g) Klein space, (h) Klein space with a horizontal flip, (i) Klein space with a vertical flip, and (j) Klein space with a half-turn. (Image courtesy of Adam Weeks Marano.)

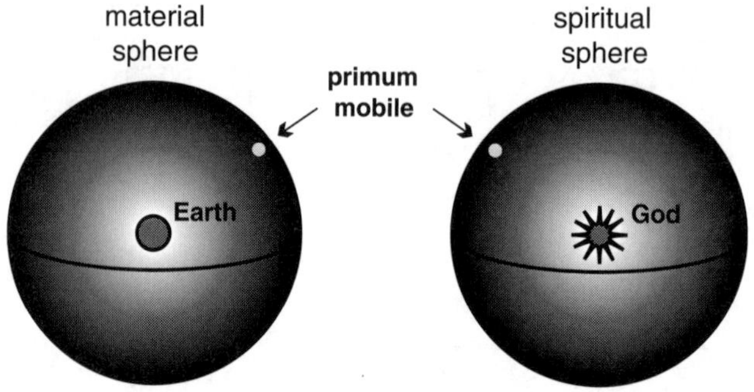

Figure 43.3. Dante's hypersphere. According to Dante, the cosmos was divided into two spheres, the material sphere and the spiritual sphere. The first starts from the center of the Earth and lifts progressively through the successive spheres of the Moon, the Sun, the planets, and the fixed stars. Beyond this sphere of the visible world is another sphere, Empyrean, the habitation of angels that gravitate like planets around a divine center. The poet explained how, at the spherical surface of the material universe (the *primum mobile*), he simultaneously sees a point of the corresponding sphere in the spiritual universe. Dante's cosmos therefore coincides with the Riemann hypersphere S^3.

three-dimensional space, that a little bit of exercise suffices to render familiar to our thinking.

There are many ways to visualize the hypersphere. One of them consists in imagining the points of the hypersphere as those of a family of two-dimensional spheres that grow in radius from 0 to a maximal value R, then shrink from R back to 0 (in the same way that a sphere can be cut into planar slices that are circles of varying radius). Another possibility is to view the hypersphere as composed of two spherical balls embedded in Euclidean space, glued along their boundaries in such a way that each point on the boundary of one ball is the same as the corresponding point on the other ball. A similar construction was imagined in the Middle Ages by Dante (1265–1321) in *The Divine Comedy* (Figure 43.3). In addition to simply-connected spherical space, there is an infinite series of multiply-connected
←73 spherical spaces, all of them closed (of finite volume) and orientable. The projective space P^3, obtained by identifying the antipodal points of the hypersphere,
←280 is the simplest of them.

The holonomies that preserve the metric of the hypersphere belong to three categories:

- the cyclic group, made up of rotations by an angle $2\pi/p$ around a given axis, where p is an arbitrary whole number;

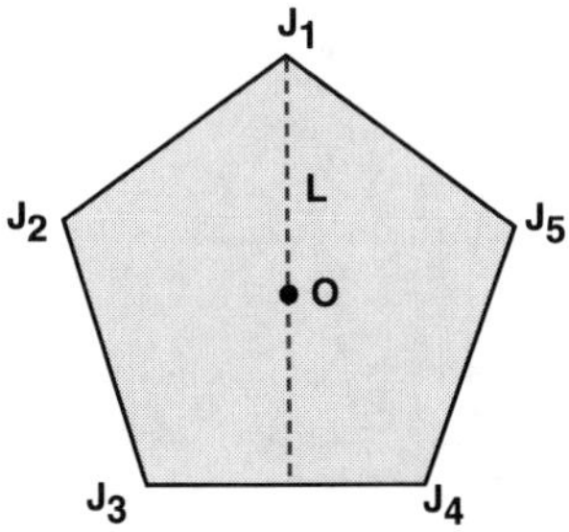

Figure 43.4. A dihedral group. The dihedral group D_5 is made up of rotations by an angle of $2\pi/5$ around the line L, which leaves the pentagon $J_1J_2J_3J_4J_5$ invariant.

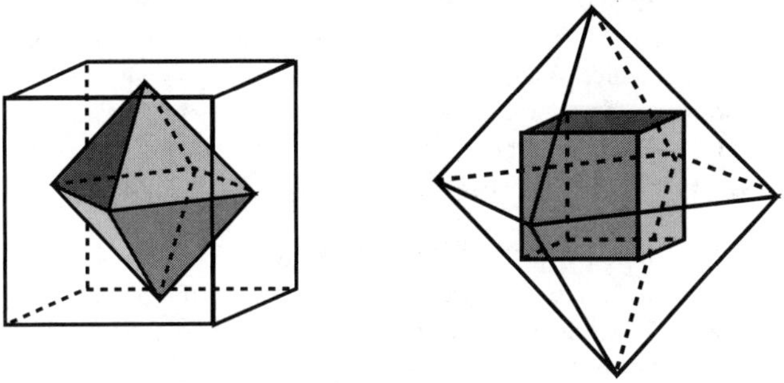

Figure 43.5. Duality of polyhedra. The cube has six faces and eight vertices, the octahedron 8 faces and 6 vertices. The cube is in fact dual to the octahedron in the sense that the centers of the faces of one become the vertices of the other, and vice versa. In the same way, the icosahedron (20 faces and 12 vertices) and the dodecahedron (12 faces and 20 vertices) are dual.

- the dihedral group, which is the symmetry group of a regular plane polygon of m sides (Figure 43.4); and
- the polyhedral groups, which preserve the shapes of the regular polyhedra. Within the polyhedral groups, the group T preserves the tetrahedron, the group O preserves the octahedron and the cube, and the group I preserves the icosahedron and the dodecahedron. There are only three distinct polyhedral groups for the five polyhedra. The reason for this is the fact that the cube and octahedron on the one hand, and the icosahedron and the dodecahedron on the other hand, are duals in the sense of Figure 43.5. As a consequence, their symmetry groups are the same.

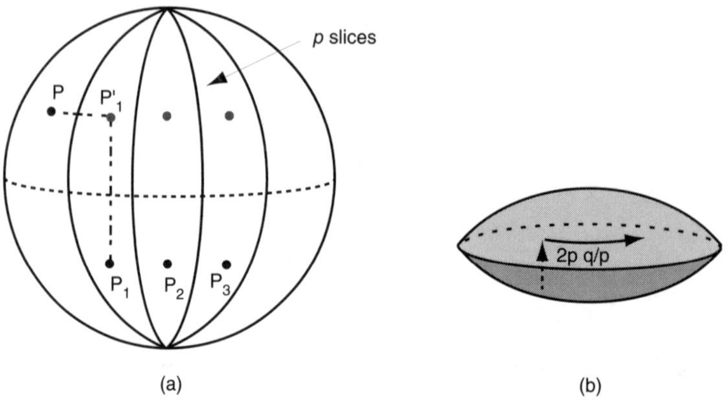

Figure 43.6. Construction of a spherical lens space $L(p, q)$. (a) We take the surface of a sphere cut into p equal slices (think of a melon) and a point P. We consider the point P_1' situated in the adjacent slice, obtained starting from P by a rotation through an angle of $2\pi/p$, and the point P_1 symmetric with P_1' with respect to the equatorial plane. We glue all the points P on the sphere to their homologous points P_1 constructed in this way, and we obtain a representation of the lens space $L(p, 1)$. If one performs the same operation by gluing the points P to their homologues P_2 situated two slices farther on, we obtain another lens space, denoted $L(p, 2)$, and so on. (b) In fact, this construction can not be done in two dimensions (there is no lens surface other than the projective plane), but in three dimensions, the lens spaces $L(p, q)$ can be obtained from a fundamental domain in the shape of a lens. They form multiply connected variants of the hypersphere S^3, with a volume p times smaller.

If one identifies the points of the hypersphere by holonomies belonging to one of these groups, the resulting space is spherical and multiply connected. There are an infinite number of these, because of the whole numbers p and m that parametrize the cyclic and dihedral groups. The larger p and m are, the smaller the volume of the corresponding spaces.

The spaces with cyclic group are called *lens spaces*, denoted thus because their fundamental polyhedra have the shapes of lenses. Figure 43.6 attempts to give a two-dimensional analog.

The spaces with dihedral group are called *prism spaces*, because of the shape of their fundamental polyhedra (Figure 43.7). Finally, the spaces with polyhedral groups are called *polyhedral spaces*. Among them, Poincaré's dodecahedral space (Figure 14.7), whose volume is 120 times smaller than that of the hypersphere with the same radius of curvature, is of particular interest for cosmology and gives rise to fascinating topological mirages.[1]

[1] See the afterword for a visualization of Poincaré's dodecahedral space.

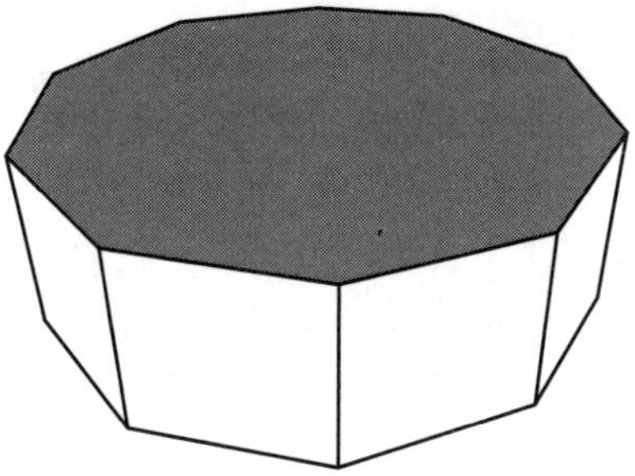

Figure 43.7. Fundamental polyhedron of a spherical prism space.

Hyperbolic Spaces (Negative Curvature)

Of all the hyperbolic spaces, only one, denoted H^3, is simply connected. It is the three-dimensional analog of the Lobachevsky plane H^2, and extends to infinity in every direction.

There is no general theorem allowing for the classification of three-dimensional hyperbolic spaces. Only some examples are known, of which some arehorn 86→ open (of infinite volume) and others closed (of finite volume). The hyperbolic is open, the Best, Seifert-Weber, and Weeks spaces are closed. 73→

All of these have interesting properties and can serve as geometric models for physical space if it should prove to be of negative curvature. Their most remarkable property comes from the theorem of rigidity. In simple terms, this means that if one fixes a hyperbolic topology, there is only a single metric compatible with this topology. From this, it follows that the volume of space is fixed by its topology—which contrasts singularly with the Euclidean case, where one can construct hypertori as small or as large as one likes (to do so, it suffices to change the size of the fundamental parallelepiped).

Thanks to this rigidity theorem, it is possible to classify the closed hyperbolic spaces by increasing volume, starting from the smallest one that is presently known, the Weeks space.

Further Reading (for informed readers)

William Thurston. *Three-Dimensional Geometry and Topology.* Princeton, NJ: Princeton University Press, 1997.

After Newton, the savants knew
How to render more and more attractive
The large
And small spaces

—André Verdet, *Hommages*

44
Topos and Cosmos

"One could imagine that as a result of enormously extended astronomical experience, the entire Universe consists of countless identical copies of our Milky Way, that the infinite space can be partitioned into cubes, each containing an exactly identical copy of our Milky Way. Would we really cling on to the assumption of infinitely many identical repetitions of the same world? In order to see how absurd this is consider the implication that we ourselves as observing subjects would have to be present in infinitely many copies. We would be much happier with the view that these repetitions are illusory, that in reality space has peculiar connection properties so that if we leave any one cube through a side, then we immediately reenter it through the opposite side" [Schwarzschild 98].

This description, that one will recognize as that of a hypertoric space, dates from almost exactly a century ago. It figures into the postscript of an article published in 1900 by the German astronomer Karl Schwarzschild (see Figure 20.1). ←105 The anticipation was all the more extraordinary in that, at the beginning of the twentieth century, the spiral nebulae had not yet been identified as galaxies entirely apart, exterior to our own.

Schwarzschild was one of those visionary astronomers who were also mathematically sophisticated. Since the foundational work of Riemann, the mathematicians of the nineteenth century had begun to discover examples of finite spaces ←161 that nevertheless had no boundaries. Among them was the hypertorus, made up of a single rectangular block of Euclidean space that is reconnected to itself, because of which everything that passes through one face reappears at a point on the opposite face. However, everyone saw this closed space as a pure abstraction, without any relation to physical space. Schwarzschild was the first to take the opposite view, to mix topos and cosmos, place and totality.

Topos and cosmos[1] have hardly mixed well throughout the development of relativistic cosmology. When the father of the discipline, Albert Einstein, introduced into his cosmological model of 1917 a three-dimensional space of positive curvature, the hypersphere, his principal motivation was to provide a model of finite space. This solution so ingeniously resolved all the paradoxes connected to infinite Newtonian space that cosmologists rapidly adhered to this new idea, at the expense of other possibilities. Einstein also thought that the hypersphere determined not only the metric of space, meaning its local geometric properties, but also its global structure, its topology; for example, the fact that the volume of space is finite and equal to $2\pi^2 R^3$, where R is the radius of the hypersphere. Now, questions about the volume, the global shape of space, and more generally of its finite or infinite nature, are not only a matter of the metric. They are above all a matter of topology. In this regard, they require a supplementary approach beyond that of Riemannian differential geometry, which serves as the mathematical basis for general relativity.

Topology was not one of Einstein's major preoccupations. In a letter to Willem de Sitter, dated March 12, 1917, he jokingly wrote: "We philosophize [whether the Universe] extends infinitely, or has a finite size and is a closed unit. Heine has provided the answer in a poem: 'And a fool waits for an answer.'" Einstein meanwhile claimed to give a definitive response in favor of a closed Universe with his hypersphere model.

Some of his colleagues, who were more fully informed about recent developments in topology, remarked to him about the arbitrary character of his choice. De Sitter, for example, noted that Einstein's cosmological solution admitted a different form of spherical space: projective space, also known as elliptic space, constructed from the hypersphere by identifying two by two all points situated at antipodes to each other. Elliptical space has the same metric as the hypersphere, 272→
but a different topology; in particular its volume is smaller by half.

De Sitter preferred the elliptic topology for a problem connected to the circumnavigation of light. The models of Einstein and de Sitter, which are both finite, were also assumed to be static, meaning that their radii of curvature remain fixed. Now, in a few billion years, a ray of light leaving a star has the time to

[1] One could also consider these terms as nicknames for topologists and cosmologists!

converge anew toward its point of departure, and create a phantom star at the position where the star was at the moment of emission.

This possibility was judged to be sufficiently paradoxical for de Sitter and, following him, other cosmologists like Arthur Eddington, to propose elliptic space in place of the hypersphere for modeling finite space. In spherical space, at each point there corresponds a single antipodal point, analogous to the pair of North and South poles on the surface of the Earth. In elliptic space, the topology is such that the maximal separation from a given point no longer corresponds to a unique point situated at the antipode, but to an infinity of points situated in a region analogous to an equator. This comes from the fact that, in spherical space, straight lines intersect twice, while in elliptic space they only intersect once, since a point and its antipode are considered as identical.

De Sitter was not the only one to react against Einstein's arbitrary choice. Two of the greatest mathematicians of the epoch, Hermann Weyl and Felix Klein, wrote to Einstein to alert him to the topological question. A supreme expert in geometry and in group theory, Weyl occupied the mathematics chair at the Zurich Polytechnic Institute—where Einstein had carried out rather mediocre studies. "My work has always sought to unite truth and beauty; but when I am forced to make the choice between truth and beauty, I always choose beauty," he would later write. As for Felix Klein, no one was more capable than he to discuss the topology of space: in 1872, at the age of only 23 years, he had presented for his academic entry into the University of Erlangen a dissertation on geometry that included the seed of a program of study aimed to unite the geometric disciplines that had until then been dispersed. This was a major moment in the history of mathematics—the origin of an integration of geometry into a unified view of mathematics founded on symmetry, transformation groups, and topology!

Einstein replied to Weyl in a letter[2] dated June 1918: "I nevertheless have an obscure feeling that makes me prefer the spherical [model]. I have the feeling that the manifolds in which every closed curve can be continuously contracted to a point are the simplest. Other people must equally well have this feeling, because otherwise, in astronomy, one would also have doubtless taken into consideration the case that our space could also be Euclidean and finite. Two-dimensional Euclidean space would then have the connective properties of an annular surface. It is a Euclidean plane in which every phenomenon is doubly periodic, where points that are found on the same periodic mesh are identified. In finite Euclidean space there would be three types of closed curves which are not continually reducible to a point. Analogously, elliptic space possesses, in contrast to spherical space, a class of curves which are not continuously reducible to a point; this is why it pleases me

[2]Einstein Archives, Princeton University.

less than spherical space. Could one prove that elliptic space is the only variant of spherical space? It seems so to me."

Einstein reaffirmed this belief in a postcard dated April 16, 1919, addressed this time to Felix Klein. The terms chosen (*obscure feeling*, *prefer*, *please*, etc.) clearly indicate that Einstein had no physical arguments that would enforce a decision in favor of the simply-connected character of space; he was guided by an aesthetic prejudice.

In his response to Weyl, Einstein was also mistaken on the final point: in three dimensions, in addition to elliptic (or projective) space, there are an infinite number of topological variants of spherical space, all closed, including in particular the 264→ so-called lens and polyhedral spaces—whereas for two-dimensional surfaces of positive curvature, there are in fact only two distinct topological types, the sphere 272→ and the elliptic plane.

However, this was not yet known. The topological classification of spaces had hardly begun. The study of different forms of Euclidean space had begun with the work of crystallography. In 1885, Fedorov had classified the 17 symmetry 250→ groups for crystalline structures, but it was only in 1934 that Werner Nowacki would demonstrate how the crystallographic groups give rise to the topologies of Euclidean space. The case of spaces of positive curvature was posed by Felix Klein in 1890 and by Wilhelm Killing in 1891, and it was known in Einstein's time under the name of the Clifford-Klein problem. This problem would not be exhaustively resolved until 1932 by William Threlfall and Herbert Seifert. As for the problem of hyperbolic spaces, their classification would not be seriously undertaken until the 1970s under the leadership of William Thurston,[3] and it is today still the subject of intensive research.

The historic relation between topology and cosmology therefore contrasts with the relation between non-Euclidean geometry and general relativity. Non-Euclidean geometry and tensor calculus in Riemannian space were well developed at the moment when physicists had need of it for modeling gravitation; on the other hand, the pioneers of relativistic cosmology who desired to take account of topology in their models for the universe did not have much to glean from the mathematical side.

We have seen elsewhere how the discovery of non-static model universes, by Friedmann in 1922 and then by Lemaître in 1927, had definitively changed the face of cosmology. What is much less known, is that their articles also contained deep discussions of the global structure of the Universe. Both quickly took notice of the incompleteness of the theory of general relativity concerning the topological question.

---

[3] A luminous synthesis is presented in [Thurston 97]. Thurston received the Fields medal in 1982.

Cosmology	Topology
1687 Newton: $\mathbf{R}^3$	
	1885 Fedorov: $\mathbf{R}^3$ groups
	1890 Clifford-Klein: $\mathbf{S}^3$ groups
1917 Einstein: $\mathbf{S}^3$	
1917 de Sitter: $\mathbf{P}^3$	
1922 Friedmann: $\mathbf{S}^3$	
1924 Friedmann: $\mathbf{H}^3$, $\mathbf{H}^3/\mathbf{G}$	
1927 Lemaitre: $\mathbf{S}^3$-$\mathbf{P}^3$, $\mathbf{H}^3$	
1929 Robertson-Walker: $\mathbf{S}^3$-$\mathbf{R}^3$-$\mathbf{H}^3$	
1931 Einstein-de Sitter: $\mathbf{R}^3$	
	1932 Threlfall-Seifert: $\mathbf{S}^3/\mathbf{G}$
	1934 Nowacki: $\mathbf{R}^3/\mathbf{G}$
1958 Lemaitre: $\mathbf{H}^3/\mathbf{G}$	
	1978 Thurston, Fomenko, Weeks . . . $\mathbf{H}^3/\mathbf{G}$

Figure 44.1. Parallel history of the topological classification of spaces and of their application to cosmology. The symbols $\mathbf{R}^3$, $\mathbf{S}^3$, and $\mathbf{H}^3$ designate respectively the simply connected Euclidean (infinite), hyperspherical (finite), and hyperbolic (infinite) spaces. The symbol $\mathbf{P}^3$ designates projective space, the simplest multiply connected variant of the hypersphere. The symbols $\mathbf{R}^3/\mathbf{G}$, $\mathbf{S}^3/\mathbf{G}$, and $\mathbf{H}^3/\mathbf{G}$ designate multiply connected variants of Euclidean, spherical, and hyperbolic spaces, obtained by identifying points by transformations of a group $\mathbf{G}$.

The discussion that Friedmann gave can be found in the article from 1924 where he discovered eternally expanding hyperbolic models for the universe. In the final paragraph, he clearly defined the fundamental limitations of the cosmological theory founded on general relativity: "Einstein's world equations, without additional assumptions, do not allow us to decide the question of the finiteness of our world." He undertook to define how space could become finite if one identified different points (which, in topological language, renders the space multiply connected). He also foresaw how this possibility allowed for the existence of phantoms, in the sense that at a single point an object and its proper images coexist. "A space of positive curvature is always finite," he added, but mathematical knowledge "does not allow us to settle the question of the finiteness of a space of negative curvature."

In contrast to Einstein, Friedmann had no prejudice in favor of a simply connected topology. Nevertheless, he believed, following the example of most physicists of the time, that only spaces of finite volume were physically admissible for describing real space. The universe models proposed by Einstein and de Sitter in 1917, and by Friedmann in 1922, all had a positive spatial curvature and therefore satisfied the criterion of finiteness. With models of negative curvature, the problem was more arduous: hyperbolic space, in its simplest version (with a simply connected topology), extends to infinity. Therefore, even at the moment of "creation" of the Universe, space was itself infinite; in other words, the Universe was not born at a point, rather the expansion started at every point of a preexisting infinite space. Aware of this difficulty, Friedmann saw a loophole in the fact that Einstein's equations were not sufficient for deciding if space was finite or infinite, even if the curvature is negative or zero. One must make additional hypotheses to specify the conditions at the limits, in particular, whether certain points in space are identified with others or not. The whole problem of cosmic topology was thus posed, but Friedmann did not have sufficient mathematical tools at his disposal to pursue the discussion.

Lemaître plainly shared the common penchant in favor of a finite space. He spoke of Riemannian geometry as that which had "dispelled the nightmare of infinite space" [Lemaître 78]. All the cosmological models he constructed starting in 1927 made the hypothesis of a space of positive curvature, which is necessarily finite, but having the topology of projective space and not that of the hypersphere. This did not stop him from recognizing the other possibilities at all. He was, it seems, the first cosmologist to remark that metrics of negative curvature admitted topologies of finite volume: "It is true that a locally hyperbolic space is not necessarily open. It is possible to construct such spaces having a finite volume. This is even true for Euclidean space" [Lemaître 58].

Unfortunately, the cosmologists of the first half of the twentieth century had no experimental means at their disposal to measure the topology of the Universe, in the same way that Gauss and Lobachevsky, in the middle of the nineteenth century, had no way of measuring the curvature of space. The majority of them therefore lost all interest in the question. In 1932, when Einstein and de Sitter proposed the perpetually expanding Euclidean model, they did away with any- 195→
thing that would go against maximal simplicity; they assumed zero curvature, a zero cosmological constant, and did not even take pains to specify that the topology was simply connected.

Nevertheless, it is this highly particular solution that served as the standard cosmological model for the 60 years that followed. Those years were therefore the dark ages of cosmic topology. Almost all cosmology texts, specialized or not, omitted any mention of the topological question, reducing the problem of the finite or infinite character of space to its curvature alone. All the efforts of observational and theoretical cosmology were concerned with the determination of the rate of expansion and the curvature of space. The two most illustrious examples were a specialized text by Steven Weinberg [Weinberg 72], which constituted a sort of bible for all the students of my generation, and George Gamow's popularization books [Gamow 93]. Through his talents as a pedagogue and humorist, Gamow planted seeds in the imaginations of two generations of young readers, to the point of converting some of them to the profession of cosmology. The flip side of the coin is that, by affirming again and again that Euclidean and hyperbolic spaces were necessarily infinite, Gamow was in part responsible for the lack of interest shown by the profession for cosmic topology!

Luckily, this diktat was not followed by everyone. Certain rebels affirmed themselves, first and foremost among them the theorists of quantum cosmology, for whom the question of the topology of the Universe was completely natural. Quantum cosmology seeks, in fact, to understand the mechanism, assumed to be quantum in nature, by which the Universe would have come into existence. Theorists of quantum cosmology therefore attempt to extend the history of the Universe, as told by general relativity and the big-bang models, back to an epoch so far back that it is certain that general relativity breaks down and a quantum generalization becomes necessary. In the 1960s, the physicist John Wheeler was the first to suggest that the topology of space-time could fluctuate at the microscopic level. Scenarios were then born where the big bang was replaced by a spontaneous birth of the Universe, starting from random quantum fluctuations. Since the ultimate ambition of these new approaches was to calculate all the properties of the Universe thus created, and to reconcile these predicted properties with observable reality, one of the properties that one could hope to predict was the topology of space. No one knew how to lead such a program to completion, but at least interest in topology within the framework of classical cosmology was revived.

In 1971, the English cosmologist George Ellis published an important article taking stock of recent mathematical developments concerning the classification of spaces and their possible application to cosmology. While he was at it, a revival of interest for the subject ensued, under the lead of theorists like the Russian Dimitri Sokoloff, the Brazilian Helio Fagundes, and the Chinese Fang Lizhi (who was moreover a political dissident who had to emigrate to the United States). An observational program was started up in the Soviet Union, under the direction of Victor Shvartsman. With the six meter diameter telescope newly installed at Zelenchuk, in the Caucasus, the phantom sources of which Friedmann had spoken in 1924, meaning multiple images of the same galaxy, were sought. All these tests failed: no ghost image of the Milky Way or of a nearby galaxy cluster was recognized. This negative result allowed for the fixing of some constraints on the minimal size of a multiply-connected space, but it hardly encouraged researchers to pursue this type of investigation. Interest again subsided.

For my own modest part, the idea of a topologically folded space began to titillate me in my student years. An excellent popular-level book by the French cosmologist Jean Heidmann [Heidmann 89], which I devoured while I was finishing a master's degree in mathematics, not only convinced me to throw myself professionally into cosmology, but also brought my attention to the extraordinary richness of possible shapes that topology could bring to the description of physical space. However, my first research was essentially dedicated to black holes—a subject that already offered original insights into the structure of space and time. I momentarily put the topological question into a drawer of my brain, contenting myself to write, in 1987, a paragraph on cosmic topology in an encyclopedia article dedicated to the geometry of the Universe [Luminet 87]. After ten years of research on black holes, these speculative objects lost a bit of their mystery in my eyes, as their real existence became more and more assured by astronomical observations. I then reopened the topology drawer and threw myself into a deeper bibliographic search, after which I was able to take notice that nearly everything remained to be done. During my stay at the prestigious astronomy department at Berkeley, in 1989–1990, my discussions about cosmic topology only gave rise to polite interest. To suggest to astrophysicists that the quasars and distant galaxy clusters they had observed for so many years were perhaps no more than phantoms was perhaps not the most diplomatic way to interest them in the subject. It was also necessary to combat the prejudice that astronomy only pushes back the observable frontiers of a Universe that, if not infinite, is at least forever inaccessible as a whole. To believe the inverse, namely that the real Universe is perhaps smaller than the observed Universe, clearly goes in opposition to common sense. But this pattern is not sufficient for the idea to be dismissed—quite the contrary, as the history of science has often shown.

It was only upon my return to France that I truly put myself to work on the question, in the company of Marc Lachièze-Rey of the Service d'astrophysique of Saclay. Doing so, we were taking a certain professional risk, since by pioneering a nearly unexplored subject, our annual rate of publication would necessarily decrease. Fortunately, the system of French research within the *Centre national de la recherche scientifique* (CNRS) research organization has the advantage of tolerating this type of situation. Outside of France, the frequency of publication governs the fate of young researchers: publish or perish. In these conditions, they are tempted to interest themselves only in minor, fashionable problems, guaranteeing rapid publications, to the detriment of profound conceptual questions. The English mathematician Andrew Wiles, who became famous in 1995 for proving Fermat's theorem, had nearly lost his academic position because he had not published anything for several years!

It was only in 1995 that our voluminous article, entitled "Cosmic Topology," saw the light of day in the American review *Physics Reports*. There we made a census of everything that researchers had published on the question since 1917 (about 50 articles in total) and indicated new research tracks that seemed interesting to us, like the method of cosmic crystallography.[4] We expected renewed interest at the international level for this manifestly neglected subject, and called upon the joint competences of mathematicians, cosmological theorists, and astronomers. The fact is that the time was ripe for this: cosmologists finally had mathematical and experimental tools at their disposal for testing the topology of the Universe, and the astronomical satellite COBE had just scrutinized the structure of the cosmic microwave background in detail, allowing us to probe the most distant past of the Universe.

By a curious irony, two American articles appeared the same year, moving in complete opposition to our hopes: their authors declared that the concept of a small wraparound universe was excluded by the data from the COBE satellite and that it was useless to work any more on the question. The articles came from the Berkeley astronomy department! My Californian "ravings" did not therefore fall completely on deaf ears, but they had not been taken as I had hoped.

Our first article, detailing the method of cosmic crystallography and written with Roland Lehoucq, was even rejected by the *Astrophysical Journal*. The referees abruptly responded that the question was closed by the negative data from COBE. After a brief time of discouragement, we submitted our article to the European journal *Astronomy and Astrophysics*, which accepted it. We perceived also that the interpretation of the COBE data made by our American colleagues was a little bit biased; their analysis drew a generally negative conclusion exclusively from some extremely particular cases—exclusively Euclidean, in fact—that they were able to treat.

In fact, that which we hoped for occurred: we finally saw a blossoming of worldwide interest in cosmic topology, as much from the theoretical point of view as the observational. In the five years that followed—which brings us to 2001—the number of articles appearing on cosmic topology far surpassed all those published during the eighty previous years. Today, the shape of space has become one of the most exciting problems in cosmology. Several dozen researchers throughout the world devote themselves to it, and the first conferences dedicated exclusively to the subject have been organized. And I am persuaded that the career of cosmic topology has only just begun: if not, I would not have written this book!

[4]We had not yet discovered that Schwarzschild, in 1900, had already proposed the idea of a crystallographic universe.

Some people believe football is a matter of life or death.
I can assure you it's much more important than that.

—Bill Shankly, manager of the Liverpool team

45 Afterword: Listening to the Cosmic Drum

The Music of the Universe

231→

We have seen how cosmologists hoped to "listen to the shape of space," that is to say to find its topology, by analyzing in detail the temperature fluctuations of the cosmic microwave background. For the early universe "rang" like a musical instrument: before the separation of matter and light, particles were subject to two opposing forces, the pressure of gravitation and that of radiation. The primordial soup, pulled by both, began to oscillate and vibrate. When the photons finally escaped, after 350,000 years, and our microwave telescopes captured them about 14 billion years later, they brought with them this vibrational signature. The little cool and warm spots that appear in the maps of the background radiation are like grains of sand on the surface of a drum: their arabesques reveal the way in which sound propagates.

The detailed study of these patterns reveals a wealth of information on the size and shape of the drum, on the elasticity and the physical makeup of its skin, and so on. In particular, it is easy to see that the statistical distribution of the grains depends on four factors:

- It depends on the initial vibration of the drum. Depending on whether you set it vibrating with a strong hit of the drumstick in the center or by rubbing with a violin bow along the edge, the arabesques will be different; for the universe, the hit of the drumstick translates into the initial fluctuations, for example, fluctuations of quantum origin.

- It depends on its material composition. Under the same given impulse, an animal skin and a metallic membrane do not vibrate in the same way; for the universe, the forms of matter and energy that it contains dictate in part its manner of vibration.

- It depends on its curvature. The patterns will be different according to whether the surface of the drum is flat (which is generally the case), lightly concave, or lightly convex. For the universe, the value of the spatial curvature, according to whether it is zero, negative, or positive, is coded in the size of the lumps in the background radiation.

- It depends on the overall shape of the drum, that is to say on the *boundary conditions.* The waves reflect and recombine differently according to whether the boundary of the drum is a circle, an ellipse, a square, or another shape. Of course, the universe-drum does not have a boundary, but its topology plays the same mathematical role as boundary conditions do.

How does one analyze the cosmic vibrations in order to draw out all this information? In the same way that the vibration of a drum can be expressed as a combination of its fundamental harmonics, the temperature fluctuations of the primordial universe can be expressed as combinations of the fundamental vibrational modes of space.

←105 Beginning in 2001, after having finished our program of study of the topology through the methods of cosmic crystallography, we devoted ourselves entirely to the study of topological signatures inscribed in the cosmic microwave background. Alain Riazuelo, who had just completed his doctoral thesis by revealing himself to be one of the best specialists worldwide in the harmonic analysis of the fossil radiation, had joined our little group. Starting then, we were able to develop a very technical model for studying the vibrations of the Universe. To the ends of comparing with the observational data to come from the WMAP satellite, we had simulated high resolution maps of the fossil radiation on a computer, calculated theoretically using a large class of spaces with multiply connected topologies (Figures 45.1 and 45.2).

←116 We verified that the method of matched circle pairs in fact offers a well-defined topological signature. In simple terms, we showed how the shape of

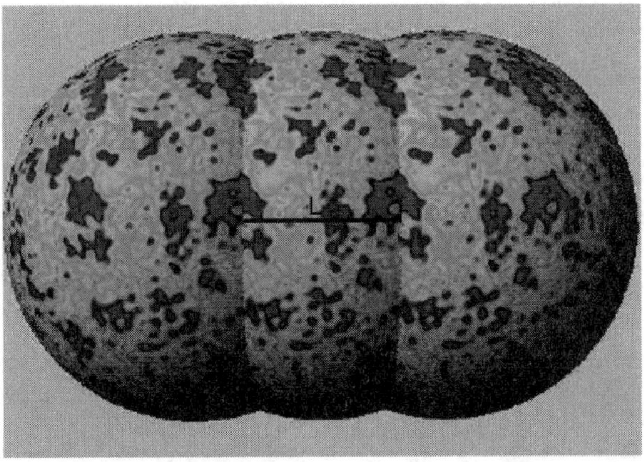

Figure 45.1. Simulated map of the cosmic microwave background in a small hypertorus. This map of the fossil radiation was calculated by Alain Riazuelo for a flat multiply-connected space, more precisely a cubic hypertorus whose size is 3.17 times smaller than the diameter of the cosmological horizon. As a consequence, the surface of last scattering self-intersects along pairs of matched circles, related to each other by translation. Here, two copies of the last scattering surface are shown in the universal covering space, so as to clearly show a pair of circles along which the temperature values are the same. (See Plate XV. Image courtesy of Alain Riazuelo.)

space could be "heard" in a unique way. For this, one must calculate the harmonics (called eigenmodes of the Laplacian) for the various possible topological models. Then, starting from initial conditions that fix the manner in which the Universe was initially set in motion, one must calculate the time evolution of these harmonics, so as to simulate realistic and precise maps of the cosmic microwave background on a computer for a great number of possible configurations. For Euclidean topologies, it was found that the proper modes are already known mathematically, which greatly simplified the problem. On the other hand, original work was needed to calculate the proper modes of spherical spaces, a task belonging more to mathematical work than to astronomical work! It is here that the interdisciplinary composition of our little group proved to be extremely useful. As for the hyperbolic spaces, in the majority of cases the proper modes are not calculable in any other way than by complicated and unreliable numerical approximations.

The balloon observations made on the cosmic microwave background between 1999 and 2002 (the Maxima, Boomerang, Dasi, and Archeops experiments) had imposed certain limits on the curvature of space, constraining its value

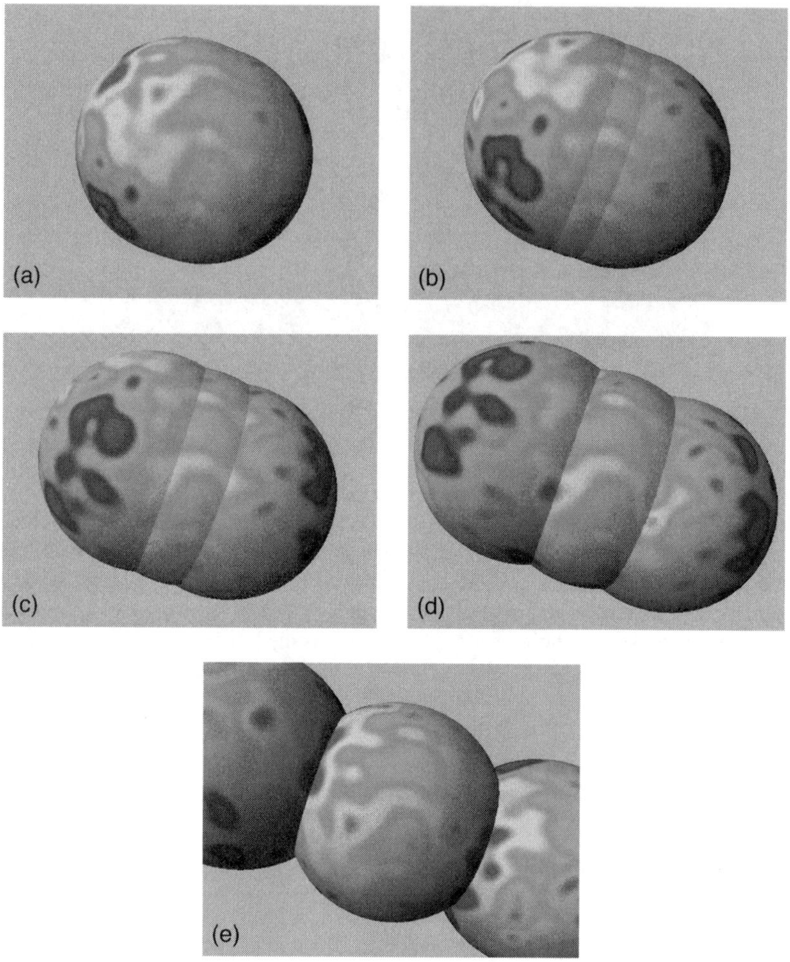

Figure 45.2. Simulated map of the cosmic microwave background in a small lens space. When space has a non-zero curvature, the signature of a multiply-connected topology can manifest itself in an unexpected way. For example, when the space is spherical, the last scattering surface can repeat itself after rotation of one of its hemispheres, as shown by the five figures above. These maps of the fossil radiation are calculated for a multiply connected spherical space of lens type $L(12, 1)$, whose volume is 12 times smaller than that of the usual hypersphere, and $\Omega_0 = 1.3$. Here again, two copies of the surface of last scattering are shown in the universal covering space, to make the five pairs of circles predicted by this model appear. From illustrations (a) to (e), the homologous circles are separated respectively by $2\pi n R_c/12$, where R_c is the radius of curvature, and matched together after a rotation by an angle of $2\pi n/12 (n = 1, 2, \ldots, 5)$. (See Plate XVI. Image courtesy of Alain Riazuelo.)

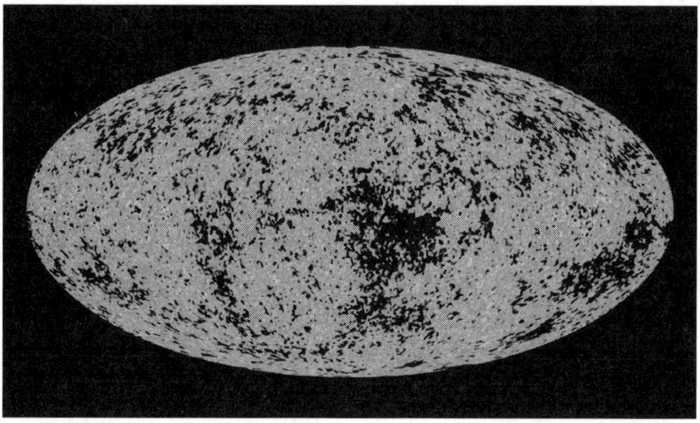

Figure 45.3. WMAP map of temperature fluctuations of the fossil radiation. The entire sky is represented here in the microwave range, and shows the residual light that the young universe emitted after 380,000 years of expansion. The minuscule temperature fluctuations are encoded by colors. They correspond to the infinitesimal lumps of density that, after condensing, created the first galaxies. The age of the universe, its geometry, its composition, and its fate are inscribed in the statistical distribution of lumps. (See Plate XVII. Image from NASA/WMAP Science Team.)

within a narrow interval around zero (corresponding to a "flat" model universe); nevertheless they have not allowed one to determine the exact sign of this curvature, nor above all to know if the topology of space is simply connected (for example that of infinite flat space) or multiply connected (for example that of a flat and finite hypertorus, or that of a dodecahedral spherical space). The reason is that these measurements only covered a very small fraction of the celestial sphere, and therefore could not be used to test the large scale geometric properties of space. For this, it was necessary to wait for the maps of the cosmic microwave background acquired by the WMAP satellite, whose launch was predicted for 2001 and the first data for 2003.

The WMAP Revolution

Finally launched on June 30, 2001, the WMAP satellite has meticulously completed, replaced, and improved the data collected ten years previously by its ancestor COBE (Figure 45.3). As hoped, the first results, published in February of 2003, have allowed one to decode to a good extent the vibrations of the cosmic drum and to draw from them information of an until now unequaled precision on the values of the cosmological parameters. These govern the age, evolution, fate, composition, and shape of the universe.

←188 The age of the universe is now fixed at 13.7 billion years, within 1%. The epoch of emission of the first light—that which we see in the cosmic microwave background—is dated to 380,000 years after the big bang.

←184 The present rate of expansion is 71 km/s per megaparsec, within 5%.

←217 The composition of the Universe is 4% ordinary atomic matter (called baryonic), 23% exotic (nonbaryonic) matter, and 73% dark energy. The nature of the latter remains to be elucidated. If it is a constant term of the cosmological constant type or quantum vacuum energy, this repulsive energy definitively dominates the ←207 future dynamics of the universe and confers upon it a perpetual accelerated expansion. In 2003, independent measurements made on supernovae have reinforced this result and even given the epoch at which the accelerated expansion would have begun: about six billion years ago. Nevertheless, since the exact nature of the dark energy remains hypothetical, certain models (for example, those baptized quintessence models) admit the possibility of time variation in the dark energy; if therefore it were to decrease again in the future instead of remaining constant, then the attractive matter density would prevail anew, and the acceleration of the universe would give place to a deceleration, which could lead to a closed universe with, at the end, a big crunch.

WMAP has also brought unexpected and deeply interesting information on the formation of galaxies: the epoch of appearance of the first stars in the universe appears to be 200 million years after the big bang (or 500 million years earlier than previous estimates). From this one deduces that it is cold dark matter that prevails over hot dark matter. The fact that the first stars had already been formed within a time as short as 200 million years clearly favors this hypothesis. In fact, in a universe dominated by hot dark matter (for example, by neutrinos), gravity would have had a much harder time producing a condensation of gas and the first stars would have appeared much later. The formation of structures in the universe therefore probably began with little galaxies, which little by little agglomerated to form more massive galaxies, then clusters and superclusters.

Let us now move on to the curvature of space. Recall that this is encoded in the normalized value of the energy density parameter Ω_0: $\Omega_0 < 1$ corresponds to a space of negative curvature (hyperbolic geometry), $\Omega_0 = 1$ to a space of zero curvature, also called a flat space (Euclidean geometry), and $\Omega_0 > 1$ to a space of positive curvature (spherical geometry). Now, combined with other measurements, the WMAP data give $1.00 \leq \Omega_0 \leq 1.04$. This result is marginally compatible with the model of flat space until now favored by consensus and by the theory of inflation. Nevertheless, the balance leans clearly in favor of a space of positive curvature, and therefore a finite space, and seems to eliminate the case of a hyperbolic space!

In a certain fashion, this result was comforting for our team. We had in fact demonstrated on the one hand that the hyperbolic topologies would in practice be undetectable, and on the other hand that the spherical topologies would be more easily detectable than the Euclidean topologies (for reasons too complex to be discussed here). Therefore, we had focused our theoretical work on the case of spherical spaces, which have remained unexplored as possible models for the physical universe. However that may be, the two hypotheses—Euclidean space or spherical space—offer, mathematically, a great diversity of topological structures: 272→ 18 in the case of a Euclidean space, an infinite number if it is spherical. Each form has its special features: ghost galaxies in greater or lesser number, lumps of the fossil radiation that repeat in such or such way, universes that are smaller than what we see with our telescopes.

Therefore, what does it say about the question that is the principal object of this book: the topology of the universe? Here things become truly interesting. There is an intriguing feature in the cosmic maps from the WMAP experiment:[1] the infant universe does not ring completely, as it would do if space were Euclidean and infinite. It is in some way mute at large wavelengths. Let us look into this more precisely.

The temperature fluctuations in the fossil radiation can be decomposed into a combination of spherical harmonics, just like the sound produced by a musical instrument can be decomposed into ordinary harmonics. The fundamental harmonic fixes the basic pitch of the sound (for example the A of a tuning fork has a frequency of 440 hertz) while the relative amplitude of each harmonic determines the timbre (the A on a piano is different from one played on a harpsichord). In the case of the fossil radiation, the relative amplitude of each spherical harmonic fixes the *power spectrum* (see Figures 38.5 and 38.6), containing a signature of the geometry of the universe and of the conditions that prevailed at the moment of emission of the radiation. Recall that the power spectrum exhibits a collection of peaks when the difference is measured between regions of the sky at small and medium dimensions, which is to say, separated by small angles. In the harmonic analysis of WMAP (Figure 45.4), these peaks conform to what is predicted by the standard model (infinite Euclidean space). On the other hand, on large angular scales (for regions typically separated by more than 60°), there is a notable loss of power that does not agree with the predictions of the standard model.

The first observable harmonic is the *quadrupole* (whose wave number is $l = 2$). It corresponds to an angle of observation of 90°—a quarter of a circle, from whence the term. Now, WMAP has observed a quadrupole 7 times weaker than

[1]This was already present in the earlier data of COBE, but at a level of precision that was not statistically significant.

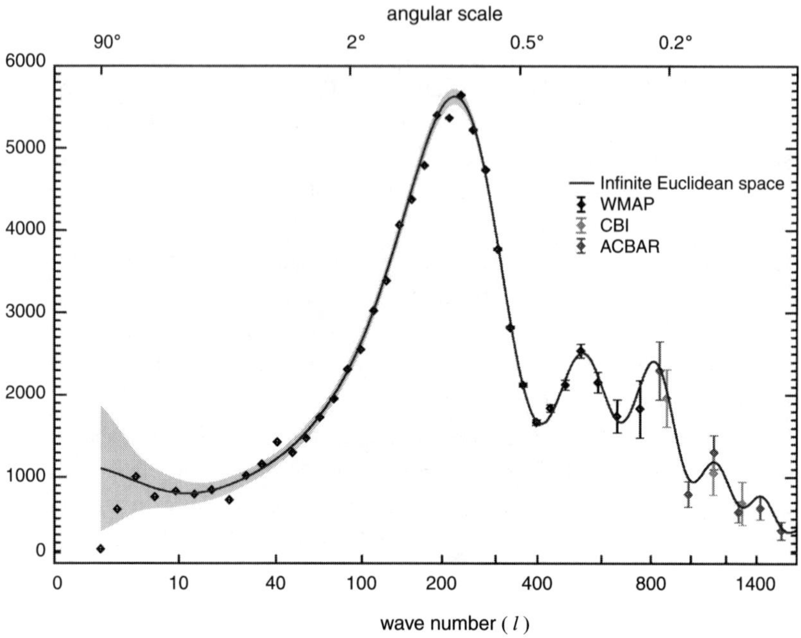

Figure 45.4. Power spectrum of the cosmic microwave background. The data from the WMAP satellite (black diamonds) have improved the precision of the spectrum far beyond that obtained by earlier measurements. The diagram reflects the small differences in temperature in the fossil radiation. The spectrum exhibits a series of peaks corresponding to small angular separations (or, equivalently, large wave numbers l). The main peak corresponds to the angular size of the most frequent lumps from Figure 45.3. The position and amplitude of this peak in particular allow one to constrain the curvature of space. At larger angular separations, this structure disappears and the data predicted by the standard model are expected to follow a plateau. However, the WMAP measurements fall well below the plateau in the range of the lowest wave numbers, named the quadrupole ($l = 2$) and the octopole ($l = 3$). The infinite, Euclidean cosmological model (giving the theoretical curve) cannot explain this, while well-proportioned finite models for the universe with multiply-connected topologies explain it very well. (Image from NASA/WMAP Science Team.)

what is expected in an infinite Euclidean space. The probability that this difference is produced by chance is estimated at 0.2%. The next harmonic, the *octopole* (whose wave number is $l = 3$), is also weaker than the theoretical prediction, at 72% of the expected value. For larger values of the wave number, up to $l = 900$ (which corresponds to temperature fluctuations on very small angular scales), the observations are, on the contrary, well explained by the standard model.

The weak value of the quadrupole signifies that the long wavelengths are missing. Some cosmologists have proposed attributing this anomaly to as yet undiscovered physical laws that would have governed the primordial Universe. Another explanation of this phenomenon, which appeared more natural to us, since it was purely geometric (which is to say, more in the spirit of the theory of general relativity), was founded on a model of finite space, in which the size of the space imposes a maximal value on authorized wavelengths.

In fact, a large and a small drum do not sound similarly. No low notes will come from the small one, since the wavelength must be smaller than the diameter of the instrument. The same thing holds for the Universe. If its size is infinite, or at least much larger than the cosmological horizon, all wavelengths are allowed and fluctuations should be present on all scales. On the contrary, if its size is finite and smaller than that of the horizon, then very long wavelengths are forbidden. In this type of small wraparound universe, there must therefore be a natural length scale above which the Universe ceases to vibrate, and this translates into a loss of power in the spectrum of the fossil radiation on angular scales greater than this maximum. Exactly this is observed in the WMAP data.

A Well-Proportioned Space

One might have believed intuitively that any multiply-connected topology, as soon as it introduces a spatial finiteness in at least one direction, must lower the power spectrum at large wavelengths. We have examined the question and demonstrated that in reality, certain multiply-connected finite topologies indeed lower the quadrupole but that others augment it. The large wavelength vibratory modes only tend to be relatively weakened in a family of finite multiply-connected spaces known as *well-proportioned spaces*. Very generally, among those spaces whose dimensions are comparable to the radius of the surface of last scattering (the necessary condition for topological effects on the power spectrum to be observable), the spaces whose three dimensions are all of the same order suppress the quadrupole. As soon as one of the dimensions becomes significantly smaller or larger than the others, the quadrupole is augmented. The demonstration is geometric, but we have also tested it numerically in the case of Euclidean tori, whose proportions we have varied (the cubic torus reduces the quadrupole while the oblate or prolate torus increases it), as well as for spherical spaces (the polyhedral spaces weaken the quadrupole while the lens spaces of high order, which are strongly anisotropic, reinforce it).

The well-proportioned spaces therefore represent the WMAP data better than the standard model does (Figure 45.5). The result was important, and I thought right away to write a "letter" destined for the British journal *Nature*.

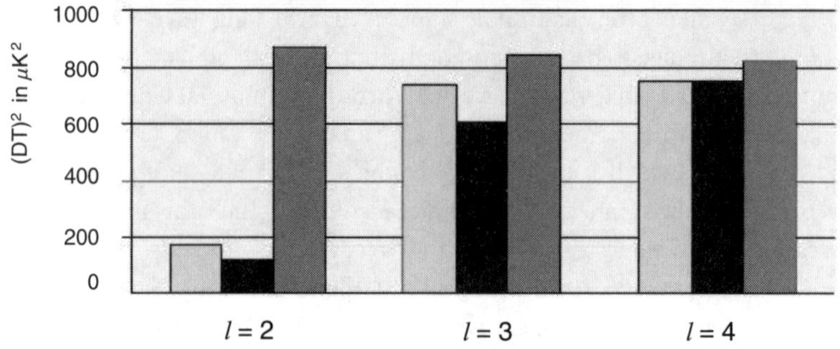

Figure 45.5. Fluctuation amplitudes for low wave numbers. This graph is a comparison of the amplitude of fluctuations observed by WMAP (in black) with those calculated for an infinite Euclidean space (dark gray). The calculation has been performed for a matter density of 0.28 and a cosmological constant of 0.74. The disagreement is large for the first wave number ($l = 2$), called the quadrupole, and less for the second wave number ($l = 3$), called the octopole. The agreement is much better for a particular well-proportioned space, the dodecahedral Poincaré space (light gray).

This scientific weekly, the most prestigious in the world, only publishes short articles of relatively nontechnical nature, since they are meant to be understood by the whole of the scientific community: biologists must follow the arguments of astrophysicists, and vice versa. Because of the flood of communications that the journal receives in all scientific domains, the choice is obviously very selective. Any submitted manuscript has to make it clear that the new model proposed is the best available; as a consequence, a number of issues do not contain even a single article on astrophysics, much less on cosmology.

In June 2003, we nevertheless contacted the editor of the journal in order to submit to him a preliminary version of our work. He replied to us:

> We get numerous papers in cosmology submitted simply for the purpose of putting forward a new idea for further discussion. Such papers normally are rejected without review, because we do not see our role in that way. Rather, that it is what the many fine specialist journals are for. All of the above is not necessarily supposed to be discouraging, but I do think you should be aware of the huge potential barrier against publishing in *Nature*. We publish less than 10% of what is submitted to us (and that does not include the cranks), and only about 25% of submissions are even sent to referees (we get over 10,000 submissions per year). As written, I might not send either of the draft manuscripts out to referees, though I might well consult informally with an expert in the field to explore whether the work showed sufficient potential to warrant sending out to referees.

At that moment, a wind of disappointment blew upon us. Jeff Weeks wrote us by email, "Here is the reply from *Nature*. As you will see, they discourage the proposed letter. So ... maybe the best thing is to forget it. ... I am sorry to be a pessimist, but in light of what the editor said (about being the 'best available model,' not just an interesting possibility) it seems the chances that *Nature* would accept this letter are close to zero."

There was a more lugubrious clang of the bell with my French collaborators. "I think that *Nature* is right in its judgment and this confirms what Alain was saying. I do not think that in the present state of things we are in a position to publish anything whatsoever in *Nature*, nor even in a specialized journal like *Physical Review Letters* or *Astrophysical Journal*," Jean-Philippe Uzan commented.

As for myself, I ruminated over this seeming setback for an entire weekend. A few years before, I had succeeded in publishing an innovative article in *Nature* on *stellar pancakes*, which was a theoretical description of the way in which entire stars could be destroyed by the tidal forces of giant black holes; see [Luminet 92]. It was in fact, at the time, the best available model of this astrophysical phenomenon. What should we do now to transform our idea of a well-proportioned universe into that of the "best available model"? There was no possibility of sifting through a vast ensemble of candidate shapes in order to select, after long months of fastidious calculations, the most appropriate.

Dodecahedral Space

The illumination came to me while working ... in the garden of my country house! The mind always works in the background when the body is occupied with very physical and down-to-earth tasks; a number of ideas have come to me while I was cultivating my garden or taking long hikes in the mountains.

On that day, in the spring of 2003, the image of a particularly well-proportioned space sprang to mind: the spherical dodecahedral Poincaré space. 73 → Why? It is difficult a posteriori to decode the mental images that serve to advance us in fundamental research. To rationalize the thing, I would say that intuitively, we needed a model of physical space that was neither too small with respect to the observable universe (otherwise the observed discrepancies with the theoretical power spectrum calculated for a very large universe would not be restricted to large angles only), nor too large (if not no discrepancy would have been observed at the horizon scale). At first sight, the dodecahedral model fit: its volume is 120 times smaller than that of the hypersphere having the same radius of curvature. Now, this is perfectly calculable if one is given the three cosmological parameters: the Hubble parameter H_0, the energy density Ω_m in the form of gravitating matter (both visible and dark), and the anti-gravitating energy density Ω_Λ (in the

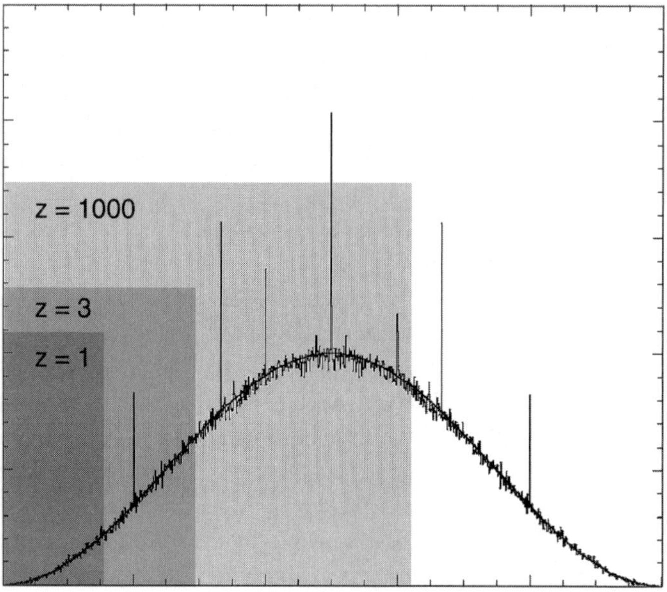

Figure 45.6. Cosmic crystallography in the Poincaré space. The figure shows a histogram of pair separations in the spherical Poincaré space, calculated for $\Omega_0 = 1.1$. Although the main spike is situated at a spectral shift too large to be detectable, the first spike is at $z \sim 1.5$, accessible to the deep observational surveys expected in this decade.

form of a cosmological constant or other more exotic energy fields). For plausible values, according to the WMAP data, the calculation gives 140 billion light-years for the radius of curvature of the simply-connected hypersphere, 53 billion light-years for the radius of the observable universe,[2] and 45 billion light-years for that of the Poincaré dodecahedral space. In volume, this corresponds to a hypersphere about 100 times larger than the observable universe. Therefore, the dodecahedral space, 120 times smaller than the hypersphere, would have a volume of about 80% of the observable universe. There is no spectacular topological mirage with hundreds of ghost galaxies in this case: we had calculated two years earlier the ←105 histogram of pair separations in such a space, and showed that the topological signature in the form of multiple images of galaxies would be difficult to detect (Figure 45.6). However, with a relatively large space, there remained the hope of a small detectable effect on large angular scales of the fossil radiation.

[2]Although the photons of the fossil radiation have only voyaged for 13.7 billion years, the expansion of space has lengthened their route, because of which the starting point of a photon reaching us today after travelling for 13.7 billion years is now 53 billion light-years away.

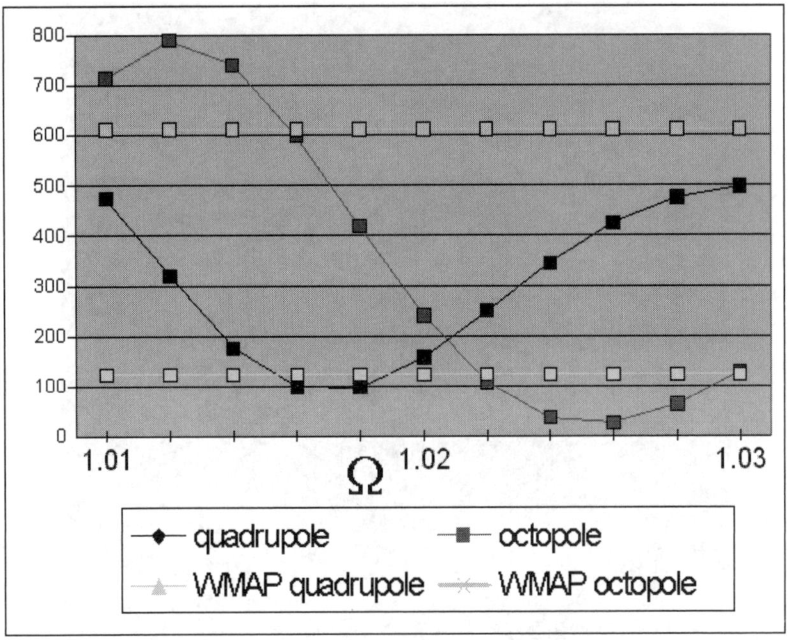

Figure 45.7. Comparison of predictions from the Poincaré space model with WMAP data. Values of the mass-energy parameter for which the Poincaré dodecahedral space (black curve for the quadrupole and gray curve for the octopole) agrees with the WMAP observations (lower line for the quadrupole and upper line for the octopole). The agreement is optimal for a value of the mass-energy density parameter somewhere between 1.012 and 1.020.

After the dream, the work. The key to the problem consisted in demonstrating that a spherical dodecahedral drum theoretically vibrated at large wavelengths exactly like the real universe as observed by WMAP vibrated. Jeff Weeks had begun the calculation and did not tarry in sending us this email: "The results are absolutely gorgeous! What a joy! Well, there is no doubt that the Poincaré dodecahedral space can be the star of the show in our *Nature* article (and also no doubt that this will be a super article)."

We therefore wrote a short letter in which, by assuming that space possesses this dodecahedral topology (allowing us to account for the quadrupole and octopole observed by WMAP), we deduced very precisely its radius of curvature and its size. Since this space is of spherical type, the density parameter of the universe must be constrained by $\Omega > 1.01$, with the value $\Omega_0 = 1.018$ as the best fit to the data on the quadrupole and octopole (Figure 45.7). Our model is

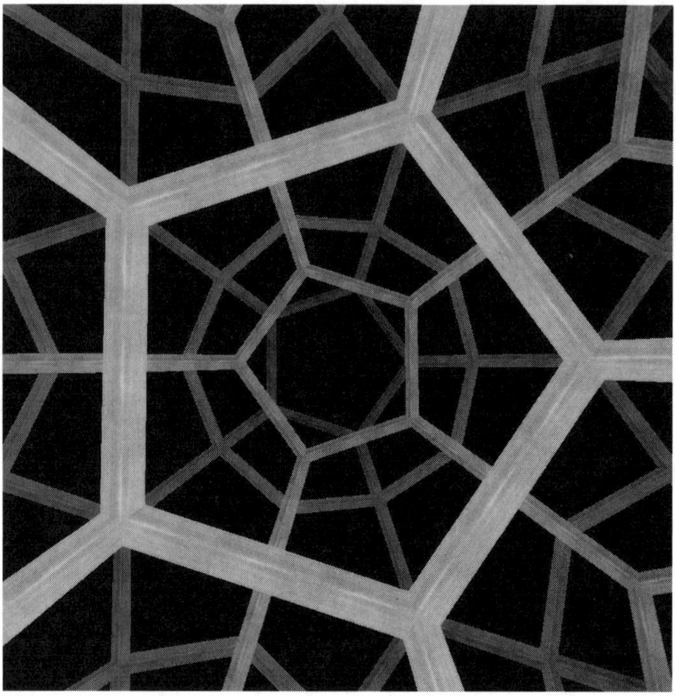

Figure 45.8. Topological mirages in the Poincaré space. The walls of the dodecahedral space are not really walls. It would be better to speak of "magic portals": when a space vessel—or, more realistically, a light ray—tries to exit by one face, it immediately reenters the dodecahedron by the opposite face, after having turned by 36°. In other words, in this finite space, a ray of light would voyage indefinitely, since each light ray exiting from one face would come back by the opposite face after having made a tenth of a turn. The consequence: this light ray needs to make ten exits and entrances before ending up in its initial state, which explains this image showing part of the topological mirage created when one looks perpendicularly to a pentagonal face of the dodecahedron: ten stacked pentagons, each shifted by 36°. Since the dodecahedron has 12 faces, the overall mirage gives the illusion of 120 dodecahedra tiling a hypersphere! (See Plate XVIII. Image courtesy of Jeffrey Weeks.)

therefore refutable, since the data from the future European PLANCK Surveyor satellite (expected to launch in 2008) should allow the determination of this density parameter up to one percent. A value smaller than 1.01 would eliminate the Poincaré space as a physical model, in the sense that the size of the dodecahedron would become greater than that of the observable universe and would leave no detectable signature on the quadrupole.

Not only is the model of spherical dodecahedral space refutable, it is also provable (in contrast, for example, with the model of infinite Euclidean space). 116→ We have seen in fact that, if space is multiply connected, there must be particular correlations in the fossil radiation taking the form of pairs of circles along which the temperature fluctuations are identical, up to rotation. Now, the dodecahedral model predicts the existence of six pairs of circles, diametrically opposite to each other in the sky (corresponding to the six pairs of faces identified in the dodecahedron), whose angular radius would lie between ten degrees and forty degrees (values that depend sensitively on the cosmological parameters).

Written thus, our article obeyed the pure rules of the scientific approach: a new observation (the deficit of acoustic vibrations in the WMAP data) raises a theoretical problem, an argued explanation (a particular multiply connected universe that accounts for the data) is proposed, which moreover makes two predictions (the precise value of Ω_0 and six pairs of matched circles in the fossil radiation), whose confirmation or invalidation could prove or refute the model.

Our article was therefore accepted without any difficulty by *Nature* in mid-July 2003. There is nevertheless a price to pay for publishing in this journal: strict secrecy. Normally, any article accepted for publication in a scientific journal is immediately placed on electronic archives, which considerably accelerates its diffusion into the community. However, *Nature* forbids that any article accepted by the journal be circulated in any way before the day of its paper publication: neither on the internet, nor to colleagues in the form of preprints, nor in the form of communication at a colloquium! One must therefore champ at the bit while waiting for its appearance, hoping that no other team will meanwhile complete an analogous work which, submitted to an ordinary journal, will have been posted earlier on the electronic archives and will take precedence!

We were expecting the article's appearance at the end of August. Publication was then delayed, probably because of summer. On September 30, 2003, the suspense came to an end, along with some consolation: *Nature* informed us that not only would the article appear in the October 9th issue, but it would also make the cover of the magazine, accompanied by a spectacular color representation of the Poincaré dodecahedron as seen from the interior (Figure 45.8).

On the big day, a press communication written without our consultation by a skillful journalist from *Nature* made a sensation: it announced the discovery of a model for the universe in the shape of a soccer ball!

We had not at all expected the comparison. However, it is true that, in order to explain to the readers of the journal how the ordinary hypersphere could be tiled by 120 stacked dodecahedra, we had given the simplified two-dimensional image of the surface of a sphere, which could be tiled by 12 curved pentagons (Figures 45.9 to 45.11). This image evokes a soccer ball.

Figure 45.9. The Poincaré dodecahedral space. The Poincaré space can be described as the interior of a sphere whose surface is tiled by 12 curved regular pentagons.

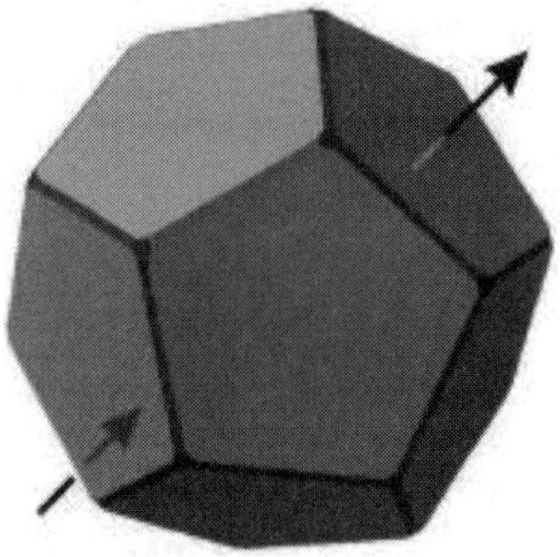

Figure 45.10. Connectedness of the Poincaré space. There is a big difference between the Poincaré space and the soccer ball that Figure 45.9 may remind one of: when one leaves through a pentagonal face, one returns to the ball through the opposite face, after having turned by 36°.

←280 More seriously, the cosmologist George Ellis, from the University of Cape Town in South Africa and one of the pioneers of cosmic topology, commented on our work in a short preliminary article for *Nature*: "An analysis of astronomical data suggests not only that the Universe is finite, but also that it has a specific topology. If confirmed, this is a major discovery about the nature of the Universe."

Figure 45.11. Universal covering of the Poincaré space. This space is finite but without boundaries or limits, therefore one can voyage through it indefinitely. Because of this, one has the impression of living in a space 120 times larger, paved with dodecahedra that multiply images like a hall of mirrors. The return of light rays that cross the walls produces optical mirages: a single object has several images. (See Plate XIX. Image courtesy of Jeffrey Weeks.)

The Controversy

To make the cover of *Nature* with an article on theoretical cosmology is not so common, and the scientific milieu is not always tender. In the midst of geopolitical controversy on the military intervention in Iraq, a portion of the general North American press made some unkind comments on our model—without, of course, understanding a single word of it. The *New York Times* began its article thus: "Cosmic soccer ball? Theory already takes sharp kicks. A revolutionary new model of the universe, as a soccer ball, arrives on astronomers' desks this morning at least slightly deflated." The rest is in the same vein.[3]

[3]The article in the *Economist* was more sober: under the title of "Platonic Truths," it recalled that the dodecahedron had already been used by Plato to geometrically model the constitutive element of the sky, the *quintessence*.

It is true that these cutting judgments were based on the precipitous declarations of two teams of American researchers. One of them declared that they had already looked in vain for pairs of opposed circles in the sky, and had published their results in mid-August. In reality, they had only looked for circles in the very particular topology of the hypertorus, and their work had no relation to ours. They recognized their error with us, as they were unfamiliar with spherical topologies, but it was too late to rectify the negative impression given in the North American media. Another team, directed by Neil Cornish (the very one who had developed the beautiful method of correlated circles), had also declared to the journalists that they had searched in vain for correlated circles in the WMAP data, but they had not published anything! The discussions that followed between our two groups were lively but fruitful. By computer, Alain Riazuelo had fabricated artificial maps of the fossil radiation founded on a topology that was kept secret, and that led to a large number of pairs of correlated circles; after a few months of calculation, our American colleagues announced that they had detected our artificial circles and found the hidden shape, which was a promising sign for the validity of their method. Reassured by this result, they improved their program for the real data and persisted in their refutation of the dodecahedral model, excluding in particular a size of the universe smaller than the cosmological horizon.

However, in parallel, a team of Polish cosmologists led by Boud Roukema (a former collaborator of mine) published an article in early 2004 in which, analyzing the same WMAP data, they declared that they had found the six pairs of matched circles predicted by the dodecahedral model. One year later, a third team led by Frank Steiner, from Germany, performed a very careful search for matched circles and found that the putative topological signal in the WMAP data was considerably degraded by various effects, so that the dodecahedral space model could be neither confirmed nor rejected. This shows in passing how delicate the statistical analysis of observational data is, since different analyses of the same data can lead to radically opposed conclusions! The controversy went up a notch when the first (American) team claimed in a new article that their negative analysis was not disputable, and that accordingly, not only the dodecahedral hypothesis was excluded, but also any multiply-connected topology on a scale smaller than the horizon radius! Because of such an argument from authority, a fair portion of the U.S. academic community believes that the WMAP data has ruled out multiply-connected models. However, the second part of the claim is wrong. This is because they searched only for antipodal or nearly-antipodal matched circles. But in the meanwhile, my team had shown that for generic multiply-connected topologies (including the well-proportioned ones, those that are good candidates for explaining the WMAP power spectrum), the matched circles are generally not antipodal; moreover, the positions of the matched circles in the sky depend on the observers position within the fundamental polyhedron. The corresponding larger number of degrees of freedom involved in the circles search in the WMAP data generates a dramatic increase of the computer time, up to values that are out of reach of present facilities. Thus, at the time that I write these additional lines (January 2007), the controversy over the shape of space has not been settled. The fascinating mathematical properties of the polyhedral spaces have been investigated in more detail by various groups outside the United States. Both analytical and massive computer calculations were able to reconstruct the full power spectrum for the spherical dodecahedral space and found that the fit with the cosmic microwave background observations was still better than what we originally proposed in our *Nature* letter. In March 2006, a new release of WMAP data, integrating two additional years of observation with reduced uncertainty, strengthened the evidence for an abnormally low quadrupole and other features that do not match with the infinite flat space model. (This explains the unexpected delay in the delivery of this second release, originally announced for February 2004.) Since some of these anomalies are one of the possible signatures of a finite and multiply-connected universe, there is sill a continued interest in the Poincaré dodecahedral space and related finite universe models. And even if the particular dodecahedral space is eventually ruled out by future improved experiments, all of the other models of well-proportioned spaces will not be eliminated as such. In addition, our numerical simulations show that, even if the size of a multiply-connected space is larger than that of the observable universe, we could all the same discover an imprint in the fossil radiation, even while no pair of circles, much less ghost galaxy images, would remain. The topology of the universe could therefore provide information on what happens outside of the cosmological horizon! But this is a search for the next decade.

Consequences for the Physics of the Primordial Universe

The models of finite and well-proportioned universes, and in particular the Poincaré dodecahedral space, are a veritable Pandora's box for the physics of the early universe. In fact, the standard model predicts that in its first fractions of a second of existence, the expansion of the universe would have been governed by a phase of exponential expansion called inflation, due to a spontaneous breaking of symmetry at the time of decoupling of the fundamental interactions. Before even speaking of topology, it is good to note that the theory of inflation leads to some difficulties. In its simplest version, inflation predicts a universe much larger than the observable universe. The simple fact that the curvature is positive (which is the case even if $\Omega_0 = 1.000000001$) would already pose a problem for inflation.

It is, nevertheless, possible to construct models where the inflationary phase is shorter and allows for an observable curvature. Therefore, if space is not flat, the existence of a multiply connected topology is not in flagrant contradiction with inflation. The latter should be able to accommodate itself since the theory has many free parameters. However, no explicit such model has yet been formulated.

In modern relativistic cosmology, it is generally assumed that spatial homogeneity remains valid beyond the cosmological horizon. In the model of chaotic inflation—a variant of the inflationary model—the Universe would become very heterogeneous on scales much larger than what we can observe, and we would therefore be in a homogeneous expanding bubble among innumerable other bubbles. Now, our model is at first sight incompatible with that of chaotic inflation: it only concerns a single expanding bubble of universe, which is, moreover, sufficiently small for us to be able to see it in its entirety.

In 1917, Einstein had already stressed that spatially finite universes presented the advantage of suppressing the problem of boundary conditions. A small universe in which we can see most of what exists is even more advantageous. It is in fact the only type of universe in which we could predict the astronomical future with certainty—the return of Halley's comet, for example—because it is only in this type of universe that we have all the data at our disposal to make such predictions. This argument is not a gauge of validity, but an interesting philosophical feature. For the moment, the WMAP data suggest that we may in fact live in a finite universe, one that is sufficiently small to create a cosmic mirage.

For the (necessarily provisional) final word, I close with a poem:

> The universe is made up of degrees
> arranged along exact mathematical proportions
> the sun applies itself to ripen each seed
>
> The sidereal spheres are woven of immaculate fire
> Numbers and figures are their spiritual objects
>
> We weave real objects from thread of the stuff
> of which mathematical dreams are made
> The universe is cabled by arithmetic
> trace a circle and pi surges forth
> Enter a new solar system
> and Kepler's laws await you
> hiding under the black velvet cape
> of space-time

(J.-P. Luminet, *Itinéraire céleste*)

Paris, February 2007

Bibliography

[Berthoz 02] A. Berthoz. *The Brain's Sense of Movement*. Cambridge, MA: Harvard University Press, 2002.

[Biggs et al. 86] N. L. Biggs, E. K. Lloyd, and R. Wilson. *Graph Theory, 1736–1936*. Oxford, UK: Clarendon Press, 1986.

[Borges 00] Jorge Luis Borges. *Selected Non-Fictions*. New York: Penguin, 2000.

[Brickman 43] R. Brickman. "Francesco Patrizi on Physical Space." *Journal of the History of Ideas* 4 (1943), 224–245.

[Clifford 73] William Clifford. *The Common Sense of the Exact Sciences*. Freeport, NY: Books for Libraries Press, 1973.

[Davis and Hersh 81] Philip J. Davis and Reuben Hersh. *The Mathematical Experience*. Boston: Birkhauser, 1981.

[Eliade 54] Mircea Eliade. *The Myth of the Eternal Return: Or, Cosmos and History*. Princeton, NJ: Princeton University Press, 1954. Translated by Willard R. Trask.

[Ernst 76] Bruno Ernst. *The Magic Mirror of M. C. Escher*. New York: Random House, 1976.

[Escher and Locher 71] M. C. Escher and J. L. Locher. *The World of M. C. Escher*. New York: H. N. Abrams, 1971.

[Galilei 90] Galileo Galilei. *The Assayer*. New York: Anchor Books, 1990. In *Discoveries and Opinions of Galileo*, translated by Stillman Drake.

[Gamow 93] George Gamow. *Mr. Tompkins in Paperback*. Cambridge, UK: Cambridge University Press, 1993.

[Heidmann 89] Jean Heidmann. *Cosmic Odyssey*. Cambridge, UK: Cambridge University Press, 1989. Translated by Simon Mitton.

[Kepler 65] Johannes Kepler. *Kepler's Conversation with Galileo's Sidereal Messenger.* New York: Johnson Reprint Corp., 1965. Translated by Edward Rosen.

[Kepler 66] Johannes Kepler. *The Six-Cornered Snowflake.* Oxford, UK: Clarendon Press, 1966. Translated by Colin Hardie.

[Koyré 57] Alexandre Koyré. *From the Closed World to the Infinite Universe.* Baltimore: Johns Hopkins Press, 1957.

[Lachièze-Rey and Luminet 01] Marc Lachièze-Rey and Jean-Pierre Luminet. *Celestial Treasury.* Cambridge, UK: Cambridge University Press, 2001. Translated by Joe Laredo.

[Lemaître 58] Georges Lemaître. "The Primeval Atom Hypothesis and the Problem of the Clusters of Galaxies." In *La Structure et l'évolution de l'univers.* Brussels: Coudenberg, 1958.

[Lemaître 60] Georges Lemaître. "L'Étrangeté de l'univers." *Revue générale belge* June (1960), 1–14.

[Lemaître 78] Georges Lemaître. "L'Univers, problème accessible à la science humaine." *Revue d'histoire scientifique* 31 (1978), 345–359.

[Luminet 87] Jean-Pierre Luminet. "Géometries de la variété univers." In *Aux confins de l'Univers*, edited by J. Schneider. Paris: Fayard: Fondation Diderot, 1987.

[Luminet 92] Jean-Pierre Luminet. *Black Holes.* Cambridge, UK: Cambridge University Press, 1992. Translated by Alison Bullough and Andrew King.

[of Cusa 85] Nicholas of Cusa. *On Learned Ignorance.* Minneapolis: A. J. Banning Press, 1985. Translated by J. Hopkins.

[Olbers 83] Heinrich Olbers. "La transparence de l'espace cosmique." In *La Science de l'Univers à l'âge du positivisme*, p. 321. Paris: Vrin, 1983. Translated by J. Merleau-Ponty.

[Poe 48] Edgar Allan Poe. *Eureka: A Prose Poem.* New York: G. P. Putnam, 1848.

[Riemann 73] Bernhard Riemann. "On the Hypotheses which Lie at the Bases of Geometry." *Nature* 8 (1873), 14–17. Translated by W. K. Clifford.

[Salat 02] Serge Salat. *Les labyrinthes de l'éternité.* Paris: Hermann, 2002.

[Schwarzschild 98] Karl Schwarzschild. "On the Permissible Curvature of Space." *Class. Quantum Grav.* 15 (1998), 2539–2544. Translated by J. Stewart and M. Stewart. First published in *Viert. d. Astron. Ges.* 35 (1900), 337.

[Sokal and Bricmont 03] Alan Sokal and Jean Bricmont. *Intellectual Impostures*, Second edition. London: Profile Books, 2003.

[Thurston 97] William Thurston. *Three-Dimensional Geometry and Topology.* Princeton, NJ: Princeton University Press, 1997.

[Valéry 60] Paul Valéry. *Œuvres*, Vol. 2. Paris: Gallimard, 1957–1960.

[Weinberg 72] Steven Weinberg. *Gravitation and Cosmology.* New York: Wiley, 1972.

Index of Names

Subject Index